ANALOG
LINE-OF-SIGHT
RADIO LINKS
A Test Manual

ANALOG
LINE-OF-SIGHT
RADIO LINKS
A Test Manual

A. A. R. TOWNSEND

 **Prentice/Hall International**

Englewood Cliffs, NJ   London   Mexico   New Delhi
Rio de Janeiro   Singapore   Sydney   Tokyo   Toronto

To my wife Ngozi

Library of Congress Cataloging in Publication Data

Townsend,A. A. R., 1944-
  Analog line of sight radio links.

  Bibliography: p.
  Includes index.
  1. Microwave communication systems--Testing.
I.Title.
TK7876.T66   1987   621.38'0413   86-17076
ISBN 0-13-032707-7

British Library Cataloguing in Publication Data

Townsend, A. A. R.
  Analog line of sight radio links: a test
  manual.
  1. Mobile communication systems
  2. Microwave communication systems
  I. Title
  621.3841'65   TK6570.M6
  ISBN 0-13-032707-7

PRENTICE-HALL INC., *Englewood Cliffs, New Jersey*
PRENTICE-HALL INTERNATIONAL (UK) LTD, *London*
PRENTICE-HALL OF AUSTRALIA PTY LTD, *Sydney*
PRENTICE-HALL CANADA INC., *Toronto*
PRENTICE-HALL HISPANOAMERICANA S.A., *Mexico*
PRENTICE-HALL OF INDIA PRIVATE LTD, *New Delhi*
PRENTICE-HALL OF JAPAN INC., *Tokyo*
PRENTICE-HALL OF SOUTHEAST ASIA PTE LTD, *Singapore*
EDITORA PRENTICE-HALL DO BRASIL LTDA, *Rio de Janeiro*

Printed and bound in Great Britain for
Prentice-Hall International (UK) Ltd,
66 Wood Lane End, Hemel Hempstead, Hertfordshire, HP2 4RG
at the University Press, Cambridge.

1 2 3 4 5   90 89 88 87 86

ISBN 0-13-032707-7

CONTENTS

PREFACE

The existing use of analog radio communications for spur and for backbone links, and the installation of new systems or the upgrading or expansion of existing systems, mean that analog radio will remain an important means of communicating in many developing and developed countries throughout the world for at least another two decades. This book provides the engineer with an account of the testing procedures used, the theory behind the testing, the equipment used for testing, and typical results, for a complete radio system.

To do this the parameters which are basic to the communications system are first defined. These parameters are then developed to a stage where they come together to provide an in-depth understanding of the whole system and of the required equipment and system tests. Where necessary, worked examples are given to assist in understanding the theory. Interwoven into the development of the system and equipment parameters are chapters on typical systems and their associated subsystems, together with subjects pertinent to radio link system design engineering. I believe that this will provide a realistic framework for the theory and consequent tests.

As the subsystems placed on either spur or backbone links are an important consideration for the practicing engineer, the theory, equipment and testing are also included for a number of different subsystems. Furthermore, frequency-division multiplex (FDM) theory, testing and equipment are described to bridge the gap between the radio system and the subsystem or telephone channel.

Consideration is first given to telephony requirements and to frequency-division multiplex. This is then followed by the subsystems and the various signaling methods used. The remaining chapters cover those areas important to the radio link, such as noise, multichannel traffic loading, system factors, modulation, antenna systems and link planning considerations. Radio and multiplex equipment block diagrams are extensively used for different manufacturers' equipment to demonstrate the levels required during testing, and the functions and operation of interfacing and monitoring equipment used in radio relay systems are also included.

This book is intended to fill the gaps between equipment, system design, and practice as encountered in the field. It is written primarily for practicing engineers, but should also prove a handy reference book for both engineering managers and field technicians, as well as students studying communications engineering. For those interested in setting up corrective maintenance and preventive maintenance procedures, the layout of the test procedures given should prove to be invaluable. Emphasis has been placed on American equipment and terminology, and on CCITT and CCIR recommendations. Extensive references are used to permit further and more detailed reading.

A.A.R.T

ACKNOWLEDGEMENTS

To Messrs G. Perzel, S. Johns and P. Woolner of Harris RF Communications Inc., Rochester, New York, Mr A. A. Ajilogba of Team Technik (Nigeria) Ltd, Mr G. Mouvet of BTMC (Antwerp) and Mr Jorma Poranto, Farinon, Canada.

INTRODUCTION

The method of testing microwave links is similar to the philosophy of experimentation. Thus a working hypothesis is established, known as the *Null hypothesis*. Suppose this hypothesis to be 'that the microwave radio and ancillary equipment are operating correctly'. The null hypothesis cannot be proved, nor can it be directly proved that the equipment is not working correctly. However, if the null hypothesis can be *rejected*, by saying 'the hypothesis that the equipment is operating correctly is rejected', its alternative can be asserted. Note that the support of the alternative hypothesis is *always* indirect, and to support the statement that the radio and ancillary equipment are not working, the hypothesis that it was had to be rejected.

Since the alternative hypothesis can neither be proved nor disproved directly, we can *never prove* the null hypothesis by rejecting the alternative hypothesis. The strongest statement we are entitled to make in this respect is that we *failed to reject* the statement that 'the microwave radio and ancillary equipment are operating correctly', or that we *failed to reject* the null hypothesis. Hence every test that we perform on a microwave link, and on its radio and ancillary equipment, is done *only in order to give the facts a chance of disproving that the link and equipment are operating correctly*, i.e. disproving the null hypothesis.

One purpose of this book is to lead the way to obtaining these facts by field tests. Before these tests can be performed, however, the purpose of the theory behind the tests, the equipment used to perform them, and of course the equipment on which the tests are performed, need to be known. Finally, how the equipment and system operate, as a single link, and as part of a larger system interacting with other subsystems, also needs to be known. In order to reach the point where we can intelligently gather the facts which either disprove or otherwise that the equipment and the system (in part or in total), are operational, we must start a long way back.

In this book some of the parameters that will be used throughout the text, at the telephone handset and throughout the system, will first be briefly considered. In order to place the microwave link and multiplex equipment in its proper perspective, how they fit into a large communications system and the role which they play (Chapter 2) will also be discussed. Chapter 3 moves away from the handset and into the next stage, where FDM is reviewed. After developing the theory and multiplex plans, and after describing the equipment, the tests required to be carried out on the equipment are dealt with, together with the test equipment necessary to perform these tests. Chapter 4 considers the subject of signaling. This is especially important to the transmission engineer in the field, for an understanding of it can save many hours of frustration during the end-to-end or loop-back tests. Chapter 5 is a theoretical treatment of thermal noise at the equipment level. This chapter is a necessity for later treatment of the subject. Chapter 6 considers how the output power levels and upper baseband frequencies of the multiplex equipment can be simulated, and Chapter 7 discusses the equipment and the measuring techniques used in this baseband power level simulation. Chapter 8 sets out the principles of line-of-sight (LOS) microwave radio link engineering. Chapter 9 deals with the antenna and with antenna systems. In Chapter 10, the theory of frequency modulation is developed and applied to radio equipment. There is some repetition in this chapter of points raised in Chapters 6 and 7. This is necessary to consolidate the information, and to permit Chapter 10 to stand alone, without continual reference to other chapters. Chapter 11 brings together all the previous chapters and deals also with the testing of the microwave radio system.

The radio and multiplex in-station and end-to-end tests, together with end-to-end system tests, produce facts. These facts disprove or otherwise that the equipment, subsystem, link or system is operating. These facts, or test results, must be considered against a reference before a judgment can be made on the equipment, etc. The references used, and which have been included in part throughout this book are the CCIR, CCITT and manufacturers' factory test results, together with results of experience on different systems. Obviously, in order not to reject the hypothesis that the equipment and link are operational, the subscriber must be able to use the system with a high degree of satisfaction.

Corrective maintenance also plays an important role in the field of microwave radio link engineering, as does preventive maintenance. The following is an algorithm, which possibly could assist in locating and rectifying any problem encountered.

1. A problem is recognized and defined, and from this an objective is formulated.

2. All the relevant data and information are collected and collated. This can be from preliminary tests, or from observation.

3. A working hypothesis or model is formulated from step 2.

4. Deductions and/or further tests are drawn from the model or hypothesis, so that the model or hypothesis can be further modified.

5. The deductions or tests are tried in the field, or if practicable in the laboratory.

6. Depending on the results, the model or working hypothesis is tentatively accepted, modified or discarded.

7. If the working hypothesis is tentatively accepted, further experiments are designed, or further deductions drawn, and these are then tested by trial in the field in order to increase the generality of the model or hypothesis.

8. The process of step 7 is continued until such time as the model permits the solution to be obtained, which satisfies the original objective. The additional information gathered may then be used to formulate further experimentation, applications, hypothesis, etc.

9. If the model or working hypothesis has to be modified, a return to step 5 is in order, and the process continued from there.

10. If the model or working hypothesis is discarded, then a return to step 3 is in order, and the process repeated and continued.

1 INTRODUCTION TO TRANSMISSION FOR TELEPHONY

INTRODUCTION

In speech communications the basic building block is the telephone channel. It is important that this channel provides an acceptable alternative to direct face-to-face conversation by conveying shades of meaning through variations in voice amplitude and inflection. The ideal telephone circuit is that which conveys the feeling of presence that exists in direct face-to-face conversation. Thus, for voice transmission through a transmission medium, the factors which must be considered are:

1. Frequency range
2. Audio dynamic range
3. Speech or signal power-level variations
4. Distortion
5. Frequency stability
6. Noise and signal-to-noise ratio

1.1 FREQUENCY RANGE AND AUDIO DYNAMIC RANGE

The frequency components of speech for a range of individuals may vary from as low as 50 Hz up to 20 kHz, while the intensity range (audio dynamic range) used by these different people may be as great as 60 dB – from the weakest syllable of a soft speaker to the loudest syllable of a loud talker. Within these ranges, it would be difficult to construct a multichannel communications system without restricting the ultimate channel capacity. As long ago as 1931 it was found[1] that most of the speech energy is concentrated in the lower frequencies (vowels), up to 700 Hz, and the higher frequencies (consonants) contribute mostly to the intelligibility, 500 Hz to 6 kHz (approximately). This means that any voice circuit should include both the low frequencies and the high frequencies. In practice a compromise is necessary because of the limitations on bandwidth. The CCITT defines[2] the standard bandwidth of a voice channel as 3.1 kHz, occupying the frequency band 300 Hz to 3.4 kHz. The working dynamic range is about 50 dB. This is sufficient, for in a normal conversation the dynamic range for the average person is about 30 dB.

In order to provide a guard band between voice channels during the frequency stacking process (see Chapter 3), and to permit out-of-band signaling to occur within each individual voice channel, the telephone channel or *normal* voice channel occupies the band 0 Hz to 4 kHz.

1.2 SPEECH OR SIGNAL POWER VARIATIONS

During transmission, owing to the transmission medium, the loss in signal power is not constant, but varies with time. This can be due to changes in circuit loss caused by temperature variations, power supply fluctuations, etc. Unless some means is used to control the variations in level, there will be signal level variations at the receiver. It is desirable to keep level variations occurring over a short time interval to a minimum.

1.3 DISTORTION

Distortion is the general term used to describe any change in the output waveform relative to the input waveform as it passes through a transmission medium. Distortion may reduce the intelligibility and identification of the speaker. The three basic types of distortion are:

> Frequency distortion
> Non-linear distortion
> Delay distortion

1.3.1 Frequency Distortion

This type of distortion is also commonly known as *attenuation distortion*. It is the selective attenuation of some frequencies or some bands of frequencies over the whole of the spectrum in use. If all the frequencies over the whole passband suffered the same attenuation or gain this form of distortion would not be apparent. If frequency distortion appears within a voice channel and is excessive, the effect becomes noticeable. Where low frequencies are greatly attenuated, the resulting speech will sound 'tinny', whereas if the higher frequencies are greatly attenuated, the speech will sound 'drummy'. Frequency distortion is measured against a reference frequency, which is 800 Hz for CCITT circuits, and 1 kHz for US circuits. It is one of the more important forms of distortion to consider when testing end-to-end baseband signals of a microwave link. As the baseband bandwidth is usually far greater than that of a voice channel, since it carries the output signal from the multiplex equipment, its reference frequency is not 1 kHz, but usually the pre-emphasis crossover frequency. Hence 0 dB is set as the reference level at this crossover frequency. Select frequencies are then chosen, and one by one sent over the link. The receiving end of the link then records the level with respect to the reference. The difference is recorded in decibels and if the signal level is higher than the reference, the difference has a positive value – if lower, a negative value. Usually there is a requirement set by the manufacturer, or some authority, e.g. for frequencies not greater than 4 kHz, the difference should be in the range +0.5 dB to −3 dB. For frequencies 12 kHz and above, the difference should be in the range +0.5 dB to −0.5 dB. If the link does not reach these specifications, then it means that one of the voice channels when demodulated will suffer degradation in signal quality, and may not be usable.

1.3.2 Non-linear Distortion

This is caused by non-linearities in the circuits. It is characterized by the generation of harmonics which are multiples of the speech frequencies being transmitted. In addition, these harmonics may mix with themselves, or with the fundamentals, causing intermodulation products. Because of the possible interfering effects of these new frequencies with other frequencies in use, non-linear distortion must be kept to a minimum in radio transmitter design. When this form of distortion occurs in the multiplex, or baseband of the radio, the resulting frequencies cause inter-channel interference, known as intermodulation noise. Later when dealing with the subject of noise power ratio (NPR), it will be seen how to measure this form of distortion, and isolate the equipment which may produce it.

1.3.3 Delay Distortion

This form of distortion arises as the result of differences in the velocity of propagation of the various frequencies which comprise a complex wave. As a phase shift occurs, delay distortion is also termed *phase distortion*. In a band-pass filter the effect of changing the velocity of propagation through it is more pronounced at the filter edges than at the center frequency of the filter. Thus phase distortion is more pronounced at frequencies close to the edge frequencies of the filter. When the design of the RF bandwidth is considered in Chapter 11, it will be seen that the IF filter is designed so that its bandwidth is greater than the expected RF bandwidth. This has been done to reduce the effect of delay distortion on IF frequencies which would have been close to the IF filter edge.

For a speech circuit, delay distortion is not a problem because the ear is relatively insensitive to phase variation. However on voice-frequency circuits used for high-speed data transmission, as in the computer aided operations (CAO) circuits used in monitoring equipment at outstations, delay distortion is important, if the highest possible bit rate is to be realized. As delay distortion varies over the band, due to the non-linear phase-frequency relationship, it is best measured by a

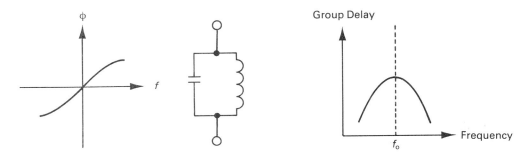

Figure 1.1   Phase Characteristic and Group Delay for a Single Tuned Circuit

parameter known as *envelope delay distortion* (EDD), or *group delay distortion* (GDD). Mathematically, envelope delay (otherwise known as 'group delay') is the derivative of the phase shift with respect to frequency. This derivative may be plotted against frequency as shown in Figure 1.1. The difference between the group delay at any specified frequency, and the group delay at the frequency where the group delay is a minimum is the EDD. Therefore EDD is always the maximum difference between the envelope delay at one frequency and that at another frequency where the envelope delay is a minimum in a passband. The EDD unit is the milli-# or micro-second. Hence, mathematically;

GD (Group delay) = ED (Envelope delay) = $d\phi/df$

$$= \frac{\text{Instantaneous phase difference}}{\text{Instantaneous frequency difference}}\ \mu s \qquad (1.1)$$

EDD (Envelope delay distortion) = $[d(GD)]_{max}\ \mu s \qquad (1.2)$

The effect of excessive amounts of group delay distortion at the upper and lower edges of the transmitted band can be described as 'ringing' and 'blurred speech', respectively. In the absence of noise or attenuation distortion the effect is conspicuous throughout the entire range of typical loudness loss values. However, the effect in a typical four-wire circuit chain is usually not serious since the group delay distortion is normally accompanied by closely related attenuation distortion, which tends to reduce the effect. The current performance objectives for group delay distortion for a world-wide chain of 12 circuits are given in CCITT Recommendation G133.

The envelope delay is produced by IF and RF circuits, such as transformers, filters, cavities, mixers and convertors, i.e. by all circuits which are bandwidth-limited. Unequalized group delay characteristics tend to be predominantly linear positive or negative, or parabolic positive or negative. Figure 1.2 shows a positive parabolic group-delay equalizer and Figure 1.3 shows group delay. This 'all pass network' would be used to equalize a network which has a negative parabolic group

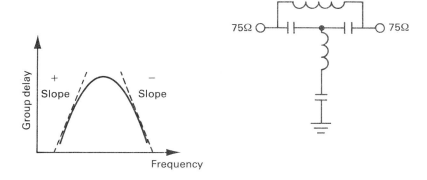

Figure 1.2   Positive Parabolic Group – Delay Equalizer

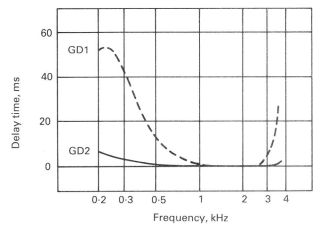

GD1 – Approximated to 12-circuit chain (95 per cent values)
GD3 – Approximated to typical modern one circuit value

Figure 1.3   Group delay (Courtesy of CCITT. Figure B1/P11 from Reference 5)

delay. This network can also be used to equalize a linear positive or negative group delay by expanding the response, and choosing the appropriate position of the curve.

1.3.3.1 Measurement Techniques for Non-Linearity and Group Delay
A dynamic sweep method is used to test linearity and group delay. For example, a 50 Hz signal whose amplitude is proportional to the deviation desired in the microwave system is applied to the modulator of the transmitter. At the receiving site, or from the receiver if in the loop-back test mode, the signal is obtained at the receiver demodulator output. The dynamic linearity test curve is obtained by using a low-frequency sweep and a high frequency superimposed on it. This is shown in Figure

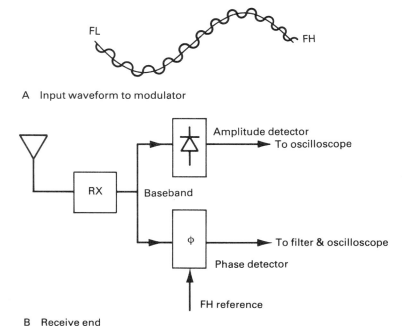

Figure 1.4   Basic Test Set-up for Measuring Linearity and Group Delay

1.4. The low frequency FL (50 Hz) is used to sweep the frequency of the modulator unit under test over the required peak-to-peak deviation. A high frequency FH (200 kHz) is superimposed on the low frequency. FH also modulates the carrier. The amplitude of FH undergoes non-linear distortion, depending upon the non-linearity of the modulator and demodulator. At the output receiver demodulator, FL is discarded, and the amplitude of FH is measured over the maximum-to-minimum range of the FL sweep. This amplitude is displayed in decibels on an oscilloscope and compared with the input signal.* The difference is a measure of the non-linearity of the transmission medium. By changing the value of FH to various frequencies, the system performance can be evaluated and linearities at various modulating frequencies can be determined.

Group delay is measured by passing the demodulated 200 kHz signal through a phase detector. The phase is compared with a reference 200 kHz phase, resulting in a DC voltage after filtering, which is a measure of the phase difference. This output DC signal can then be displayed on an oscilloscope and calibrated in microseconds. The experiment is repeated using a different FH each time until a group delay over the passband of the transmitter and receiver is determined. Note that the 50 Hz and 200 kHz signals are chosen to demonstrate the principle only. FL will depend on the lowest baseband signal the radio can take, and FH will range up to the highest. Figure 1.4 shows the test set-up.

---

* For a discussion of the decibel see Section 1.7.1

1.4 FREQUENCY STABILITY

The use of oscillators in the channel and group modems of multiplex equipment introduces the problem of frequency stability. In telephone circuits the frequency stability that is of concern is the overall change in frequency which occurs between the transmitting and receiving ends of the circuit. The amount of frequency change tolerable is directly related to the amount of change discernible to the ear, or the amount of change sufficient to impair operation of signaling, or data transmission. For voice circuits, a frequency stability (end-to-end) of approximately $\pm$ 2 Hz provides a very good circuit. In actual operation, frequency shifts approaching ±15 Hz may occasionally occur without impairing seriously the quality of a voice circuit. However the use of signaling equipment and data transmission prevents such excursions.

1.5 NOISE AND SIGNAL-TO-NOISE RATIO

Without noise in a communications circuit or system, there would be no challenge in circuit or system design. Noise can be considered as the unwanted transmitted or received signal occurring in a communications circuit. Thermal noise, crosstalk, and impulse noise only will be considered under this heading, and other forms of noise such as 'echo distortion' noise, and interference, will be dealt with later, as the need arises.

1.5.1 Thermal Noise

This type of noise will be considered in depth in Chapter 5, and as necessary in the following chapters. *Thermal noise* or *Johnson noise* is due to the random motion of the free electrons in a conductor caused by thermal agitation. This gives rise to a voltage at the open ends of the conductor. In most conductors, the frequency components of this noise voltage cover the complete radio-frequency spectrum uniformly. Thermal noise, due to its broad-band nature, is known as *white noise*, since white contains all of the colors of the visible spectrum. *Shot noise*, which occurs in vacuum tubes, is also due to the random manner in which electrons leave the cathode, generating a noise current and hence a noise voltage. Semiconductors also release large numbers of electrons randomly, and the resulting noise is called *flicker noise*. If a signal current in a conductor is smaller than or equal to the random noise current due to thermal agitation, the signal is masked by the noise, and no amount of amplification will separate them. Thus thermal noise is the factor which sets the lower limit for receiver sensitivity. Available thermal noise power is directly proportional to absolute temperature (kelvins), and bandwidth as given by the expression:

$$P_{nav} = kTB \qquad\qquad (1.3)$$

where P_{nav} = the available noise power in watts
$\quad\quad k$ = Boltzman's constant = 1.3803×10^{-23} J/K
$\quad\quad T$ = the absolute temperature in kelvins
$\quad\quad B$ = the noise bandwidth (usually of signal processing circuits) in Hz

Assuming that the noise bandwidth for the moment is 1 Hz, and that the temperature is 17°C, the noise power P_{nav} can be expressed in dBm/Hz of bandwidth.*

$$P_{nav} = -174 \text{ dBm/Hz} \qquad\qquad (1.4)$$

For other values of B and T, we have

$$P_{nav} = (-174 + 10 \log B + 10 \log T) \text{ dBm} \qquad\qquad (1.5)$$

Again B is in Hz, and T is in kelvins.

1.5.2 Crosstalk

Crosstalk is intelligible or unintelligible noise that is induced into a circuit from another circuit very close to it. Crosstalk may be produced by inductive and/or capacitive coupling in parallel lines. The inducing circuit is known as the *disturbing circuit*, and the circuit in which the crosstalk occurs is the *disturbed circuit*. Three types of crosstalk are considered to exist. These are:

1. Intelligible crosstalk, which is in the same frequency range, but lower in amplitude than the wanted signal. Usually crosstalk is considered intelligible when four or more words can be understood from the disturbing circuit in a seven-second period.

2. Unintelligible crosstalk, which is translated in frequency, or appears in the disturbed circuit in an inverted form.

3. Babble, which is crosstalk from a number of sources, either intelligible or unintelligible. The noise appears as if many people are talking due to a number of interfering signals, but is not intelligible. It usually occurs during busy periods.

When a signal is transmitted along a wire, a varying magnetic field exists which is induced into nearby parallel conductors causing current to flow and crosstalk to result. Similarly the electric field produced by the signal in the disturbing circuit

* For explanation of dBm and dBW see Sections 1.7.2 and 1.7.3.

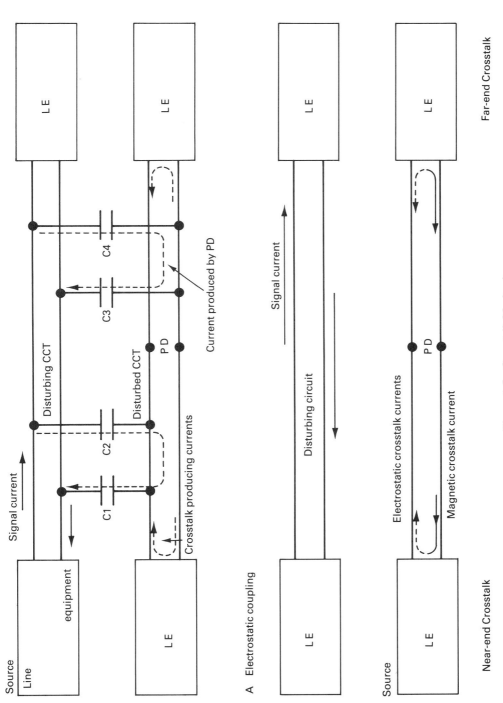

Figure 1.5 Crosstalk Mechanisms

produces electrostatic coupling, which can be represented by distributed capacitors along the length of the conductor. The size of the capacitances will vary due to the different spacings between the conductors of the disturbing and disturbed circuits. Due to the potential difference between these conductors, current will flow and the result is crosstalk. The effects of magnetic and electrostatic coupling should be combined and considered together, because the signal in the disturbing circuit produces both fields simultaneously in the form of an electromagnetic wave, which cuts the disturbed circuits. Figure 1.5 shows the induced currents in the disturbed circuit due to magnetic and electrostatic coupling. The effects of these combined currents is to produce;

1. A high level of crosstalk at one end of the disturbed circuit, because the currents at that end are in the same direction and additive.

2. A lower level of crosstalk at the other end of the disturbed circuit, because the currents are in opposition and only the difference between them produces the crosstalk. (Induced electrostatic current is in opposition to induced magnetic current.) The amount of crosstalk induced by magnetic and electrostatic coupling varies with the frequencies being transmitted over the disturbing circuit, and the spacings between the conductors of the circuits. Because crosstalk increases with frequency, transmission lines must be designed and constructed accordingly to assist in its reduction. In multipair cable, the general principle is to arrange the cables so that the induced voltages in each side of the disturbed circuit are equal and opposite, and produce a zero resultant voltage. One arrangement is to arrange each of the wires of two pairs in the 'star-quad' or 'quad' configuration. Here each wire is located at the corner of a square, and one pair of wires is located at the corner of a square, and one pair of wires for one circuit is perpendicular to the other pair used in the other circuit. A multiconductor cable contains many pairs of conductors arranged in star-quad form. Transpositions are provided to reduce crosstalk between neighbouring quads, and each layer of quads is spiraled in opposite directions around the core of the cable. Another way to overcome this problem is to use shielded twisted-pair cable, where the numbers of circuits are few.

1.5.3 Impulse Noise

Unlike thermal noise, impulse noise occurs in bursts and is not spread uniformly through the frequency spectrum. Some impulses are produced by natural causes such as lightning, etc. A large number of causes of this form of noise are man made, such as induction from traction systems, car ignition systems, power lines, etc. The major effect is on the transmission of data. Telephony effects are crackling over the line, or a single click.

1.5.4 Signal-to-Noise Ratio

Noise is often expressed as a signal-to-noise ratio S/N. This means that the difference is expressed in dB between the absolute nominal signal level (in dBm, dBW, etc.) and the noise level (also in dBm, dBW, etc.) at the same point in the circuit. Both signal and noise must of course be expressed in the same absolute units. The most important feature of signal-to-noise ratio is that it is dimensionless, and is expressed only in dB, no matter what the absolute logarithmic units are. This is because S/N is a relative level between the signal and the noise, i.e. it expresses how high the signal is above the noise. One criterion for a voice channel at its audio output is that the signal-to-noise ratio is 30 dB. If the noise level is measured to be -23 dBm, then:

$$(S/N)\text{dB} = 10\log(S/N) = 10\log S - 10\log N \qquad (1.6\text{A})$$

where S = the signal power level in milliwatts
N = the noise power level in milliwatts

If S and N are expressed in absolute units such as dBm or dBW then;

$$(S/N)\text{dB} = (S-N)\text{dBm} = (S-N)\text{dBW} \qquad (1.6\text{B})$$

This gives
$$30 = S - (-23)$$
$$\text{or } S = +7\,\text{dBm}$$

It should be pointed out, however, that the above calculation is correct only if the frequency range in which the tests are performed is specified. Outside of this range the signal-to-noise ratio may be something else. In the above example it has been implicitly assumed that it is the audio range of the voice channel, i.e. $0-4\,\text{kHz}$. Another practical point is that the measuring instrument must be designed to handle the signal frequency. If its operating frequency range is below that of the signal frequency, the results will be erroneous, because the noise level may be constant over a frequency range up to and above the signal frequency; hence as the signal level is attenuated by the measuring instrument, the noise power will not be. This is one cause of measurement error.

1.6 WEIGHTING THE VOICE CHANNEL

Up to this point it has been considered, without actually stating it, that if there was a test signal of constant level, which was sent into the voice channel, and then its frequency swept over the voice band, the signal coming out of the receiving voice channel would, although at a different level, be constant over the swept frequency. In practice, there would be small variations, according to the frequency response specifications of the channel, but, regardless of this, the response of the channel over the voice band would be considered 'flat'. Hence when the channel is referred to as being 'flat', it means that it is *not* weighted. If a telephone handset transmitter (with battery) was connected to the transmit input of the four-wire voice channel, and a telephone handset receiver connected to the four-wire output of the far-end voice channel, with a test-tone acoustic signal at a fixed level, again frequency-swept over the voice band, being sent, the response would not be flat at the receive end. The response would in fact be shaped by the handset. The frequency response reference used when performing this test is located at the minimum attenuation frequency in the voice band. For telephones designed to CCITT standards, the reference frequency used is 800 Hz, whereas for those designed in North America, and not to CCITT standards, the reference frequency is 1 kHz. Although this shaping effect does reduce the speech signal power through the telephone link, it reduces the noise power more, hence there is an improvement in signal-to-noise ratio. For all voice-channel measurements affecting speech, it must be noted that certain frequencies in the voice channel are attenuated more than others. In the testing of speech communication links, weighting networks are used to simulate these effects, and not the crude acoustically coupled test as outlined above for the purpose of explaining the concept. Different types of telephone handsets have different attenuation/frequency characteristics. The types* commonly encountered are:

144	Line weighting (North America); noise unit is the dBrn	Seldom encountered these days
FIA	Line weighting (North America); noise unit is the dBa	Now possibly phased out
C-Message	C-Message weighting (North America); noise unit is the dBrnC	Presently used
CCIR–1951	Psophometric weighting (CCITT/ CCIR); noise unit is the pWp	Presently used in countries with CCITT or CCIR standards

The reference frequency was established at the point where the reference signal level was just discernible by the ear. This level, depending upon the handset, is in the range −85 to −90 dBm. Thus the derived noise units would be positive numbers, and would depend on the zero reference and the weighting characteristics as referred to in one of the four telephone handsets mentioned above. The weighting characteristics are shown in Figure 1.6.

These derived noise units associated with weighting networks were based on two factors:

1. Subjective tests, where listeners compared the degree of interference of various frequencies in the voice band with a standard 1 kHz (or 800 Hz) tone, both with and without voice being present.

2. The frequency response of the telephone receiver being used at the time.

* For explanations of dBa, dBrnC and pWp see Sections 1.7.6, 1.7.7 & 1.7.8.

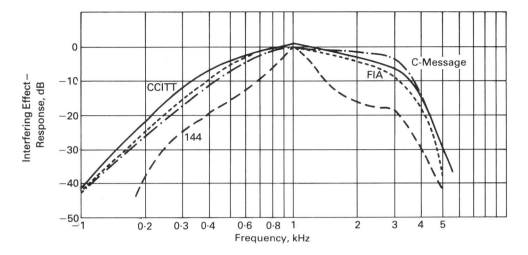

Figure 1.6   Line Weighting Curves

From the curves in Figure 1.6 it can be seen that each succeeding telephone receiver as listed above had an improved frequency response, resulting in a smaller difference between the power of a 1 kHz tone measured through a network and the power of a 3 kHz band of thermal noise measured through the same network. This can be intuitively understood if the zero dB point represents the noise power -174 dBm/Hz, and the -10 dB point the noise power -184 dBm/Hz, etc. For a 3 kHz band of noise, the power would be approximately -139 dBm. If the response of each curve is considered as a window to noise between the curve and the frequency axis the noise power can be calculated by starting at the top of the curve and determining the power level for a particular frequency range, then similarly working down the curve. For example, taking the 144 curve between 1000 and 1200 Hz the level is -174 dBm/Hz; therefore the power is $-174 + 10 \log 200 = -151$ dBm $= 8 \times 10^{-16}$W. Next consider the area bounded by -184 dBm/Hz and the frequency ranges 700 Hz to 1000 Hz, and 200 Hz to 1.2 kHz, the noise power is $-184 + 10 \log (1000 - 700) = -159.2$ dBm, and $-184 + 10 \log (1200 - 200) = -154$ dBm. To add these values we must first convert to non-logarithmic power. After doing this and adding, the resultant power is 5.18×10^{-16}W. This power is added to the first calculation to give 13.18×10^{-16}W. The process is then repeated until all the band, 180 Hz to 3 kHz, has been covered. The smaller the bands chosen the more accurate will be the final result. The total noise power through the 144 mask should be -147 dBm. The results of the quick calculations above show -148.8 dBm. The reason for the closeness is that the power is falling logarithmically as we move away from the 1 kHz frequency. If the FIA Line weighting network is chosen, the difference is 3 dB, and with the C-Message weighting network, this difference is 1.5 dB. Finally with the CCITT psophometric weighting network the result is 2.5 dB.

The first noise-measurement unit was the 'dBrn' which meant dB above reference noise. The reference was a −90 dBm, 1 kHz tone. This meant that, using this handset, a listener could just discern a 1 kHz tone at a level of −90 dBm. The second unit was the 'dBa' (or dBrn adjusted). Alongside the development of each of the telephones was the development of noise-measuring tests sets. The 2A test set incorporated the 144 line weighting and the 2B incorporated the FIA line weighting. In order to make the 2A and 2B read the same numerically for one milliwatt of thermal noise 3 kHz wide (82 dB), the reference for the dBa was set at −85 dBm for a 1 kHz tone. When the C-Message weighting handset came into being, and produced only a 1.5 dB difference between a 1 kHz test tone and 3 kHz of thermal noise, it was decided to go back to the −90 dBm, 1 kHz test tone reference, the unit being named the 'dBrnC'. Europe had however adopted a different frequency weighting curve. The measuring set was called a psophometer, with a reference frequency of 800 Hz, at −90 dBm.

To convert from flat channel noise measurements in dBm, CCITT recommendation G223 recommends that 'If uniform-spectrum random noise is measured in a 3.1 kHz band with a flat attenuation frequency characteristic, the noise level must be reduced 2.5 dB to obtain the psophometric power level. For another bandwidth B the weighting factor will be equal to

$$2.5 + 10 \log(B/3.1) \text{ dB where B is in kHz'}.$$

The relationships between each of the above mentioned noise units will be given later in the chapter.

1.7   LOGARITHMIC UNITS OF MEASURE

Assume for the moment that an audio tone with a power of ten watts from a loudspeaker can be heard. Before any increase in the intensity of the sound can be discerned, it would have to be turned up to approximately 12.5 W. Again before any change in intensity could be discerned, the power would have to be increased by approximately another 3.1 W, to bring the output power to 15.6 W. In other words the volume of a sound must change by approximately 25 per cent before the ear can detect the difference. These changes are logarithmic, hence it follows that the response of the ear is logarithmic in nature. Because communications circuits were originally developed for people, any practical unit used to express signal power gains or losses should also vary logarithmically. The transmission unit used for the comparison of electrical powers at two different points in a circuit is the *bel* – a name given in honour of Sir Alexander Graham Bell, the inventor of the telephone.

The bel is defined so that two electrical powers differing by a factor of 10 differ by one bel. In general the ratio of two powers in bels = log (power ratio) where 'log' is the logarithm taken to the base 10.

Thus, if the output power of say an amplifier is 10 W, and the input power is 1 W, the power ratio is 10, and the number of bels is one. Thus when the output power is ten times that of the input power, the output is 1 bel above the input. If the output is one-tenth of the input power, then the output is 1 bel below the input, and an attenuator has been described.

1.7.1 The Decibel (dB)

We return to the discussion on the logarithmic response of the ear in Section 1.7; the power ratio which the ear can just discern is 12.5/10, and its value in bels is given by log (12.5/10 W) = log (1.25) = 0.097 bel. This is approximately 0.1 bel. Thus it appears that to follow the logarithmic response of the ear, the bel is ten times too large. Because of this factor, the decibel came into being, which is defined as one-tenth of a bel. Hence;

Power ratio in decibels = 10 log (power ratio), or

$$N = 10 \log (P_o/P_i) \text{ dB} \tag{1.8}$$

where P_o = the output power of an amplifier or attenuator, or is an absolute level generally.
P_i = the input power of an amplifier or attenuator, or is an absolute level generally.
N = the number of decibels in the power ratio.

It is important to note that the decibel is a unit of power ratio, and does not represent a particular power level, i.e. decibels are not absolute units like dBm, or dBW, but are a measure of a relative level between two absolute units. If the value in decibels is negative, it represents an attenuation figure, which is the number of decibels below some absolute power level to where another absolute power level is to be found.

Frequently, it is convenient to know the increase or decrease in decibels, by a change in non-logarithmic units of power, such as watts, or milliwatts. The rules of thumb for quick calculations are: if the power doubles, add 3 dB; and if the power halves, subtract 3 dB.

Example applying the rules of thumb. If the power through an amplifier has increased from 8 W to 120 W, how many decibels has the power increased by?

To calculate the power ratio simply, multiply 8 repeatedly by 2, e.g. 8 × 2 = 16 (once), 16 × 2 = 32 (twice), 32 × 2 = 64 (thrice), 64 × 2 = 128 (four times). As 128 is a little above 120, because the next number is 256, it can be seen that we have doubled up 4 times, or 4 × 3 dB = 12 dB. Our answer should be just under 12 dB; consequently it could be said: '120 W represents an increase of approximately 12 dB above 8 W'. The exact ratio, found by calculator, is 11.76 dB.

Similarly, it can be found how many decibels are lost if the input power is 10 W, and the output power is 0.1 W. Thus, 0.1 × 2 = 0.2 (1), 0.2 × 2 = 0.4 (2), 0.4 × 2 = 0.8 (3), 0.8 × 2 = 1.6 (4), 1.6 × 2 = 3.2 (5), 3.2 × 2 = 6.4 (6), 6.4 × 2 = 12.8

(7). Now $12.8 > 10$, but the next number is 25.6. Mentally linearizing, it is probably about 0.5 dB too high; hence $7 \times 3 - 0.5 = 20.5$ dB. Thus, the approximate value of power that is lost is 20.5 dB. Using a calculator, the power that is actually lost is found to be 20 dB. Of course an easier way would have been to recognize the fact that there is a factor of 100 between 10 and 0.1 W which gives $10 \log 100 = 20$ dB.

1.7.2 The dBm

The dBm is the decibel referred to 1 mW. A power P_0mW is equivalent to a power of $10 \log (P_0/1)$ dBm, or:

$$\text{Power in dBm} = 10 \log (P_0/1), \text{ where } P_0 \text{ is measured in mW} \tag{1.9}$$

Adding the reference power of 1 mW makes the dBm a measure of absolute power rather than just a ratio as with the dB. The dBm is the most commonly used reference power unit in the field of telecommunications.

Example What is 22 W expressed in dBm?
22 W = 22 000 mW, and from equation 1.9,

$$22 \text{ W} \Rightarrow 10 \log (22\,000/1) = 43.42 \text{ dBm}$$

A 30 dBm (1 W) signal applied to an amplifier with 10 dB gain results in a 40 dBm output, while a standard 0 dBm (1 mW) test tone would be measured at -16 dBm after passing through a 16 dB attenuator.

Although the unit dBm expresses absolute power, it has little meaning in a transmission system unless the gain or attenuation at the measurement point is also specified. Gain and attenuation similarly have significance only if a reference level is defined for the system, allowing any point to be described in terms of dB above or below the reference. The point at which the reference appears is designated the *zero transmission level point* or 0 TLP. The 0 TLP, which is also known as a *reference transmission level point*, is defined as the point in a system at which a standard 0 dBm test tone would have an absolute value of one milliwatt. All other points throughout the system in its entirety is referred to the 0 TLP, by the algebraic summation of gains or attenuation from 0 TLP to the point being considered.

1.7.3 The dBW

The dBW is the decibel referred to 1 W.

$$\text{Power in dBW} = 10 \log (P_o/1), \text{ where } P_o \text{ is measured in watts} \tag{1.10}$$

Like the unit dBm, the unit dBW is an absolute unit of power. To convert from dBW to dBm, just add 30 to the dBW value, and the result is the value of the power in dBm:

$$\text{Power in dBm} = \text{Power in dBW} + 30 \tag{1.11}$$

Example 1 W is equivalent to $10 \log (1/1) = 0$ dBW; thus 1 W $= (0 + 30)$ dBm $= 30$ dBm.

7 W is equivalent to $10 \log (7/1) = 8.45$ dBW $= (8.45 + 30)$ dBm $= 38.45$ dBm. $(8.45 + 30) = 38.45$ dBm.

Similarly, to convert from dBm to dBW, subtract 30 from the dBm value.

1.7.4 The dBm0

As mentioned previously, all levels in a communications system are referred to the zero transmission level point. If some of these levels are at 0 dBm, then these are additional 0 TLP points. The system levels are referred to 0 TLP by algebraically adding gains or subtracting losses from this point to the point where the level is considered. The point to where the level is being considered is referred to as the *relative transmission level*, or the 'dBr point'. A 0 TLP is thus a zero dBr point. At a -30 dBr point there has been a 30 dB loss between it and the 0 TLP point; so, if the absolute level was measured, as long as the 0 TLP point was at 0 dBm, -30 dBm would be read. When 0 TLP is at 0 dBm, there are no problems, because instruments should read the dBr levels throughout the system directly in dBm. In an actual case, however, something different occurs; the managers of a communications system make a policy that, when the voice channels are being tested on a 'live' system, they should not be tested with a test tone level referred to 0 dBm0, but referred to -16 dBm0. The reasons given are; that a test tone entering the two-wire or four-wire circuits at 0 dBm0 does not represent the average speech power of an individual, and would permit crosstalk to be produced by the high level test tone on the channel under test, which would affect the other working circuits. In this case the power measured at a -30 dBr point would be $(-30 - 16)$dBm or -46 dBm.

In 1964 the CCITT (Recommendation G101 Section B) adopted the four-wire sending point level as -3.5 dBr and receiving point level as -8.5 dBr. They did not, however, define the exact location in the four-wire circuits where these levels were to appear, as they considered this unnecessary. Some systems in use adopt the sending level as $+7$ dBr and the receiving level as -16 dBr. These levels are located directly at the four-wire terminals of the multiplex equipment.

If the 0 TLP is not 0 dBm (0 dBm0) but -16 dBm (-16 dBm0), then all of the dBm levels as measured at the dBr points, or relative level points throughout the system, also have been dropped by this amount. Hence the input level to the transmit of the four-wire channel modem will be $+7 - 16 = -9$ dBm, and the level coming out of the receive of the channel modem will be $-16 - 16 = -32$ dBm.

The 0 in dBm0 means a power level referred to zero transmission level point. Thus;

$$dBm = dBm0 + dBr \qquad\qquad (1.12)$$

where dBm is the actual measured power using instruments, at the dBr points.

dBm0 is the actual measured power level using instruments at the 0 TLP point.

dBr is the relative level at any point in the system, referred to the 0 TLP point.

The dBm0 unit is particularly useful where levels other than telephonic levels are to be used. Because of the high power content of a digital signal, it may be prudent not to send it over a telephone circuit at 0 dBm0 or test tone levels, but to send it at, say, −15 dBm0. Knowing the value of the dBm0 level means that a check can be made throughout the complete system, at dBr points for the correct level, to ensure the circuit is operating as it should.

The 0 occurs on other units besides the dBm. It can be used with the same inference on noise units such as the dBrnC, as dBrnC0 (pronounced dabrenko); dBa, as dBa0; and as pWp0 or dBmp0.

(For the sake of completeness, and easy reference, the most commonly occurring noise units are defined below, in addition to the units defined in Sections 1.7.1 – 1.7.4.)

1.7.5 The dBrn

A unit of measurement for noise interference was first developed by taking into account the capabilities of the ear and the telephone. A 'weighting curve' was thus produced by graphing the test results from the ear and telephone combination; let us consider the 144 curve as shown in Figure 1.6. This weighting curve indicates how much each individual frequency in the voice band will interfere with a conversation, compared with how much a reference frequency of 1 kHz will interfere . The reason why 1 kHz was chosen was because it was found to produce more interference to a conversation than any other voice-band frequency. The 144 line weighting curve shown in Figure 1.6 indicates that at 700 Hz the response is approximately − 10 dB, or 10 dB below that at 1 kHz. This means that, if the level of 700 Hz was at the same level as 1 kHz, it would produce 10 dB less interference to a conversation than the 1 kHz tone would.

As any tone or noise combined with a conversation has an interfering effect, the level at which this interfering effect becomes negligible must be determined. The level of −90 dBm was decided as the level at which an interfering frequency of 1 kHz (worst case) became negligible. This level of −90 dBm thus became the noise reference. Any noise power capable of producing interference then would have to be above −90 dBm, or the noise reference, and could therefore be expresssed in 'dB above reference noise', or 'dBrn'. For example, if a noise measurement showed that the level of noise was 45 dBrn, this would indicate that the level of noise was 45 dB higher than the reference, or at a power level of −90 + 45 = −45 dBm. Hence:

$$PdBrn = (P-90) \text{ dBm at 1 kHz} \qquad\qquad (1.13A)$$

$$PdBrn = (P-82) \text{ dBm 3 kHz noise bandwidth} \qquad\qquad (1.13B)$$

Points to Note
1. The North American Western Electric 144 handset was used to determine the 144 line weighting curve shown in Figure 1.6.
2. The reference frequency in North America is 1 kHz.
3. The reference level as determined from the 144 handset was −90 dBm.
4. dBrn means dB above reference noise level, and is given by equations 1.13.
5. If a noise band of 3 kHz is used, the reference level is −82 dBm.

1.7.6 The dBa

After the 144 handset was in use, another handset was developed in North America. This is called the FIA. The listener tests were repeated until a 1 kHz tone was barely discernible, and so on for other frequencies in the voice band. The results of these tests produced the 'FIA line weighting' curve shown in Figure 1.6. It was found that the level at which the 1 kHz tone was barely audible was not at −90 dBm, as in the case of the 144 line weighting network, but at −85 dBm, implying a 5 dB difference between the two telephone handsets. Rather than change the existing standard from −90 dBm to −85 dBm, it was decided to establish a new unit of measure, the reference level being −85 dBm. The new unit of noise measure was the 'dBa' or 'decibels above reference noise adjusted', the reference frequency again being 1 kHz. Thus if a noise measurement was found to be 45 dBa, it meant that the noise level tone of 1 kHz was 45 dB above noise adjusted, or 45 dB above −85 dBm, giving a level of −40 dBm. Hence:

$$q \text{ dBa} = (q - 85) \text{ dBm at 1 kHz} \tag{1.14A}$$

$$q \text{ dBa} = (q - 82) \text{ dBm 3 kHz noise bandwidth} \tag{1.14B}$$

Points to Note
1. The North American Western Electric FIA handset was used to determine the FIA line weighting curve shown in Figure 1.6.
2. The reference frequency is 1 kHz.
3. The reference level at 1 kHz is determined from the FIA handset to be −85 dBm.
4. dBa means decibels above noise reference level adjusted, and is given by equations 1.14.
5. If a 3 kHz band of white noise is used, the reference level becomes −82 dBm.

1.7.7 The dBrnC

In 1960 another handset was put into service. This was called the 500-type telephone handset. The resulting weighting curve, based again on listener tests, was called the 'C-Message weighting'. Since there had been an improvement in this handset resulting in only a difference of 1.5 dB between a 1 kHz tone and a 3 kHz band of noise, rather than create a new reference, the old reference level of −90 dBm at a 1 kHz test-tone reference was again chosen. The new measurement unit was called the 'dBrnC', or 'decibels above reference noise, C-Message weighted'. If then a noise measurement resulted in a value of 22 dBrnC, this would mean that the noise level as measured through a C-Message weighting network in the measuring instrument, and thus referred to a 1 kHz test tone, would be 22 dB above reference noise, equivalent to $(22 - 90) = -68$ dBm.

Note that in the examples given in Sections 1.7.5, 1.7.6 and 1.7.7, it has been assumed that the noise measurement was made through the appropriate weighting network. In the testing of multiplex voice channels the tests are usually done out of a 'flat' channel, and unless the measuring instrument reads directly into dBa or dBrnC, the instrument will not have the appropriate weighting network, and the measurement will be 'flat'.

Because the 1 kHz reference is a single tone, it has all its power concentrated at one point in the voice band. Thus a 0 dBm 1 kHz tone will produce an interference level of 90 dBrn, 85 dBa, or 90 dBrnC. With a 3 kHz band of white noise, the noise power is evenly distributed throughout the 3 kHz voice band, and the level of interference is reduced. The interfering level for a 3 kHz band of white noise in the voice band, at a level of 0 dBm becomes: 82 dBrn, 82 dBa, or 88.5 dBrnC. Equations 1.14B, 1.15B and 1.16B reflect this by the change in the reference level of each unit. Hence if measurements are made using a 1 kHz test tone the reference level is -90 dBm for the dBrn and dBrnC, and -85 dBm for the dBa. If the measurements are made out of a voice channel with 3 kHz of thermal noise, the reference levels become -82 dBm for the dBrn and dBa, and -88.5 for the dBrnC. Thus when measurements are taken in white noise circuits, or flat channels, the reference noise power must be changed accordingly.

$$r \text{ dBrnC} = (r - 90) \text{ dBm at 1 kHz} \tag{1.15A}$$

$$r \text{ dBrnC} = (r - 88.5) \text{ dBm 3 kHz noise band} \tag{1.15B}$$

Remembering that the difference in noise reference power in a 3 kHz band between the dBrnC (-88.5), and the dBa (-82), is 6.5 dB, a conversion factor between dBrnC and dBa can be established. The generally accepted conversion formula is;

$$r \text{ dBrnC} = (r - 6) \text{ dBa} \tag{1.16}$$

Points to Note
1. The North American 500-type handset was used to determine the C-Message weighting curve as shown in Figure 1.6.
2. The reference frequency is 1 kHz.
3. The 1 kHz reference level as accepted for the C-Message weighting network is -90 dBm.
4. dBrnC means decibels above reference noise, C-Message weighted, and is given by equations 1.15.
5. If a 3 kHz band of white noise is used, the reference level becomes -88.5 dBm.
6. The conversion between dBrnC and dBa is given by equation 1.16.

1.7.8 Psophometric Weighting

The CCITT unit which is used in Europe and in other parts of the world, and which is generally accepted as the international standard unit of noise measurement, is the 'pWp' or, 'picowatt psophometric'. The reference level is -90 dBm at a reference frequency of 800 Hz. A noise level measured through the 'CCITT psophometric weighting' network may be expressed in pWp, or in dBmp. Thus 1 nanowatt of noise would represent 1000 pWp, or 1000 picowatts above a reference of 1 picowatt psophometrically weighted, at a reference frequency of 800 Hz. Alternatively, as -90 dBmp is the reference level, 1000 picowatts of noise $= 10 \log [(1000$ pW$)/(1$ milliwatt$)] = -60$ dBmp referred to a reference frequency of 800 Hz. Hence

$$-90 \text{ dBmp} = 1 \text{ pWp at a reference frequency of 800 Hz} \qquad (1.17)$$

$$\text{Noise power in dBmp} = 10 \log (\text{pWp in milliwatts}/1 \text{ milliwatt}) \qquad (1.18)$$

If a 3.1 kHz band of noise is used instead of 800 Hz, the reference level becomes -87.5 dBm. If a 1 kHz reference frequency is chosen instead of the 800 Hz frequency, the reference level becomes -90.5 dBm.

When all the noise units are referred to 0 TLP, so that anywhere in the system the expected noise level can be verified, the units will contain a '0', i.e. dBa becomes dBa0, dBrnC becomes dBrnC0, dBmp becomes dBm0p, and pWp becomes pW0p. If all noise units are referred to 0 TLP, the conversion factors between the weighted noise units dBa, dBrnC and dBmp become;

$\underline{-90 \text{ dBm0p}}$	$= -5.5 \text{ dBa0}$	$= 0.5 \text{ dBrnC0}$
-90.5 dBm0p	$= -6 \text{ dBa0}$	$\underline{= 0 \text{ dBrnC0}}$
-84.5 dBm0p	$\underline{= 0 \text{ dBa0}}$	$= +6 \text{ dBrnC0}$

The CCITT Volume IV Supplement 1.2 shows a rounding-off of the above to permit the conversion formulae as shown below. These formulae are correct for any tone or a 3 kHz band of noise, due to the rounding-off,

$$\text{dBmp} = \text{dBa} - 84 \qquad (1.19)$$

$$\text{dBa} = -6 + 10 \log \text{pWp} \qquad (1.20)$$

$$\text{dBrnC} = 10 \log \text{pWp} \qquad (1.21)$$

$$\text{dBrnC} = \text{dBmp} + 90 \qquad (1.22)$$

In using the above conversion formulae the following applies:

1 mW unweighted 3 kHz white noise reads 82 dBa
= 88.5 dBrnC, rounded off to 88.0 dBrnC

1 mW into 600 ohms = 775 mV = 0 dBm = 0 dBmp

Readings of noise measuring test sets when calibrated on 1 mW of test tone:

> FIA line weighting; at 1 kHz reads 85 dBa
> C-Message; at 1 kHz reads 90 dBrnC
> Psophometer (1951); at 800 Hz reads 0 dBm

Points to Note
1. The CCITT weighting curve is shown in Figure 1.6.
2. The CCITT reference frequency is 800 Hz.
3. The CCITT reference level is −90 dBm, or 1 picowatt measured with a 800 Hz tone to produce pWp.
4. dBm0p means the absolute noise level in dBm referred to 0 TLP. The reference level at 800 Hz is −90 dBm0p.
5. If a 3 kHz band of white noise is used, the reference level becomes −87.5 dBm.
6. If a 1 kHz reference frequency is used, the reference level becomes −90.5 dBm.
7. If white noise is measured in a 3.1 kHz bandwidth, the noise level must be reduced 2.5 dB to obtain the psophometric power level. If another bandwidth B kHz is used, the weighting factor will be $[2.5 + 10 \log(B/3.1)]$dB.
8. Conversion factors between dBa, dBrnC, and pWp or dBmp are given in equations 1.17–1.22.

1.7.9 Other Units

1.7.9.1 Crosstalk
The two crosstalk measurements commonly made on transmission lines are *crosstalk attenuation* and *crosstalk ratio*.

Crosstalk attenuation is the logarithmic ratio of the power sent on the disturbing circuit to the power received on the disturbed circuit, i.e.

$$\text{Crosstalk attenuation} = 10 \log [(\text{Power sent on disturbing circuit})/(\text{power received on disturbed circuit})] \tag{1.23}$$

Crosstalk ratio is the logarithmic ratio of the power received by the disturbing circuit to the power received by the disturbed circuit at the point of measurement, i.e.

$$\text{Crosstalk ratio} = 10 \log [(\text{Power received by disturbing circuit})/(\text{power received by disturbed circuit})] \tag{1.24}$$

For far-end crosstalk there is a difference between crosstalk ratio and crosstalk attenuation due to the attenuation of the line.

Crosstalk coupling in telephone circuits is indicated in 'dBx', or 'decibels above reference coupling'. Reference coupling is defined to be the difference between a 90 dB loss and the amount of actual coupling. If two telephone circuits were measured to have a coupling of 30 dB, the coupling could be expressed as 30 dB above the reference coupling of a 90 dB loss, i.e. as $(90 - 30) = 60$ dBx.

$$\text{Crosstalk coupling } t = (90 - \text{actual coupling in dB}) \text{ dBx} \qquad (1.25)$$

1.7.9.2 The Volume Unit

Although most transmission measurements are made using a sine wave test tone, at times it is necessary to measure a complex speech signal. The decibel meter will give steady readings for a test tone, but will be erratic if a speech signal measurement is attempted. The volume indicator is a damped meter which is calibrated to read 0 volume units (vu) when connected across a 600 Ω line carrying a 0 dBm test tone in the frequency range 35–10000 Hz. It has a damping coefficient slightly less than one and is so designed that if a constant sine wave is placed across the terminals the needle will within 0.3s move to within 90 per cent of the steady state value and not overshoot it by more than 1.5 per cent of the steady state value.

Volume indicator uses are:

1. To indicate a suitable level of speech or music at a particular circuit point to prevent distortion in line amplifiers by overloading.

2. To measure attenuation of speech signals in circuits carrying speech traffic.

For a complex signal the following approximation holds:

$$\text{VU} - 1.4 = \text{power in dBm} \qquad (1.26)$$

There are many other units besides those mentioned in this chapter, and those which are used in the testing of LOS radio links will be defined as the need arises.

EXERCISES

1. Express 2500 watts in dBm (*Answer* 63.98 dBm)

2. Express 600 microwatts in dBW (*Answer* −32.21 dBW)

3. In a 3.1 kHz bandwidth, noise was measured to be −60 dBm. What is its equivalent value in dBrnC? (*Answer* 28.5 dBrnC)

4. What would be the theoretical thermal noise, expressed in dBm, at the input of a receiver, operating at 17°C, with a bandwidth of 1 MHz? (*Answer* −114 dBm)

5. If a signal entering the receiver of Question 4 was at a received level of −84 dBm, what would be the signal-to-noise ratio? (*Answer* 30 dB)

6. A system is carrying information at a level of −10 dBm0. If noise is measured to be −80 dBm at a −35 dBr point, what is the signal-to-noise ratio? (*Answer* 35 dB)

7. What would be the actual measured value of noise in dBm, at a dBr point of -16 dBr, if the noise value was 38.5 dBrnC0, in a 3.1 kHz slot? (*Answer* -50 dBm)

8. What would be the expected signal-to-noise ratio for Question 7? (*Answer* 34 dB)

9. If a level of -35 dBm was measured for some test-tone frequency in a radio system, at a -15 dBr point, what level in dBm0 would the test tone be inserted? (*Answer* -20 dBm0)

10. A 1 kHz interfering tone was measured in an 80 Hz wide slot by a selective level meter. The level recorded was equivalent to 30.5 dBrnC. What is the level expressed in pWp? (*Answer* 1000 pWp)

REFERENCES

1. *Bell System Technical Journal*, July 1931.

2. CCITT Recommendations G132, G151A, Orange Book, Vol III–1, *Line Transmission*, VI Plenary Assembly, 1976.

3. CCITT Recommendations G111, G132, G151, G223, G227, G232, G322, White Book, Vol. III, *Line Transmission*, IV Plenary Assembly, 1968

4. CCITT Recommendation P11 'Effects of Transmission Impairments' (Geneva 1980), Yellow Book, Volume V, Telephone Transmission Quality, 1980

5. Tant, M. J., *The White Noise Book* (Marconi Instruments, 1974)

6. Freeman, R. L., *Telecommunication Transmission Handbook* (Wiley, 1975)

7. ITT, *Reference Data for Radio Engineers*, 6th Ed. (Howard W. Sams, 1981)

8. White, R. F., *Engineering Considerations for Microwave Communications Systems*, (GTE Lenkurt, San Carlos, California, 1975)

2 AN INTEGRATED COMMUNICATIONS SYSTEM

2.1 OVERALL SYSTEM DESCRIPTION

An integrated communications system (ICS) provides voice and data communications between remote locations, which may be mobile, and a central office. Primarily it will route all remote radio and hard-wire circuits to a central switching facility. Once all communications have been routed to this single location, the calling and called parties can be either connected automatically, or connected with the assistance of an operator. Figure 2.1 is a block diagram of such a system showing the interacting subsystems.

The system configuration which will mainly be discussed in this chapter is the 'loop' system. The straight 'chain' system, which is a more simple version of the loop, but which is more often used because of geographical considerations, is also described. Rather than treat each separately, they will be discussed together to avoid repetition.

Each fixed remote location, or outstation as it will be now called, is connected to the adjacent outstation, and eventually into the central office by means of microwave links. These links then form a loop. At each location where a radio link originates or terminates, there is a transmitter and receiver pair facing in both the clockwise and anticlockwise directions. At the central office location the loop is broken at the baseband to prevent the baseband information circulating and rendering the system useless. This loop configuration differs from a straight chain of microwave links in that if one of the links becomes inoperable, a baseband switch at the central office closes, completing the radio baseband circuit. The opening in the loop is now the link which is 'down'. In the chain system, if one link fails, then from the central office end all communications from the failed link onwards are 'out'. Figures 2.2 and 2.3 show typical chain and loop configured systems.

Consider Figure 2.4. The baseband signal into the microwave transmitter at outstation 1 is transmitted in an anticlockwise direction, and received at outstation 2. As there is a four-way/four-wire bridge at this site, this baseband information is split. One direction is towards the engineers' order-wire* circuitry and the supervisory circuits, another is towards the multiplex† equipment, and the third is towards

* See Section 2.17
† See Section 3.2 for detailed explanation.

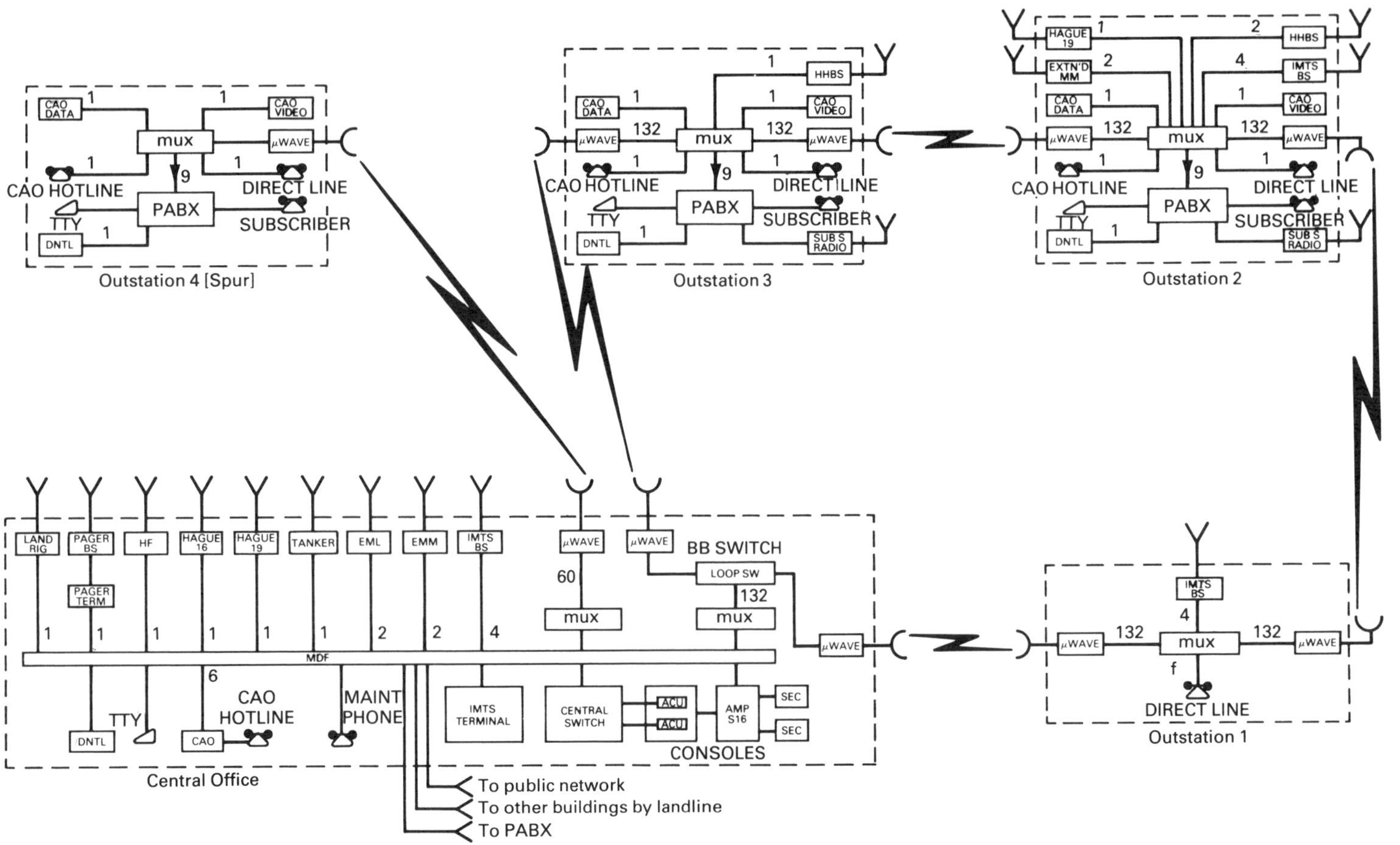

Figure 2.1   Example of an Integrated Communications System

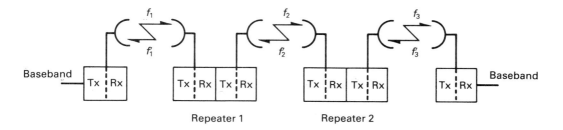

Figure 2.2   Chain Configuration

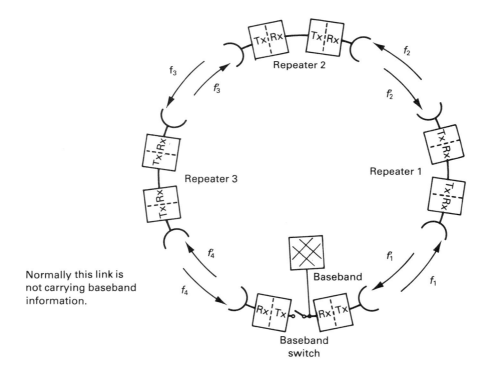

Figure 2.3   Loop Configuration

the transmitter facing outstation 3. The multiplex equipment at location 2 is equipped with only a limited number of group or channel modems, so that, if the received baseband information does not contain the frequencies of those groups or channels pertinent to that location, it cannot provide outstation 2 with any information, i.e. at this particular time no call has been placed there from anywhere in the system. The baseband signal which entered into the transmitter at location 2 is transmitted by FM to outstation 3, where it is again passed into a four-way/four-wire

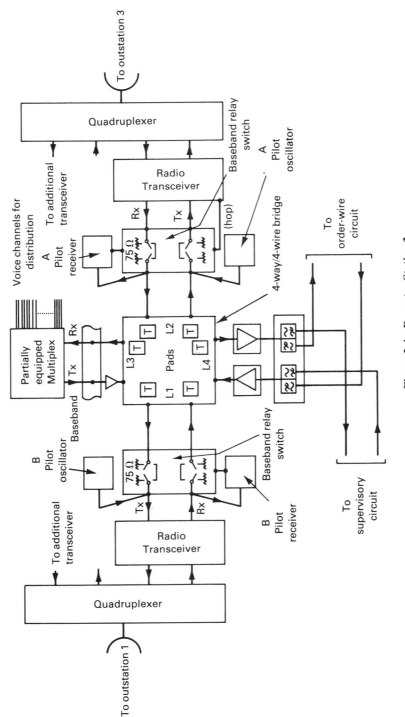

Figure 2.4   Repeater Station 2

Notes:

1. At repeater site, the pilot oscillator and receiver are *internal* to the transceiver and are *hop* pilot equipment

2. At the central office, the pilot oscillator and receiver are *external*, and are *system* pilot equipment.

3. At the central office only, normally baseband switch facing outstation 1 is open.

bridge, and associated amplifiers, etc. Three baseband paths are formed, one for the transmitter facing location 4, and the others for the multiplex equipment and order-wire/supervisory circuits. If it is assumed that, somewhere in the system, a subsystem or subscriber wishes to place a call at outstation 3, the baseband signal will contain one of the 4 kHz band of frequencies which correspond to a particular group and channel for which outstation 3 is equipped. Out of the multiplex channel modem at outstation 3, will come information to set up a call, and once the call is established, the resulting information will be speech or data. During the stage when the system was being planned, each multiplex group and channel number was allocated to a particular subsystem. Thus the final destination of the call depends on this allocation, and, for example, it may be to the station PBX (private branch exchange) if the station is so equipped. Figure 2.1 gives details of the various subsystems where this call could end up.

When a subsystem or subscriber wishes to place a call from outstation 3, the multiplex equipment will translate in frequency the relevant voice band, and stack it frequency-wise into its designated place in the baseband spectrum, where it is added to the loop baseband signal by means of the four-way/four-wire bridge. This baseband signal is then passed to the transmitter facing outstation 4, and also to the baseband feeding into the transmitter facing outstation 2. The bidirectional nature of this information is necessary since the loop is broken at the central office, and the station which may be equipped with the multiplex demodulating equipment located by approach from only one direction around the loop, as for outstation-to-outstation trunks or junctions against outstation-to-central-office trunks. Similarly for a 'chain' configuration, the required demodulating station could be in either direction from the originating location. When a station in a chain or loop microwave system can place information onto a common baseband, it is said to have 'insert' facilities. If it has equipment which can demodulate the common baseband, it has 'drop' facilities. The common baseband in the system means that multiplex equipment which is duplicated at outstations 1 and 3 may also be duplicated at outstation 4 or any other station. If, by accident, a group or channel modem was added to the multiplex equip-ment at outstation 4, which was identical to that at either outstation 1 or 3, then out-station 4 could make and receive calls to outstations 1 and 3 simultaneously, or break into an existing conversation, etc. Multiplex channelization plans are carefully designed to insure that this type of mistake does not occur, and also the plans provide for dedicated lines or trunks to compatible subsystems, according to the customer's requirements.

Each transmitter and receiver at a location which faces in more than one direction operates on a different bearer or radio carrier frequency. This is essential for the prevention of interference. Due to frequency band allocations, there are usually duplicated frequencies used throughout the system.

To control the switching in case of link failure, there are two types of pilot for the loop configuration. These are the hop pilot which monitors the performance of the link itself, and the system pilot which controls the baseband switching at the central office in case of a link failure elsewhere in the system. The hop pilot is different in frequency from either of the system pilots, and controls the local baseband switching only in one direction of transmission and reception (towards outstation 4 or outstation 2 for example). The two system pilots are individually used to control the closing of the loop at the central office location, by the failure of one from either the anticlockwise or clockwise direction. Because the hop pilots are restricted to one hop only, the receiver baseband amplifier usually has an upper band-pass cut-off frequency below that of the pilot frequency. This prevents the hop pilot received for a particular link being passed further down the line and interfering with the operation of other link receivers.

Figure 2.4 provides information on how the system pilots at the central office may be connected to the radio equipment using separate pilot oscillators and receivers. The hop pilots for each repeater station and for the central office are derived from circuits within the transmitter itself, and similarly are detected by circuits within the receiver. Each of the outstations in a chain or loop type of configuration is a center to other subsystems. These subsystems may be directly connected to the outstation via interfacing equipment, or by other radio bearers, called 'spurs'. The final result resembles nerve fibres sending and receiving signals into and from the spinal chord. This analogy is carried through to the chain of micro-wave links in the same way, and thus the main links are referred to as the 'microwave backbone'. To reduce the 'down time' due to link outages, most systems which carry a reasonable amount of traffic, or a small amount of traffic which is vital, have protection systems installed. These protection systems can take the form of frequency and space diversity to reduce the down time due to fading, or, as described above, a loop system. Other forms of protection can come from alternative routing by exchanges or baseband switching, if the system is large enough.

The use to which the four-way/four-wire bridge is put has been described above, but because transmission loss through this bridge is approximately 15 dB, some account of the level change to and from the associated equipment must be made. Usually amplifiers and/or additional 'T-pads' are added to provide the correct levels. (T-pads are attenuators, so named because of their resistive circuit construction.) Figure 2.4 indicates in principle these additions.

Other bridge configurations which are used are the six-way/four-wire, which has a two-wire input that can feed to any of the remaining five two-wire outputs. By terminating one input and one output of a leg in an impedance of 75 ohms, a six-way/four-wire bridge can be reduced to that of a five-way/four-wire bridge, if such a bridge is not readily available. These balanced bridges are important components in the design of centers which may combine the basebands of several radio links feeding into a central terminal – either a repeater station or a central office. They are especially useful when an existing system is required to be upgraded without the traffic being carried on that system being disrupted during the period that the new equipment is being installed, and also when the existing equipment has baseband levels that are non-compatible with that of the new equipment.

Figure 2.5 shows a block diagram of the combining of order-wires from several radiating spurs from a central repeater station using a six-way/four-wire bridge. The use of the 3 kHz low-pass filters are to permit only order-wire frequencies below 3 kHz to appear on any of the basebands of the spurs, and so not to permit higher order-wire frequencies from interfering with the spectrum reserved for the multiplex information. Also shown is the case where each of the radiating spurs has a different

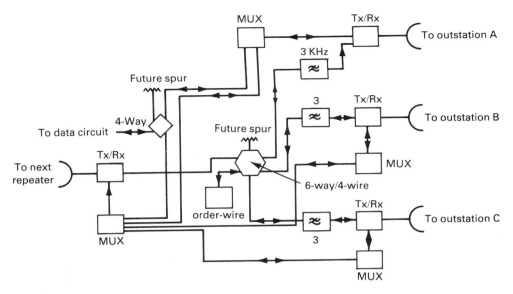

Figure 2.5 ***The Use of a Six-Way/Four-Wire Bridge at a Junction, and of Voice-Channel Coupling of Multiplex with the same Group and Channel Frequencies (23 dB Pads are not shown)***

baseband that is fed directly into the corresponding multiplex equipment. This sometimes arises when each spur has the same group and voice channels, and so the baseband frequencies are all similar and therefore non-compatible. The order-wire, due to its operation, does require the combining of similar frequency spectrums, hence the use of the six-way/four-wire bridge. To permit each of the separate central repeater station spurs to communicate to its distant terminal or repeater each spur has its multiplexed voice channels brought down to four-wire audio circuits plus E and M signaling leads. Each four-wire audio circuit is then combined with an allocated audio channel in the multiplex equipment used for the transceiver facing the next repeater and similarly its E-leads are connected through to the M-leads.* In order to make the received audio signal of the spur compatible in level with the transmitted audio signal of the repeater multiplex, usually a 23 dB pad is inserted between the two audio ports. This pad is also included in the connection of the audio receive of the repeater to the audio transmit of the spur path: an attenuation of 23 dB is chosen because the normal audio receive level out of the multiplex equipment is +7 dBm (referred to 0 TLP), and the normal input level into the multiplex channel modem audio transmit is −16 dBm. The order-wire is used to permit conversations of testing personnel to be transmitted to all stations in the system. Any person speaking at one location will be heard at all other locations simultaneously.

An alternative to the use of the four-way/four-wire bridge is the use of multi-input amplifiers. These can be an advantage if cost is most important, and flexibility is not. A system so designed may reduce the number of amplifiers and pads because the levels coming out of the multi-input amplifiers are exactly designed to feed from

* See Section 4.3

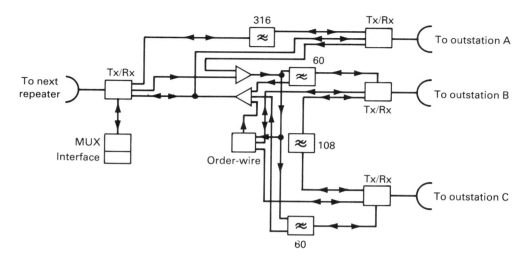

Figure 2.6   The Use of Multi-Input Amplifiers instead of Balanced Bridges

and to specific equipment. Figure 2.6 provides a block diagram of a case where multi-input baseband amplifiers are used instead of a four-way/four-wire bridge. This diagram also shows how the order-wire may be connected in such a case.

In an integrated communications system, many subsystems may exist. To monitor all of these subsystems and to monitor the many variables of a subsystem, such as power, noise, card failures, etc., an automatic monitoring system may be installed. A section of the baseband frequency spectrum may be allocated for this purpose. One such example would be above the order-wire frequency band (0–3 kHz), and below the multiplex band (below 12 kHz or, for some US manufacturers, below 8 kHz). Thus the band 4–8 kHz in the baseband spectrum may be used to permit a polling type, or some other type, of supervisory system to operate. Figure 2.4 shows how the output of the supervisory system may be connected to the repeater baseband.

Some subsystems which may be integrated into an operating system via the microwave backbone and into the central switch are described below.

2.2 SWITCHING UNITS AND OPERATOR CONSOLES

The functions of switching units and operator consoles may be best described by the placing of a call from one point in the network to another. Two calls will be placed. The first is from a mobile subscriber wishing to contact another mobile subscriber. The second is from someone on a construction site who wishes to call somebody outside of this private network.

2.2.1 Mobile-to-Mobile Call

The subsystem to be used is the mobile telephone system (Figure 2.7). The remote-end subscriber, taking his handset 'off-hook' starts a long and involved sequence of operations before he is able finally to speak to the person he wishes to call. What is of interest here are not the fine details of each mechanism of the call processing, but the type of equipment which is brought into use in the setting-up of the call, and its general function. By the subscriber going off-hook, the mobile's transmitter signals its base station. The base station is for convenience at an outstation. This base station in turn via its control unit sends audio tones into the 4 W channel modem of the multiplex equipment via level-changing pads. The two E & M signaling wires of the channel modem are not used in this type of connection since audio tones will be trans-

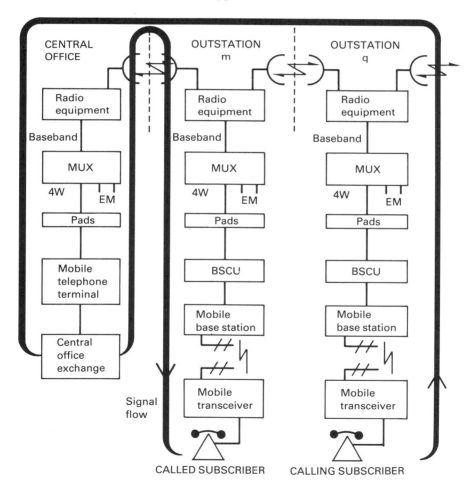

Figure 2.7   Mobile-to-Mobile Call

ferring the signaling information. The signaling tones are multiplexed and placed onto the baseband signal, and thence onto the microwave backbone. The group and channel which contain the signaling information are located at the central office. This information then appears on the four-wire receiving end of the channel modem and is passed into pads which change the signal level to make it compatible with the mobile telephone switching terminal in the central office. Once into this terminal, the information is processed and then passed to a free line connecting the terminal to the central office exchange. The exchange sends back a dial tone to the originating subscriber. The route is the same backwards as it is forwards. The subscriber, upon hearing the dial tone, enters into his keyboard the required number. Once the central office exchange receives this number, it decides whether to route the call:- to the operator console for connection to one of the HF subsystems; to the tie lines from the central office exchange to the PABX connecting the public network, for an outside call; or to any of the other compatible subsystems which are connected to the central office exchange.

The exchange recognizes the dialled number as being another mobile subscriber, and so finds a vacant tie line to the mobile telephone switching terminal. This terminal, after receiving the number, decides if the number belongs to a valid subscriber, etc., and if all is in order, the call is allowed to proceed. Assuming that this is so, the terminal looks up in its memory the geographical location of the called party. This is done by remembering where the called party was when the last call was made to or by him. The terminal, knowing the location, creates a circuit between itself and a free multiplex voice channel, again via the level-changing pads. The voice channel is selected as being one which will go to the outstation closest to the called party. The terminal now sends audio signals representing ring tone to the 4 W transmit of the selected voice channel. The signal after modulation in the multiplex equipment is then sent via the microwave backbone to be received by the multiplex equipment at the desired outstation. As the voice channel is dedicated to this subsystem, it passes directly into the base station control unit (BSCU) via the level-changing pads, and is thus transmitted by the mobile telephone base station transmitter to the called party's mobile radio receiving equipment. This equipment generates ring current which in turn rings the telephone. The called party on answering or going 'off-hook' sends signals back to the terminal, which after further processing permits the call to proceed. Hanging up or going 'on-hook' clears down the circuit in a similar fashion through the network as that of the setting-up of the call.

2.2.2 Walkie-Talkie–to–Public Network

In this type of call the subscriber switches the frequency of his walkie-talkie to a channel whose frequency permits connection to the operator console (Figure 2.8). The subscriber keys his walkie-talkie (or hand-held radio), i.e. he presses the transmit button. The hand-held base station (HHBS) at the outstation receives the carrier, and automatically opens an audio path between itself and the nominated voice channel of the outstation's multiplex equipment. Again this path contains level-changing pads to permit correct interfacing levels between the HHBS and the

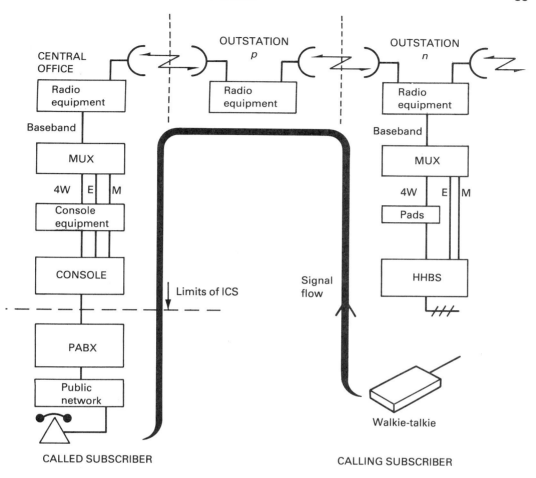

Figure 2.8   Hand-Held Radio-to-Public-Network Call, placed through the Operator Console

voice-channel 4 W transmit and receive. The signaling E- & M-wires of the multiplex channel are also connected to the HHBS, but not through the pads. When the HHBS receives the carrier, it also keys the M-lead. The keying of this M-lead is transferred through the microwave backbone, as is the audio circuit, and appears at the central office multiplex E-wire, and at the audio receiving end of the companion multiplex channel. This channel is connected to the console equipment. The activation of the E-lead alerts the operator, who in turn asks for the desired number. Upon receiving speech information from the subscriber, the operator dials for an outside line, using a code for access to the PABX connecting the system to the public network. With a dial tone he dials the public number. The called party upon answering is then connected by the operator to the calling subscriber. Clear-down is automatically completed by either party going 'on-hook'.

If a public subscriber wishes to call a person on a construction site, the process is the same, except that the holder of the walkie-talkie may not be switched to the frequency required to accept calls from the operator. Usually the holder will be using the HHBS as a repeater to talk to all other holders of walkie-talkies operating on the same frequency. To talk to a particular individual, a verbal request to that individual is made for communication.

The central office exchange switches all calls within an integrated communications system, except those calls which are dealt with by the console units. Lines which tie the exchange to the multiplex equipment are not all necessarily four-wire E & M. They may be two-wire E & M, where a hybrid is included in the multiplex channel equipment, or they may be two-wire, where the exchange senses a 600 Ω impedance bridged across the line from equipment interfacing the four-wire E & M channel to an ordinary telephone line into the exchange. There are also variations in the signaling voltage levels of the E & M leads. Some exchanges may respond by battery on the E-lead (or F-lead*), others may require a ground start. The exchange may respond in answer or handshake by placing either battery or earth on the M-lead. Later exchanges have moved away from the concept where the signaling information to the exchange is carried on the same audio trunk entering the exchange, in favour of the 'common-channel signaling' (CCS), where the signaling information for all trunks is carried on a separate single circuit. This channel carries signaling information in a digital form, in a serial binary format. The standards which have been laid down are the European CCITT No. 6, and the North American CCIS (common channel inter-office signaling). With the increase in the number of digital exchanges, this type of signaling will come more into the foreground.

The mobile telephone switching unit processes all calls to and from mobile telephones, and behaves similarly to a modern exchange. The system which is now being marketed is the cellular system, which is an outgrowth of the improved mobile telephone service established in 1964 to eliminate operator intervention in mobile calls. It is expected that the total US user base should be just over 1 million by 1990. Continental Europe, once it has settled on what type of cellular system to build, should follow the US by having a rapid growth in this area of communications in the future.[1]

2.2.3 Operator Consoles

The console units permit manual patching of calls to anywhere in the network. Access to the consoles can be made via the central office exchange or radio subsystems. One primary console and two secondary consoles will be able to satisfy the requirements of the operator-handled subsystems used in the example of an integrated communications system described in this chapter. All operator consoles are located in the central office and as close to the radio equipment or carrier room as possible. The primary console may be equipped with two operator positions so that either operator may access any of the circuits brought to the console. Each operator

---

* If tied to ground the F-lead permits the E-lead on activation to be at ground – See Section 3.7.2.

position also has an attendant console unit (ACU) to provide access through the central office exchange to all the circuits which have been designed for access to the telephone network. The consoles' interface equipment permits each of the consoles to access the microwave system via the multiplex equipment. For each of the multiplex-fed circuits, either primary console operator may:

Monitor the communications in progress using a speaker in the console itself or in the attendant console unit.

Break in on an existing call.

Establish a call on an idle channel.

Provide assistance in placing a call to or from the telephone network on those circuits so equipped.

The primary console has access to all radio subsystems which are also trunked to the central office exchange for integration into the telephone network. To place a call or connect a call to a particular radio channel, the operator simply dials the appropriate trunk via the ACU. Answering a call from a mobile is done similarly. The two secondary consoles are connected to two radio subsystems which are outside the ICS, namely the Hague Channel 16, and the SSB air-sea-rescue channel. Circuits are provided to bridge any or all of the secondary consoles across any desired circuit in the EMM or EML subsystems.* As interconnection of subsystems to the telephone network is not required, the secondary consoles are essentially radio-operator consoles, with provision for handling of calls, break-in, monitoring, and visual indicators for monitoring channel status.

2.3   THE MICROWAVE BACKBONE

The configuration of the microwave bearer system usually depends on the network philosophy adopted. The network philosophy may be that all voice channels are to be interlinked 'mesh', and connected at each point by 'node' switches, so that each exchange has a direct connection to every other exchange or PABX. If this type of arrangement is adopted, the traffic is connected over the shortest possible available route and requires a minimal number of switching stages. The disadvantages are that the total traffic originating from an exchange is split into several small groups which require a greater number of circuits than that required by a single large group, and that the exchanges are more complex due to the number of alternative routes into which the exchange is required to switch.

The alternative arrangement is the 'star' network in which all exchanges are

---

* See Sections 2.7 and 2.9.

connected via a central switching device. As switching is concentrated in the central location, the capacity required at the other exchanges is reduced. This reduction is offset by the increased cost of a large and complex central switch. The star network maintains the traffic as a single large group which needs the minimum number of circuits, but these traverse longer distances and therefore are more costly. Both types of network have advantages and disadvantages and a compromise between these two extremes that will produce a network which is less costly than either is the objective of alternatively routed networks. Figure 2.9 is a diagram of each of the two networks mentioned above.

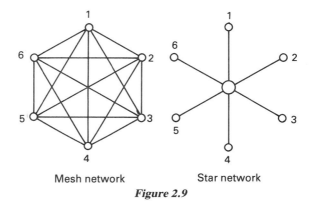

Mesh network   Star network

Figure 2.9

Once the network philosophy has been established, and the permanent sites of the exchanges have been proposed, the radio-bearer-link routing can be determined. With this tentative routing the following information is required before any further planning can be made:

Customer's concepts

The frequencies of transmit and receive for each hop

The total channel capacity and the circuit types and subsystems

The channel requirements for each link

Radio path evaluation reports based on field surveys and contour maps

k-factor* spreads throughout the year. These are obtained perhaps from the Government meteorological department or equivalent

Types of radio and multiplex equipment available and the equipment parameters

---

* Refer to Section 8.3.1 for an explanation of k-factor.

Types of power source in existence and proposed

For a new system – analogue versus digital radio considerations

Other existing and operating transceivers in the area, together with their frequencies, output power and distance from any proposed sites

Federal, State or Local Government regulations and/or specifications

With these broad headings it is possible for the planning engineer to complete the following calculations;

1. Antenna height calculations
2. Path calculations
3. Performance calculations

2.3.1 Antenna Height Calculations

To complete these calculations the following information is required:

1. Expected maximum and minimum values of the k-factor
2. Latitude and longitude of each radio terminal site
3. Obstruction data in the radio path, i.e. the obstacle distance from a terminal and its height above some reference point (e.g. LLWS)
4. Operating frequencies
5. Height of existing towers

2.3.2 Path Calculations

In addition to the information required in Section 2.3.1, the following information is required to determine the fade margin, the probability of exceeding the fade margin during the worst month, and the percentage availability:

1. Total radio bearer voice-channel capacity
2. Transmission line, jumper, branching, radome, and safety factor losses
3. Antenna gains
4. Transmitter power
5. Required receiver signal level for S/N 30 dB
6. Availability requirements

2.3.3 Performance Calculations

In addition to the information required in Sections 2.3.1 and 2.3.2, the following information is required to determine the noise performance of the link and the radio equipment:

1. IF bandwidth
2. Upper baseband frequency
3. RMS frequency deviation
4. Pre-emphasis figure
5. RF-carrier-to-noise ratio
6. Feeder line length

Thus the total noise from these calculations is added to the noise from interference and from the multiplex equipment, to estimate the total expected noise for each hop.

After these calculations have been completed, and the results indicate that a link can be designed and installed which will satisfy all requirements, a proposal is made. These calculations will be used for the testing and commissioning phase of the link after its installation, and also eventually by the system preventive and corrective maintenance personnel.

2.4 TANDEM AND SATELLITE EXCHANGES

The function of tandem networks is to connect calls between terminal exchanges in the local network and to provide an interface for trunking. Their role is to assemble small elements of traffic so it can be handled more efficiently on larger routes. Usually in a small network they provide the automatic switching of local and distant communications traffic. If the local area is small enough it is possible to use a PABX which will handle the local traffic switching within that area, as well as allowing direct dialling to any other outstation without going through the central office. This of course means that provision has been made for inter-outstation trunks within the multiplex plan, as well as the numbering plan of the PABX, and the central office exchange. The outstation besides the two switching routes mentioned above, will be able to provide the switching of traffic to the central office. The trunk lines connecting the outstations to each other are able to provide automatic route selection (ARS), if the PABXs at both outstations are provided with this capability. This means that outstation 2 may be able to dial a central-office-switched number via outstation 5. If all the trunks from outstation 2 to central office are occupied, and if all the trunks at outstation 5 are occupied, then again an alternative route will be attempted. PABXs available at the present time, such as the Harris DTS D1200 series, can provide up to 400 lines and 56 trunks. The 56 trunks can be proportioned according to the customer's requirements between the central office and different outstations.

2.5 SUBSCRIBER RADIO

Telephone or data communications to remote areas have always posed a problem of economics. To provide a service to one or two subscribers at points which are located at a remote point from the local exchange or outstation is costly in terms of the external plant requirements. The subscriber radio (Figure 2.10) to some extent provides

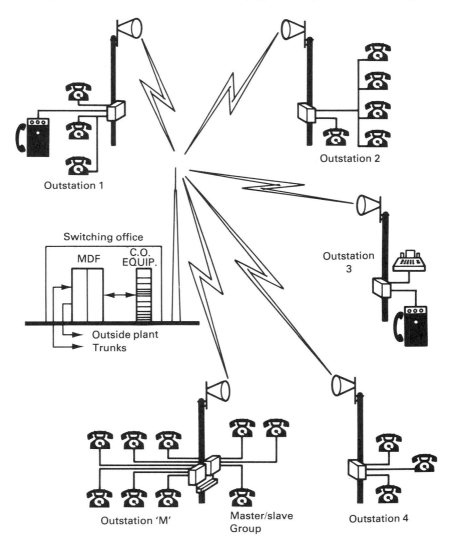

Figure 2.10 Typical SR System

a realistic way of overcoming this problem. The concept is to provide a microwave link from several subscribers near the remotely located radio to a multipoint, a central or nodal transceiver. The central or nodal location uses an omnidirectional antenna to receive and transmit to several remote locations. A single RF carrier is used with time division multiple access (TDMA) techniques which may allow up to 15 simultaneous conversations or more from any of the remote subscriber radios. The nodal location is connected via the outstation PABX to the microwave system, permitting communications to occur anywhere in the system.

2.6   IMPROVED MOBILE TELEPHONE SYSTEM[1]

Communications to any land or water mobile or stationary vehicle within the geographical area of the designed transmitter's range can best be served by the Improved Mobile Telephone System (IMTS) two-way mobile radio equipment. This equipment operates in the VHF part of the frequency spectrum as per CCIR recommendations and extends automatic dialling facilities to the telephone network. Section 2.2 describes the routing of a call using this subsystem in an integrated communications system. An alternative name presently in common use is the 'cellular telephone'.

2.7   EXTENDED MOBILE MARINE SUBSYSTEM

The extended mobile marine (EMM) subsystem allows marine craft equipped with channelized VHF/FM transceivers to call into the system telephone network with assistance from the operators at the operator consoles, and also permits telephone subscribers to call marine craft directly with open calling. Base stations for the EMM channels are usually located at strategic outstations, to permit a wide coverage. Each base station operates on a different frequency pair or channel from the others, and on receiving a call routes it to and from the operator console control centre.

2.8   HAND-HELD BASE STATIONS

The hand-held base station (HHBS) subsystem allows hand-held portable UHF radios (walkie-talkies) to call into the system telephone network with operator assistance, and telephone users to call hand-held portables which are switched to the correct frequency channel directly with open calling. This subsystem is described in

Section 2.2. Base stations for these radios are again usually located at strategic centres to permit the widest coverage possible, and route their traffic into the microwave system to the operator consoles.

2.9 EXTENDED MOBILE LAND SUBSYSTEM

The extended mobile land (EML) circuits are used for land-based vehicles around the central office area. This system may augment the IMTS (Section 2.6), but does not provide automatic switching as in the IMTS. The operating frequencies are in the VHF section of the radio frequency spectrum, and all calls into the telephone network are made through the operator consoles, using some of the channels for this purpose. Other channels are used in the open calling node with the aid of the base station located at the central office.

2.10 LONG-DISTANCE HIGH-FREQUENCY COMMUNICATIONS

The high-frequency (HF) circuits are used for telex and voice communications with sites which are so remote that they cannot be reached by any other subsystem so far described. Usually each radio independent-sideband-transmission (ISB) duplex channel uses the lower sideband for voice and the upper for facsimile (FAX) communications. These channels interface with the operator consoles and are connected to the telephone network by manual patching. As ISB may be too noisy for telex traffic, the use of single-sideband transmission (SSB) is preferred.

2.11 PAGING SYSTEM

The paging terminal permits automatic dialling to 'pagers', i.e. someone dialling into the system or dialling from within the system may gain access to any person holding a pager.

2.12 HAGUE PLAN SYSTEM

The Hague Channel 19 is a VHF frequency assigned to FM radiotelephony in the international maritime mobile service. It is not used for long-distance communications since a typical VHF transmission will involve distances of only up to 35 miles. Two frequencies are used per radio channel permitting full duplex operation between a ship and shore. Base stations are located at strategic outstations to permit the widest coverage. These base stations are tied into the outstation multiplex and microwave by the use of base station interfacing or control units. All calls to and from the Hague 19 subscribers are placed by the operators of the operator consoles. The operating frequencies are:

Transmitting ship to shore	156.950 MHz
Transmitting shore to ship	161.550 MHz

The Hague Channel 16 operates similarly to the Hague Channel 19, except that by international agreement it is used only for emergency calling and distress messages. The circuit, therefore, is monitored for radio traffic on a continuous basis by a secondary console operator, who takes the appropriate action should a message arrive. The circuit operates in simplex and at a frequency of 156.800 MHz. The Hague Channel assignments adjacent to channel 16 are reserved for guard bands, namely channels 75 and 76.

2.13 COMPUTER AIDED OPERATIONS

The computer aided operations (CAO) subsystem is used for the exchange of voice and data communications from outstations. The voice communications portion of the CAO consists of private automatic ringdown telephone circuits. When any of the telephones go off-hook, a connection is automatically made to a central CAO operator console located at the central office. The operator receives a visual indication on the operator telephone panel which indicates from where the callers come. The data portion of the CAO consists of data inputs from and to outstation locations and visual information or status reports are indicated on video display units at both the central office and outstations.

2.14 NUMBERING PLAN

2.14.1   Mobile Telephone Subsystem

On any mobile originating call, the mobile telephone switching terminal will inspect the incoming dialled digits. Based on this inspection, the terminal will determine the correct routing of the incoming call. Any first dialled digit received by the terminal within the range 0–4 will be accepted by the terminal. This digit will inform the terminal that the originating mobile is trying to access a particular number from its directory. The second digit, if in the range of 0–9, will then route the call to the central office exchange. If the terminal receives a first dialled digit within the range 5–8, the mobile will be routed to the operator console, since such digits are considered invalid. If 9 is received as the first digit, the terminal waits for the second digit, and then for all the remaining digits. The value of the second digit instructs the terminal what to do in the processing of these numbers before passing all digits to the central office.

2.14.2 Outstations

At any outstation which contains a PABX, a subscriber can make calls local to that outstation by dialling the number of the local subscriber. The first digit of the local subscriber's number is restricted to exclude that number used for gaining exit from the PABX to either another outstation or the central office. For example this digit may be 9, which will permit access to another outstation if followed by a four digit number not including 9. If the second dialled digit is 9, this would access the central office exchange.

The numbering plan is usually established long before a telephone system is implemented, and is drawn up in co-ordination with the appropriate authorities. Upgrading or expansion of an existing system requires detailed planning in order that the resulting scheme is operational under all conditions. As the numbering plans cannot be changed frequently because of major inconvenience to subscribers, great care must be taken in designing both numbering schemes and tandem networks so that long-term stability is achieved and unnecessary changes avoided.

2.15 INTRODUCTION TO MICROWAVE SYSTEMS

A microwave system is used to relay a quantity of voice-frequency (VF) channels from outlying sites or outstations to a central site or office. Each VF channel has a bandwidth of 4 kHz. The technique of placing a number of VF channels onto a single microwave RF carrier is called *multiplexing*. Each VF channel has an individual frequency-converting unit or multiplex (MUX) channel, where the 0–4 kHz audio signal is converted into a frequency band which is stacked into 4 kHz wide slots on the microwave carrier. The total sum of the 4 kHz wide slots form a signal that is used to frequency-modulate the microwave carrier. This signal is called the high frequency (HF) baseband, and its frequency spectrum commences at 0 kHz and continues in segments of 4 kHz up to the upper baseband frequency, e.g. 0–4 kHz, 4–8 kHz, 8–12 kHz, etc.

An example of a baseband signal frequency range is 0–300 kHz for 72 voice channels of multiplex. If 4 is divided into 300 kHz, we see that the expected number of voice channels is 75. The difference of three channels is due to the frequency bands 0–4 kHz, 4–8 kHz, and 8–12 kHz not being assigned to multiplex frequency plans. This is mainly due to the higher noise concentrations in the 0–12 kHz portion of the baseband, making its use less desirable for speech. The 0–12 kHz portion of the baseband is called the *sub-baseband*. The frequency band 0–4 kHz is used typically for the engineer's order-wire*, common to all microwave sites. The frequency portion 4–8 kHz is typically used for supervisory systems, which monitor equipment performance and site alarms. This information is relayed back to the central office indicating the status of the complete system. The 8–12 kHz band is used either for expansion of the supervisory system, or for additional requirements as the need arises.

After modulation of the microwave carrier with the baseband signal, the RF output is applied to a parabolic dish antenna for transmission to another line-of-site microwave station or outstation. The parabolic antenna exhibits a 'gain' since it focuses the RF power into a narrow beam. This gain is the amount of energy which is focused into a narrow beam when compared with the isotropic antenna which radiates energy spherically. The actual gain figure of a parabolic antenna increases as the dish area increases, and as the frequency is raised. As the microwave sites are usually spaced no more than 50 km apart, the microwave transmitter power is usually not excessive. Normally 10 W is sufficient. Chapter 8 deals with this subject in more detail. The RF carrier is relayed from site to site until it reaches its far terminals, or its central office. The main relay path of the microwave system is called the 'backbone', since it forms the primary route for the signals. Other locations along the way are spur links that tie into the backbone system with each spur adding its information to the MUX channel capacity.

The microwave system operates in full duplex, that is, each VF channel that is transmitted from a site has a corresponding channel that is received at the site with both operations taking place simultaneously. The received signal is converted from the microwave baseband back to an audio frequency band 0–4 kHz wide by the associated MUX equipment. Rather than have one microwave dish antenna for transmitting, and one for receiving, normally a duplexer is used. The duplexer permits the transmitter frequencies to enter into the transmission line which feeds into the antenna, without entering the receiver, and permits received signal frequencies arriving at the same antenna to feed to the receiver only. Section 9.5.3 deals briefly with the components used to make a duplexer.

If two radio systems are to feed into one antenna, that requires two different transmitters and two different receivers, the frequency path separator which permits such a configuration to occur is called a *quadruplexer*. The case where such a system may occur is when there are two loops of microwave backbone, instead of only one, as has already been described.

2.16 MICROWAVE FREQUENCY PLAN

Unless the user of an existing system is planning to have additions or modifications

---

* See Section 2.17

made to it, his main interest will be to insure that this existing system is protected against undue interference[2] by other new or modified systems. This user will possibly be approached by the designer of a new system before the designer has filed an application for new frequencies [Federal Communications Commission Rules Part 94, 15(*b*)]. The user of the existing system will review the calculations made by the designer, make his own calculations and take what action he feels is appropriate. Although the existing system user is not obliged to make any such investigation, or monitor potential interference, if he has a large or very critical system he possibly will do so, and contact the designer about any interference danger. If the matter cannot be resolved the user will file a petition to deny the application of the designer of the new system. Applicants filing for new microwave systems or additions are required under the FCC Rules and Regulations, Parts 94, 63/94.65, to do or have done for them a comprehensive frequency study in order to select frequency pairs which will not create or experience interference with existing or filed-for systems. As this process is becoming more complex due to increasing congestion, in many cases this work is either being done by the manufacturer or a specialist consulting company. The process involved is briefly described below:

1. Data on existing systems throughout the area where the new system is planned must be obtained. These data will take the form of existing frequencies, transmitted power, co-ordinates, and antenna information.

2. New system sites may be selected and the system engineered not knowing the specific frequencies, but knowing the frequency band. An attempt can then be made to find interference-free frequencies. Alternatively the site selection and frequency study can be co-ordinated to obtain a site in which the interference at a particular frequency is minimal or acceptable.

3. FCC Rules and Regulations, Part 94.15(*b*), instruct the designer of a new system to liaise with existing users in the area over potential conflicts.

4. Interference is unlikely to occur if stations using the same frequencies as those planned are at a distance of 200 kilometers or more. Hence the search is restricted to a 200 km radius for radios operating in the same band as that proposed.
    Although interference studies are now done by computers, additional work involving path profiles and manual calculations may be required because of path obstructions, etc. The practical approach is to start at one end of the proposed system, usually the most radio-congested end, and select a pair of frequencies which will be interference-free for the first hop, and a second pair which will be interference free for the second hop. The first pair of frequencies is then tried for the third hop, and the second pair for the fourth hop, etc. Additional frequency pairs would be asked for only in those cases where the selected frequencies would cause interference somewhere in the chain of links.

2.17 MICROWAVE ORDER-WIRE SUBSYSTEM

Multiplex channel assignments for the microwave baseband begin at a frequency of 12 kHz, with each multiplex channel 4 kHz in bandwidth. The portion of the baseband or sub-baseband 0–4 kHz is used for the engineer's order-wire. The order-wire is a common audio channel between all sites in the microwave system, and is open at all times, i.e. all sites can monitor order-wire traffic by simply listening to the loudspeaker on the order-wire panel, and can enter into conversations by talking into the microphone, which is also located on the order-wire panel. In addition, a signaling button may be pressed which will place a tone on the channel to alert all site personnel that speech communications is required with someone at one of the sites. The use of the 0–4 kHz portion of the sub-baseband is an effective method of using more of the baseband frequency spectrum, for this portion is usually too noisy for the assignment of a multiplex channel.

2.18 MICROWAVE SUPERVISORY SYSTEM

The supervisory system utilizes the 4–8 kHz portion of the sub-baseband to monitor alarm points and various status signals throughout the whole network. As the network expands and more alarm points and status signals grow in number, the portion 8–12 kHz can be used to accommodate the additional bandwidth necessary to relay all this information back to a central location for analysis and processing.

Supervisory systems are able to monitor at each site 30 alarm points or more using remote terminal units (RTUs) and remote terminal expander units (RTEUs). The system may operate on a polling basis with the master terminal unit (MTU) located at the central office. The MTU sends control tones to each of the RTUs, in turn and receives back binary information in separate time slots. These time slots correspond to the identification of the RTU, and control signals, while one of these slots may be dedicated specifically to the determination of any change of state which has occurred on a monitor or alarm point. The MTU on recognizing that a change of state at any of the alarm points has occurred will raise an audio and visual alarm on the control panel, indicating whether it is a major or minor alarm, and where it has occurred.

Alarm supervisory subsystems are in the main collectors of information, and do not necessarily perform any other active function, such as the turning on or off of valves, switches, etc. Typical alarms may be PABX, microwave transmitter and receiver, multiplex, baseband switch, AC power, DC power, subsystems, the opening of doors (intruder alarm). In addition to the alarms the system may have the facility for logging-in peronnel, if the system is tied into a computer. The equipment used in the system usually comes from the manufacturer with the alarm points pre-wired to a terminal block, or terminal points, which are accessible to the equipment installer. Part of the installation will be the wiring of these blocks or points to the station main distribution frame (MDF) or carrier distribution frame (CDF).

2.19 MULTIPLEX DESCRIPTION

Frequency-division multiplex (FDM) overcomes the problem of how to transmit over a radio path more than one voice channel. The manner in which this is done is by modulating the voice channel with a carrier frequency, and then choosing the upper or lower sideband. This is then used for further modulation. Schemes to organize the stacking of these voice channels for transmission have been quite numerous. Fortunately today a system has evolved which is almost universally accepted. The systematic stacking of voice channels internationally recognized is that recommended by the CCITT[3] in Recommendations G231–G235 under the heading of 'Translating Equipment used on Various Carrier-Transmission Systems' (Orange Book, Volume III, 1976).

There are still in existence, although probably being phased out, those FDM systems which are not to CCITT recommendations, and the modulation plans are necessary before any work on these systems can be attempted. Some systems although differing in the method of modulating to group and supergroup levels do end up with the same band of baseband frequencies as recommended by the CCITT. Other schemes place the channels after modulation direct-to-line (DTL)[4]. The direct-to-line multiplex may provide an economical means of dropping small numbers of channels from a high density system, since the system translates blocks of six voice channels to the HF line without using the conventional group and supergroup levels, whilst allowing end-to-end compatibility with standard CCITT high-density equipment.

2.20 CHANNELIZATION PLAN

No matter what form the channelization takes – whether it is in graphical form or tabular form – there are common characteristics. The plan must indicate for each and every channel in the system:

> The channel number
> The group number
> The supergroup number, etc.

and against each of designated groups of numbers, indicate:

> What its use is for
> Where it originates
> Where it ends
> Its carrier frequencies (channel, group, etc.)
> Baseband frequencies of each channel

With this information placed at every multiplex-equipped site, it is possible for the test personnel to see at a glance where each channel is inserted and dropped, whether duplicated channels occur, and if so where, together with information for testing each channel individually, etc. If changes or additions are to be made to the existing system involving multiplex channels, it is nearly impossible to implement them without such a plan being produced first, with the marked-up changes on it. During the design stages of a system, the plan indicates whether spurs operating with duplicated channel and group modems are required to be fully demodulated to produce the audio signal before being patched across to another multiplex bank, or whether they can be inserted directly into the baseband, etc.

REFERENCES

1. Godin, R. J., 'The Cellular Telephone goes on Line,' *Electronics*, 22 September 1983, pp. 121–129.

2. 'Interference Criteria for Microwave Systems in the Safety and Special Radio Services,' *Industrial Electronics Bulletin* No 10–C, August 1976. Electronics Industries Associative Engineering Department, 2001 Eye Street, N.W. Washington, DC 20006, USA.

3. CCITT Sixth Plenary Assembly, Orange Book, Vol. III–I, *Line Transmission*. Recommendations G231–G235 (inclusive).

4. Brown, D. C. and Ruebusch, R. R., A New Direct-to-Line Multiplex, Document 7733, Farinon Electric Co., San Carlos, California, USA.

3 THE BASICS OF FREQUENCY-DIVISION MULTIPLEX

3.1 TWO-WIRE/FOUR-WIRE TRANSMISSION – HYBRID OPERATION

Normally all multiplex channels operate on a four-wire basis, that is two wires for the transmitting circuit and two wires for the receiving circuit, unless the channel is specially set up for two-wire operation by having an internal hybrid* installed. As a telephone conversation requires transmission in both directions, if both the transmitted signal and the received signal are placed on the same pair of wires, it is called *two-wire (2W) transmission*. A radio system has one carrier frequency for transmission between two sites A and B, and a different carrier frequency for transmission between sites B and A; hence there are separate circuits for the send and receive signals, for full duplex operation. Figure 3.1 shows a block diagram of the conversion from 2 W to 4W operation using terminating sets or 'term sets'. These term sets contain a voice-frequency hybrid.

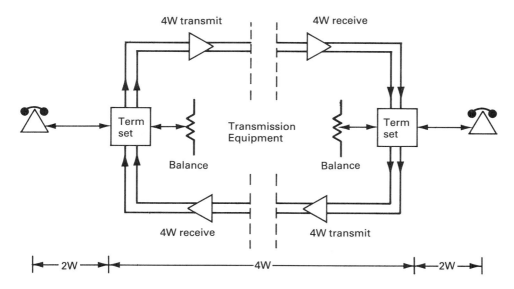

Figure 3.1   2W/4W Network

* See Section 3.1.1.

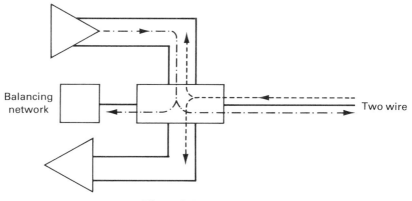

Figure 3.2   Hybrid Operation

3.1.1 Operation of a Hybrid

A hybrid is a transformer in which signal energy from the two-wire subset connection divides equally, half dissipating in the impedance of the four-wire side of the receive path (600 Ω), and the other half into the 4 W transmit path (also 600 Ω). Thus there is in the ideal case half of the speech or signal power lost by being dissipated in the 4 W receiving circuit. When a signal enters the hybrid from the 4 W receiving circuit, half the signal power enters the 2 W circuit, while the other half is dissipated in the balancing network, again ideally producing a 3 dB loss in signal power. Due to extra losses caused by the resistance of the transformer windings, balancing network imbalance, etc., in addition to the 3 dB loss, another 0.5 dB can be added. The total loss in signal power encountered by using a term set is approximately 3.5 dB. Some hybrids used on short subscriber connections purposely have higher losses to prevent singing or howling of the circuit. These higher losses are inserted by using pads which may be a part of the term-set circuits. Figure 3.3 shows a typical term-set circuit. The capacitors are *network build-out* (NBO) capacitors and are used to compensate for the capacitance of office cables or associated gain devices, and so to improve the trans-hybrid loss, or the 4 W return loss. Usually the distant outstation sends a 2 kHz test tone at a level of −16 dBm into the 4 W transmitting end of the multiplex channel. At the 4 W receiving end of the local multiplex channel a 600 Ω termination is applied and a measurement taken to insure that the received level is correct (approx. +7 dBm). The 4 W receiving end of the multiplex channel is then connected to the term set, and a measurement made at the 4 W transmit from the term set. With the 2 W circuit connected to the subscriber, the NBO capacitors are switched in until the signal measured is at a minimum. More precise adjustment may be made by varying the test tone from 2 kHz to several frequencies in the voice band. Usually all handsets are 600 Ω, but if the 2 W circuit connecting the telephone to the term set is very long, 900 Ω may be a better approximation. Switch settings on the term set allow for this so that its operation may work at 900 Ω instead of 600 Ω. Care must be taken to insert in series a 300 Ω resistor between the term set and the 4 W transmit and to

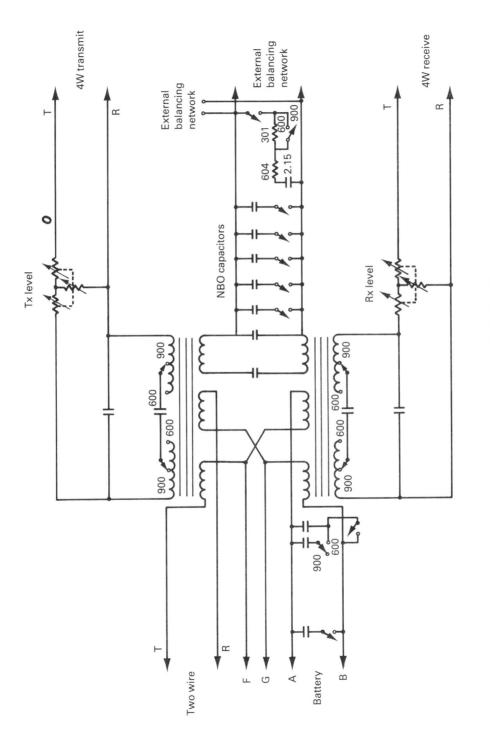

Figure 3.3 Basic Term Set

insert the appropriate network after the receiving end of a multiplex channel if it is designed to operate at 600 Ω.

In a 4 W switching system, it is often considered desirable to handle speech and signaling at lower values of absolute power through the switching equipment than is customary in 2 W systems. In 1964 the CCITT (Recommendation G101 Section B) adopted a relative level of −3.5 dBr for the sending end of a 4 W circuit, and −8.5 dBr at the receiving end, at the 'virtual' switching points. These are theoretical points; their exact location depends on national practice, and the CCITT considers it unnecessary to define them. Hence the levels of −16 dBr send and +7 dBr receive do not conflict with this requirement, since it will appear at a different point in the 4 W circuit from those in the above CCITT recommendations − a point which possibly could be immediately on the 4 W side of a 2 W/4 W hybrid. The CCITT 2 W levels are 0 dBr send, and −5 dBr receive.

3.2   GENERAL DESCRIPTION OF FDM

Frequency-Division Multiplex (FDM) is a method of allotting a unique band of frequencies into the comparatively wide-band radio-frequency spectrum. The allocation of each communications channel is done on a continuous time basis. The communications channel may be:

Voice	4 kHz wide
Telegraph	120 Hz wide
Broadcast	15 kHz wide
Data	48 kHz wide
TV	4.2 MHz wide

All multiplexing (MUX) operates on 4 W basis. Figure 3.4 shows a block diagram of the multiplex operation.

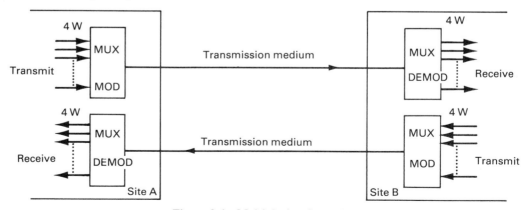

Figure 3.4   Multiplexing Operation

3.3 MIXING

The mixing or heterodyning of signals at frequencies f_1 and f_2 is shown in Figure 3.5. The output consists of the frequencies f_1, f_2, and harmonics of f_1, and f_2, together with the second-order products $f_1 + f_2$, $f_1 - f_2$, the third-order products $2f_1 + 2f_2$, $2f_2 + f_1$, $2f_1 - f_2$, $2f_2 - f_1$, fourth-order products $3f_1 + f_2$, etc. As the product order rises, the amplitude of these products falls at a rate $(n-2)$ dB faster than the falling rate of the second-order product, with decreasing input signal level to the mixer[1,2]. For multiplex applications, interest lies only in the second-order products, and furthermore in only one of the two formed. All the higher-order products, including one of the second order, are filtered out immediately after the mixing.

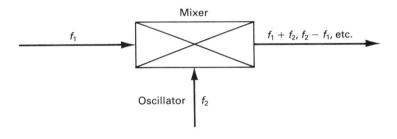

Figure 3.5   Mixing Process

Consider a 4 kHz voice channel, whose 3 dB cut-off frequencies are 0.3 kHz and 3.4 kHz. This channel is to be slotted into a basic group by pre-group modulation (double modulation), and so with a 20 kHz oscillator the resulting frequencies will be: 20 kHz ± 0.3 kHz, etc., for an 0.3 kHz input and 20 kHz ± 3.4 kHz, etc., for a 3.4 kHz input. There are a range of frequencies between 0.3 kHz and 3.4 kHz, and as each frequency in this range is modulated or mixed with 20 kHz, it is easier to represent the band 0.3 kHz to 3.4 kHz by a triangle, and the result of the mixing process or translation by triangles, one for the upper sideband and another for the lower sideband. Figure 3.6 shows these triangles. When the *sum* of 20 kHz and the band of voice frequencies is taken, and all the higher and lower harmonics and third-order and above are filtered out, as well as the lower sideband of the second-order, the upper sideband remains. This can be represented by a triangle with its right angle on the right-hand side. This triangle represents an *upright* or *erect* sideband. Similarly, if the *difference* second-order is taken, that is 20 kHz − 0.3 kHz to 20 kHz − 3.4 kHz, during translation an inversion of frequencies occurs. The higher voice-band frequency becomes the lower frequency of the translated spectrum and an *inverted* sideband is formed. This is represented by a triangle whose right angle is to the left.

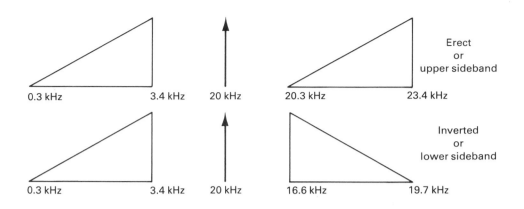

Figure 3.6   Erect and Inverted Sidebands

3.4   MULTIPLEXING SYSTEMS – MODULATION PLANS[3,4,9]

3.4.1 Formation of the Basic Group

Broad-band terminal equipment accepts a number of twelve-channel basic groups and assembles them into 60-channel units, 300-channel units, etc. Combinations of these units are arranged to produce a baseband frequency which can be applied to a coaxial cable or a microwave radio bearer. Figure 3.7 shows the relative positions for

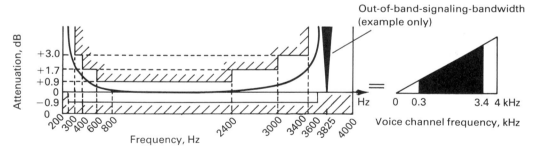

Mask. CCITT Rec. G232 - Receiving equipment of any channel of a 12-channel terminal

Figure 3.7   Voice Channel

speech and signaling bandwidths within the voice-channel band. Note that the frequency 3.825 kHz is used if 'out-of-band' signaling is required. Although the speech band is 0.3 kHz to 3.4 kHz, the actual voice band which undergoes modulation and demodulation is 0 – 4 kHz wide. The total voice band is usually considered in all of the modulation plans, since this includes the speech band, out-of-band signaling bandwidth, and guard bandwidth. The modulation of twelve separate voice channels into the basic group of 60 – 108 kHz as recommended by the CCITT recommendation G232 can be achieved in two ways:

1. A single modulation stage for each channel
2. Two modulation stages for each channel

3.4.2 Single Modulation Plan

Figures 3.8 (a) & (b) show the modulation plan for assembly of 12 channels into a

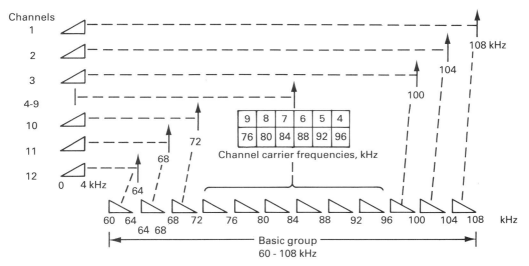

(a) Single modulation stage

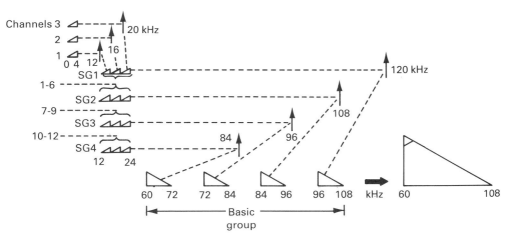

(b) Double modulation method

Figure 3.8   Basic Group – CCITT Type B

type-B group (inverted). Where a single stage of modulation is used the 12-channel carrier frequencies are 64 kHz, 68 kHz, 72 kHz, etc., in 4 kHz steps up to 108 kHz. The modulator and demodulator of a particular channel is supplied with the same carrier frequency. This plan is used for both radio links and coaxial cable carrier systems. If the type-B group is modulated with a carrier of frequency 120 kHz, the CCITT type-A (erect) group is formed.

3.4.3 Double Modulation Plan

The 12 channels are divided into a number of identical *subgroups (pre-groups)* for initial modulation, and then the frequency range output of each subgroup is further modulated to produce the frequency range 60–108 kHz. The channels are divided into four subgroups of three channels each, and then modulated with the channel carrier frequencies 12 kHz, 16 kHz, and 20 kHz applied to the modulators and demodulators. The upper sideband product of modulation is selected by the modulator band-pass filters, and the frequency translation at this stage is shown in Table 3.1. From this, four subgroups each consisting of three channels are established, which are individually modulated with the carriers, 120 kHz, 108 kHz, 96 kHz and 84 kHz. In the receiving direction, subgroup demodulators using the same carrier frequencies translate each of the bands of the incoming basic group back to the subgroup range 12 – 24 kHz, and then by individual channel demodulation to the voice frequency range.

Table 3.1

Channel numbers	Carrier frequency, kHz	Resultant frequency range, kHz
1, 4, 7, 10	12	12 – 16
2, 5, 8, 11	16	16 – 20
3, 6, 9, 12	20	20 – 24

The advantages of using two stages of modulation are:

A reduction in the number of carrier frequencies to be generated and distributed.

More simple design in the band-pass filters.

The subgroup equipment is identical, hence permitting ease in manufacture and maintenance.

The disadvantage is that there is an additional subgroup modem.

Note that for the type-B group, the translated channel 1 occupies the frequency band 104 – 108 kHz, and decreases in frequency as the channel number increases, until channel 12 occupies the frequency range 60 – 64 kHz. For the type-A group, channel 1 occupies the frequency range 12 – 16 kHz, and the frequency range increases as the channel number increases, until channel 12 occupies the frequency range 56 – 60 kHz.

3.4.4 The Group

As opposed to the 'Basic Group', the *group* is a set of 12 channels, occupying adjacent frequency bands in a bandwidth of 48 kHz. The start of the range of frequencies comprising the group, and the end frequency of the range, are defined by the group number, or where the group is slotted into the baseband spectrum. Figure 3.9 shows the assembly of groups in forming the basic supergroup.

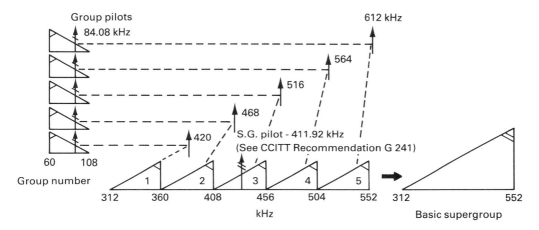

Figure 3.9   Formation of CCITT Basic Supergroup

3.4.5 The Basic Supergroup

The *basic supergroup* consists of a set of five groups (60 channels), occupying adjacent frequency bands in the frequency range 312 – 552 kHz. Due to translation, the groups in a basic supergroup are erect groups. Figure 3.9 shows the formation of the basic supergroup using five groups. Note that the CCITT symbol for an erect group is a triangle with its right angle to the right, and a single line across the top of the triangle, whereas for the basic supergroup (or supergroup) it is two parallel lines at the top of the triangle. All 60 channel units are initially assembled into the basic supergroup frequency range and then by modulation are translated to the required supergroup frequency range.

3.4.6 The Supergroup

A set of five groups (60 channels), occupying frequency bands in a bandwidth of 240 kHz, constitutes the *supergroup*. The position of a group within a supergroup is identified by a number from 1 to 5. The numbering in order of frequency is in ascending order when the groups are erect groups, and in descending order when the groups are inverted. The supergroup frequency at the beginning and the end of the range is determined by the position it takes in the baseband frequency spectrum.

3.4.7 The Basic Master-group

The *basic master-group* comprises a set of five supergroups (300 channels), occupying frequency bands separated by 8 kHz, in the frequency range of 812 – 2044 kHz. This is shown in Figure 3.10. The numbering scheme for supergroups relates to the supergroup arrangement of a standard 4 MHz coaxial cable system. All 300 channel units are initially assembled into the basic master-group frequency range, and then by modulation are translated to the required master-group frequency range. The basic master-group and master-group are both identified by three parallel lines drawn at the top of the triangle.

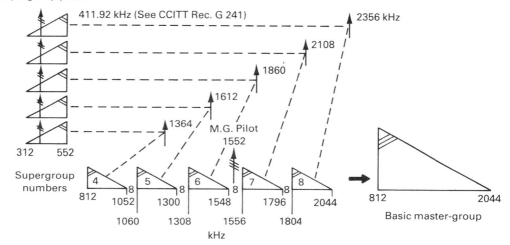

Figure 3.10   Formation of CCITT Basic Master-group

3.4.8 The Master-group

A set of five supergroups (300 channels), occupying frequency bands separated by 8 kHz, in a bandwidth of 1232 kHz. The position of a supergroup within a master-group is identified by a number from 4 to 8.

3.4.9 The Basic Super-master-group

A set of three master-groups (900 channels), separated by two free spaces of 88 kHz and occupying the frequency range 8516 – 12388 kHz. This is shown in Figure 3.11. Notice that the basic super-master-group and the super-master-group are both identified by four parallel lines at the top of the triangle.

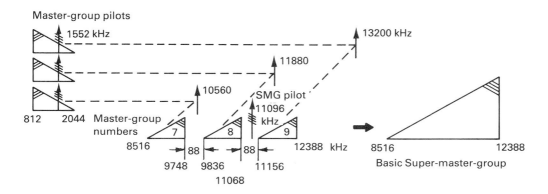

Figure 3.11 Formation of CCITT Basic Super-Master-group

3.4.10 The Super-master-group

A set of three master-groups (900 channels), separated by two free spaces of 88 kHz and occupying a total bandwidth of 3872 kHz. The position of a master-group within a super-master-group is identified by a number from 1 to 9 as shown in Figure 3.11. Typical arrangements of supergroups, master-groups, and super-master-groups for various systems are described below.

3.4.11 The Standard 4 MHz System

The arrangement of supergroups for a standard 4 MHz system, providing 960 channels, is shown in Figure 3.12(a). The frequency range used extends from 60 to 4028 kHz, and the supergroups are numbered from 1 to 16.

Figure 3.12(b) shows the arrangement of 15 supergroups for a 900-channel 4 MHz system. The frequency range extends from 312 to 4028 kHz and the super-groups number from 2 to 16. This system has the advantage that equalization for coaxial cable characteristics is simplified by avoiding the use of frequencies below 300 kHz.

3.4.12 A 6 MHz System

The arrangement of 16 supergroups and one master-group for a 6 MHz system is shown in Figure 3.13. The supergroups number from 1 to 16 and the number 4 master-group is used. A frequency range of 60 – 5564 kHz is used to provide 1260 channels. If supergroup 1 is not used, a 1200-channel system is formed.

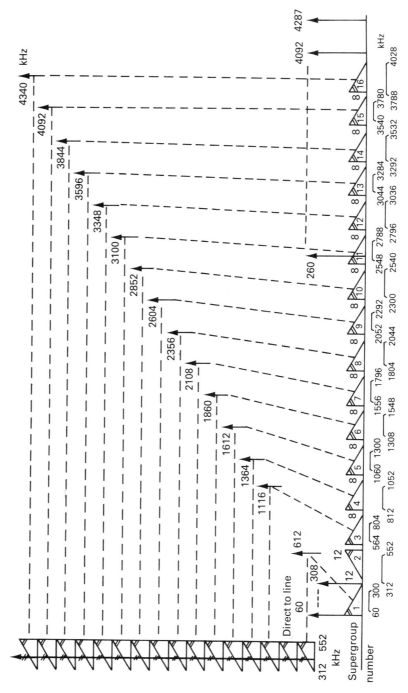

(a) 960-channel, 4 MHz system

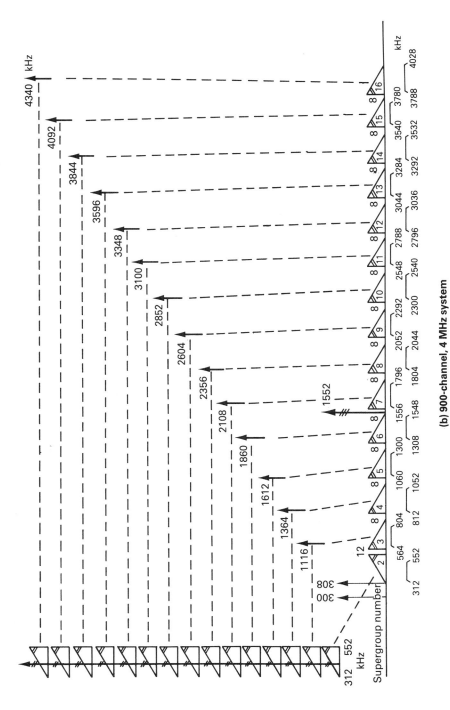

(b) 900-channel, 4 MHz system

Figure 3.12 4 MHz Systems

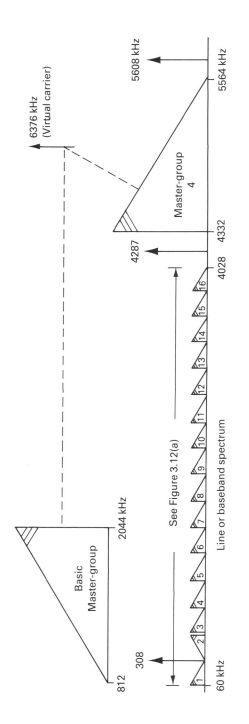

Figure 3.13   1260-Channel 6 MHz System

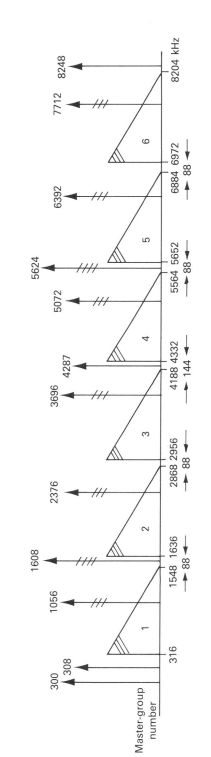

Figure 3.14   1800-Channel 8 MHz System

3.4.13 An 8 MHz System

Figure 3.14 shows the arrangement of six master-groups for an 8 MHz system. The master-groups number 1 to 6, and a frequency range of 316 – 8204 kHz is used to provide 1800 channels.

3.4.14 A 12 MHz System

Figure 3.15A shows the arrangement of 15 supergroups and 6 master-groups (2 super-master-groups) to provide a total of 2700 channels. The supergroups are numbered from 2 to 16 and the master-groups from 4 to 9. A frequency range of 312 – 12388 kHz is used. An alternative arrangement is shown in Figure 3.15(b). This arrangement provides 2700 channels using a total of nine master-groups or three super-master-groups, which occupy a frequency range of 316 – 12388 kHz.

3.4.15 Pilots

A pilot is a single frequency which is transmitted with the various groups, super-group, etc., of the multiplex information, or which is transmitted as a part of the microwave radio bearer system.

The purposes of these pilots are:

1. To control level
2. To indicate system continuity
3. To provide alarms
4. To permit in some cases, frequency synchronization of the carrier equipment

The pilots which are used in an FDM microwave radio-relay system can be divided into two groups: radio-relay system pilots and FDM carrier equipment pilots.

3.4.15.1 Radio-relay system pilot [5]
This pilot is required on radio-relay systems to indicate the continuity of the circuit. Typically, the frequency used is ten per cent higher than the upper limit of the transmitted baseband. The discrete frequency chosen so that intelligible crosstalk may be reduced is $(4n-1)$ kHz, where n is an integer. Table 10.1 of Section 10.1.5 provides additional information on this type of pilot.

3.4.15.2 FDM carrier equipment pilots [6]
These pilots are required to be transmitted with groups, supergroups, master-groups, etc. They automatically regulate or monitor the level of each group, etc., as it is presented to the transmit baseband. The output level due to the action of these pilots remains within ±5 dB of the nominal level set for the baseband. If there is a deviation greater than ±5 dB, then an alarm is raised. Figures 3.9–3.15(b) inclusive show the various pilot positions in each of the modulation plans. The symbol for a

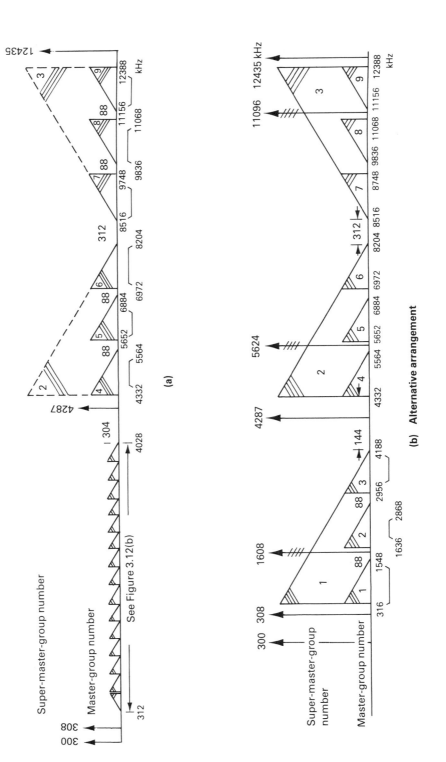

Figure 3.15 *2700-Channel 12 MHz System*

pilot is a vertical line with an empty arrowhead at the top. For a group pilot this is represented by one slash across the vertical line. The supergroup pilot has two slashes, the master-group three, and the super-master-group four. These pilots are called the reference pilots and used to monitor the levels associated with the particular group, supergroup, master-group or super-master-group. Line regulating pilots are those pilots which are inserted into the baseband in between the various groupings of voice channels, and are used where automatic gain control is required. The pilot frequencies are chosen so that interference with any part of the modulated voice spectrum is kept to a low level. The CCITT recommended maximum level of interference is −73 dBm0p. If the incoming level of a pilot frequency to the demodulating equipment falls outside of ±4 dB of the nominal received baseband level, an alarm is raised. The receive pilot filters are designed to permit the pilot to have a bandwidth of 50 Hz.

The group regulation pilots assigned by the CCITT are:

84.080 kHz at a level of −20 dBm0
84.140 kHz at a level of −25 dBm0

The basic supergroup, master-group, super-master-group, and 900 channel 4 MHz system pilots as given in Figures 3.9–3.12(b) are all transmitted at a level of −20 dBm0. These are the pilots which are transmitted in the guard band between voice channels.

3.4.16 The Farinon LD–G Multiplex Modulation Plan[7]

This plan arrives at the same frequencies for Type B CCITT group, and CCITT

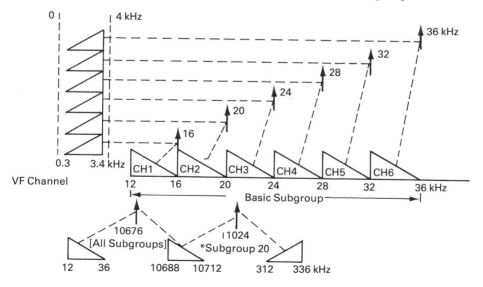

Figure 3.16   LD–G MUX, Basic Subgroup, and Subgroup

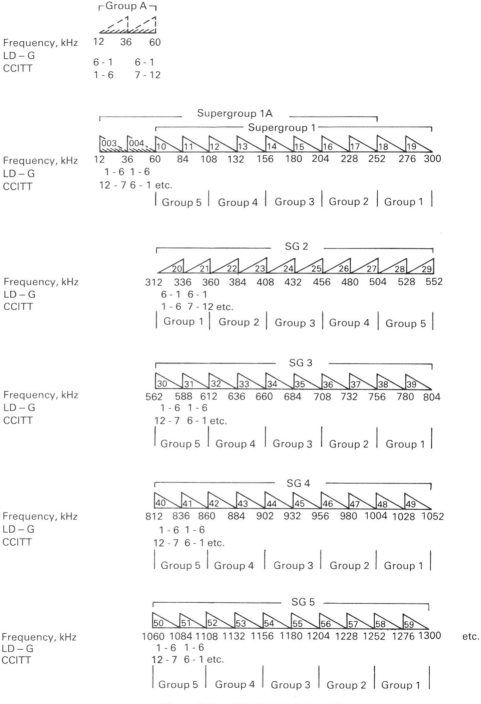

Figure 3.17   LD–G Modulation Plan

supergroups, without using the same carrier frequencies. Thus the transmission of the baseband into CCITT baseband designed transceivers and filters is possible, but the modulating and demodulating equipment would not be compatible. The basic building block is the voice channel with a frequency range 0 – 4 kHz, and is similar in all respects to the CCITT voice channel (unweighted). Six of these voice channels comprise a subgroup, which is the next building block. These subgroups then are assembled to form the CCITT supergroup frequency ranges. Each basic subgroup, as formed by the six different channel modems, is again translated by each different subgroup modem. The baseband constitutes the placing direct to line of the output of each of the subgroup modems. Each subgroup modem is different from the next, and sequentially numbered from 010 to 109. Numbers 010 to 019 are frequency-range-compatible with CCITT supergroup 1, 020 to 029 with supergroup 2, etc., up to 100 to 109 being used for supergroup 10, giving a total channel capacity of 600, within the CCITT frequency range (60 – 2540 kHz). Figure 3.16 shows the formation of the basic group, and the use of two translations to arrive at subgroup 20, whereas Figure 3.17 shows the LD–G modulation plan.

3.5 ELECTRICAL DESCRIPTION – TRANSMIT AND RECEIVE PATHS, SIGNALING, SYNCHRONIZATION

Figure 3.18 is a block diagram of the equipment which is used to show the translation of voice channels through the channel modems, to form the basic group, and the translation of these basic groups in the group modems to form the basic supergroup.

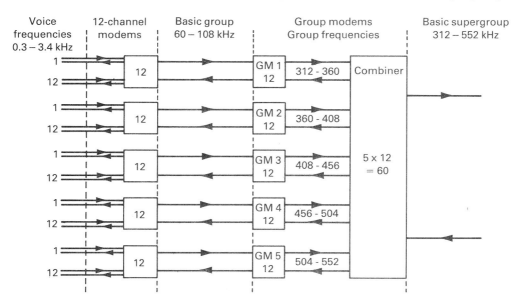

Figure 3.18   Translation to Basic Supergroup

Figure 3.19 takes the process further by using a block diagram to show the equipment which would be used to translate 960 channels to line. If the radio equipment is not located in the same building as the multiplex equipment, the baseband would be transmitted to the radio equipment by coaxial cable.

The channel modem equipment operates on a four-wire basis and accepts the voice band on the transmit path into the modem. Signaling information if applied by keying the M-lead of the channel modem either applies or removes a 3825 Hz 'out-of-band' signaling tone to be transmitted along with the speech band (300 – 3400 Hz). In the channel modem the voice band (0 – 4 kHz) is modulated by the channel oscillator, the oscillator frequency being different for each of the twelve channels. After modulation each of the twelve channels is combined to form the 'basic group'.

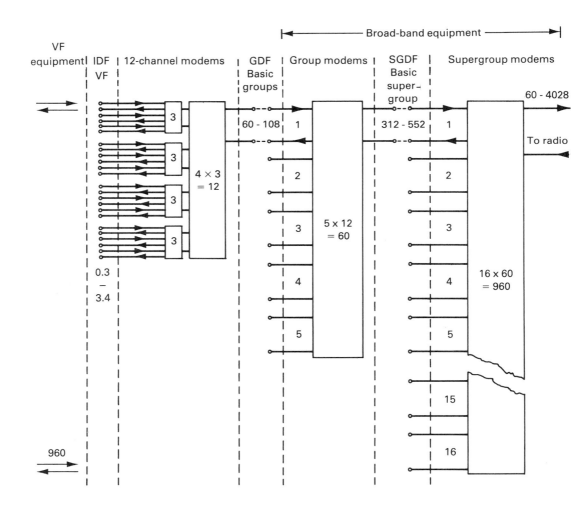

Figure 3.19   Translation of 960 Channels

The remaining four twelve-channel modems are used to form four more basic groups. The total of five basic groups are each transmitted to five separate group modems, where the basic groups are translated by individual group modem oscillators. The five different groups are then combined to form the 'basic super-group'. In a similar fashion fifteen other basic supergroups are formed, and each one applied to sixteen different supergroup modems for translation according to the supergroup modem oscillator frequency. The resulting sixteen supergroups are then combined and transmitted to line to constitute the baseband of a 960-channel system. This baseband is then entered into the baseband input of the radio transmitter and frequency-modulates the transmitter carrier. The far-end radio receiver demodulates the frequency-modulated carrier to form the receiver baseband. This baseband signal then enters into the supergroup modems where sixteen different supergroups are formed from the composite baseband signal. All the supergroups reconstituted by mixing with the supergroup carriers and filtering enter into their respective group modems. Similarly each of the basic supergroups enter into the group modems where they are mixed with each of the group carriers and filtered to form the basic groups. Each basic group is applied to the 12-channel modems associated with that basic group, and by means of channel carriers and filters separated into the 12 separate voice channels. The speech signal will emerge from the channel modem on the 'receive' pair, and the signaling information contained by the 3825 Hz tone is detected in the channel modem to activate the E-lead.

Figure 3.20 shows a block diagram of the equipment involved to implement the Farinon LD–G modulation plan. In this case, six-channel modems individually take the voice band and translate each voice channel, including the 3825 Hz signaling tone which is E-lead-activated or deactivated, into the basic subgroup. Each separate subgroup by double translation, translates each basic subgroup direct to line. The first subgroup translation involves a 10676 kHz oscillator which is common to each of the group modems. The second oscillator is pertinent to the subgroup number. Each subgroup modem up to a maximum of fifty enters into a 'combiner', which combines and adjusts the level of the composite signal before presenting it to the radio as the composite baseband signal. After being received and demodulated by the far-end radio receiver, the baseband signal enters into the combiner and is presented to all the subgroup modems after adjustment of its level. Each subgroup combiner, again by double translation and filtering, selects its own part of the baseband signal and presents this to the six-channel modems associated with it. The channel modems then individually demodulate according to their respective oscillator frequencies, and present at the receive output the speech band. The received out-of-band signaling frequency 1825 Hz again activates the circuitry associated with the E-lead, and thus the E-lead, itself.

As will be discussed in Chapter 4, if additional equipment is added to the 4 W channel modem, 'in-band' signaling may be used. The standard frequency for in-band signaling is 2600 Hz.

At times it is necessary to synchronize the near- and far-end channel oscillators, especially if data are being transmitted over the channel. This is possible if the 'slave' terminal has the provision for injection locking. The out-of-band signaling tone from the 'master' terminal, if activated during transmission, injection locks the slave

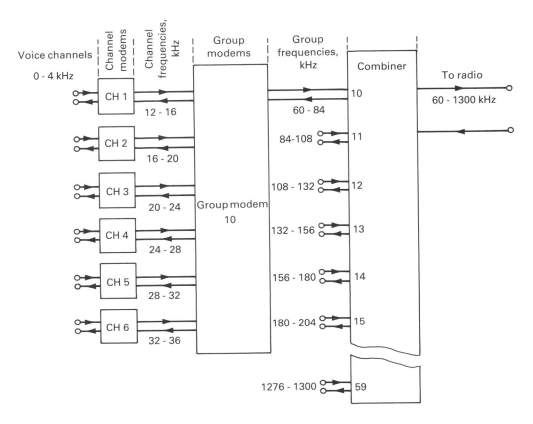

Figure 3.20   LD–G Equipment

terminal channel oscillator, and permits any frequency variations from the master terminal to be automatically followed at the slave terminal, thus maintaining continuous end-to-end synchronization.

3.6 SPECIFICATIONS

Below are the specifications for the Farinon LD–G multiplex modulation plan. These specifications have been included in this chapter since they give an indication of how multiplex specifications are written, and they also assist in providing the results to be expected from the tests performed on the equipment.

Multiplex method   Frequency division
Type of modulation   Single-sideband suppressed carrier.
Number of channels   1 to 312 (See Note 1)

HF interface

Line frequency range	4 to 2788 kHz
Transmit level (0 dB Pad)	−21.5 dBm (strappable −27.5 to −37 dBm)
Receive level (0 dB Pad)	−28.5 dBm (strappable −13 to −28.5 dBm)
Built-in pad range	0 dB to 15.5 dB in 0.5 dB steps (external pads available)
Impedance	75 Ω, unbalanced
Return loss (4 to 2788 kHz)	
60 channels	26 dB minimum, for all levels.
312 channels	20 dB minimum, for 0 dB of HF pad; 26 dB minimum, for −25 dBm HF transmitter and receiver

Carrier leak

Channels	−50 dBm0, typical (transmitter mute disabled)
	−45 dBm0, maximum (transmitter mute disabled)
	−70 dBm0, typical (transmitter mute enabled)
	−65 dBm0, maximum (transmitter mute enabled)
Groups	−45 dBm0, typical
	−40 dBm0, maximum

VF Interface (4 W)

Transmit level	−16 dBm
Receive level	+7 dBm, adjustable ±4 dB
Impedance	600 ohms, balanced
Return loss	26 dB minimum (0.3 to 3.4 kHz)
Longitudinal balance	50 Hz to 3.4 kHz; ≥75 dB Tx; ≥65 dB Rx

VF Interface (2 W resistance hybrid terminated in 600 Ω and 2.15 μF; see Note 2)

Transmit level	0 dBm
Receive level	−4 dBm, adjustable
Impedance	600 Ω, balanced
Return loss	
Echo return loss	20 dBm minimum with weighted noise signal
1 kHz return loss	25 dBm minimum
Singing margin	12 dB minimum
Trans-hybrid balance (0.3 − 3.4 kHz)	45 dB minimum

Transmission Performance

	TYPICAL	MAXIMUM
Frequency response (end-to-end)		
0.3 − 3.4 kHz	0.75 − 2.5 dB	0.9 − 3.0 dB
0.4 − 3.0 kHz	0.75 − 1.2 dB	0.9 − 1.7 dB
0.6 − 2.4 kHz	0.75 − 0.75 dB	0.9 − 0.9 dB
Delay		
Absolute delay (relative to 2 kHz)	1200 μs	1500 μs
Envelope delay (relative to 2 kHz)		
0.5 − 3.3 kHz	1350 μs	1500 μs
0.6 − 3.2 kHz	900 μs	1100 μs
0.8 − 2.8 kHz	500 μs	550 μs
1.0 − 2.6 kHz	300 μs	350 μs

Frequency stability	Individual channel synchronization for zero frequency error end-to-end	
Phase jitter (channels synchronized)	$\pm 1°$ (peak-to-peak)	
Harmonic distortion		
0 dBm, 0.3 – 3.4 kHz	0.25 per cent	1 per cent
Linearity	0.3 dB maximum, at +3.5 dBm0	
Limiting	+9 dBm0 maximum, at +20 dBm	
Level stability (end-to-end)	± 0.5 dB for 6 months	
Inter-channel crosstalk		
Intelligible		
(using 1 kHz tone, 0 dBm)	< -65 dBm0	
Unintelligible		
(using 1 kHz tone, 0 dBm)		
Adjacent channel	20 dBrnC0	25 dBrnC0
Non-adjacent channel	20 dBrnC0	25 dBrnC0
Inter-channel crosstalk (side tone)		
Near end (0.3 – 3.4 kHz tone at 0 dBm0, HF terminated)		
60 Channels	18 dBrnC0	20 dBrnC0
312 Channels	22 dBrnC0	33 dBrnC0
Far end (4 W VF terminated, 4 – 5600 kHz)		
60 Channels	16 dBrnC0	18 dBrnC0
312 Channels	21 dBrnC0	25 dBrnC0
(Loaded condition – idle condition for 60 channels included)		
	CCITT (Note 4)	-8 dBm0 (Note 3)
60 Channels	18 dBrnC0	20 dBrnC0
120 Channels	18 dBrnC0	20 dBrnC0
312 Channels	19 dBrnC0	25 dBrnC0

Signaling

Type	Out of band
Frequency	Channel carrier reinserted
Level	-16 dBm0
Speed	8 – 14 pulses/second at 30 – 80 per cent break
Pulse distortion	± 5 per cent maximum for ± 3 dB level variation
Contact rating	100 VA (with internal contact protection)

Power requirements (for each six-channel group; Note 1)

	NOMINAL VOLTAGE (DC)	OPERATIONAL VOLTAGE	CURRENT
-24 V option	-25.8 V	-21 to 36	1.0 A max
-48 V option	-51.6 V	-42 to 56	1.0 A max

Environmental Properties

Temperature ranges	
Meets all specifications	0 to 50°C
Operational	-10 to +60°C
Storage	-40 to +65°C
Humidity	95% at 40°C
Altitude	15 000 feet, maximum
Mounting	Attaches to standard 483 mm (19 inch) equipment rack.
12-channel assembly	Occupies 3 vertical spaces (133 mm, 5.25 in)
24-channel assembly	Occupies 6 vertical spaces (267 mm, 10.5 in)
36-channel assembly	Occupies 9 vertical spaces (400 mm, 15.75 in)
48-channel assembly	Occupies 12 vertical spaces (533 mm, 21 in)
60-channel assembly	Occupies 15 vertical spaces (667 mm, 26.25 in)

Weight

12-channel assembly	15 kg (33 lb)
24-channel assembly	27.5 kg (60.6 lb)
36-channel assembly	40 kg (88 lb)
48-channel assembly	52.6 kg (116 lb)
60-channel assembly	65 kg (143.4 lb)

Notes

1. Standard assemblies are available for up to 120 channels, at -24 or -48 VDC.
2. A 4W – 2 W resistance hybrid is optionally equipped within the channel unit. External 2 W equipment is also available with different specifications.
3. For 312-channel systems, those channels in the 12–36 kHz spectrum will experience loaded noise of up to 25 dBrnC0, and for 120 channels, 20 dBrnC0.
4. Adjacent channels are loaded at -15 dBm0.

3.7 CHANNEL MODEM

3.7.1 CCITT Modem

3.7.1.1 Single Modulation Stage
In the transmitting direction of a single modulation stage (Figure 3.21(a)), each of the 12 voice-frequency channels modulates a different carrier frequency, and the lower sideband is selected from the products of modulation by a band-pass crystal filter. In the receiving direction a frequency band of 60 – 108 kHz is divided by means of 12 different band-pass crystal filters into 12 channels, and the same carrier frequencies as those used in the transmitting path are then used to translate these channels into voice frequency bands of 300 – 3400 Hz.

3.7.1.2 Double Modulation Method
Figure 3.21(b) is a block diagram of the channel modem equipment for two modulation stages. It shows the stages by which the 12 voice-frequency channels are assembled into four 3-channel subgroups each producing 12 – 24 kHz. The four subgroups are separately applied to subgroup modulators and translated to the basic group frequency range of 60 – 108 kHz. In the receiving direction the basic group frequencies are separated into four subgroups and then brought back to voice frequencies by two stages of demodulation. Out-of-band signaling equipment is shown, but the four-wire terminating equipment which optionally converts four wires to two wires has been omitted as has the VF jack-field usually associated with this equipment for testing purposes.

In the transmitting direction, the voice frequencies are applied to the VF input circuit where a voltage limiter restricts any peaks of incoming speech to a suitable level. A low-pass filter establishes an upper limit of the speech band at 3.4 kHz to allow combining of the out-of-band signaling frequencies at this point. The signaling pulses of 3825 Hz are applied together with the voice frequencies to the channel modulator. The channel modulator band-pass filter selects the upper sideband of modulation: 12–16 kHz for channel 1, 16–20 kHz for channel 2, and 20–24 kHz for

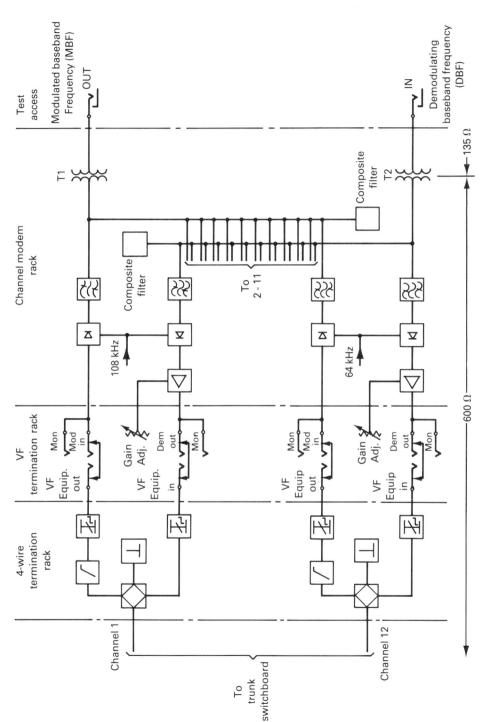

Figure 3.21 *Channel Equipment*

(a) **Single stage**

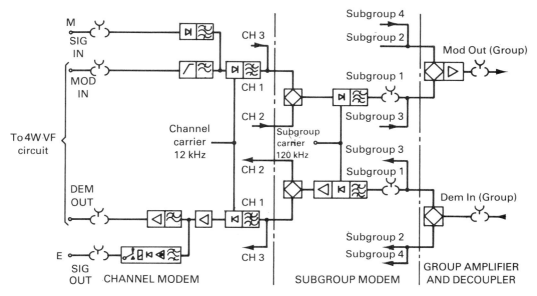

(b) Double stage

Figure 3.21   Channel Equipment (continued)

channel 3. The three channels are combined via a hybrid decoupler and, as a subgroup, are applied to the subgroup modulator for translation with the subgroup carrier of 120 kHz. The subgroup modulator band-pass filter selects the lower sideband, 96–108 kHz. Three similar circuits serve to translate channels 4, 5 and 6 to 84–96 kHz, channels 7, 8 and 9 to 72–84 kHz, and channels 10, 11 and 12 to 60–72 kHz. The four subgroups are subsequently combined into the basic group, 60–108 kHz, in the hybrid decoupler circuit incorporated with the group amplifier.

In the receive direction the basic group incoming from the hybrid decoupler is directed in subgroups to the appropriate subgroup demodulators which translate them to the subgroup position. The subgroup amplifier raises the power level to maintain a suitable margin between signal level and noise level. Each channel demodulator band-pass filter selects the appropriate band from the subgroup; this is further translated in the channel demodulator and amplified in the first stage of the channel amplifier. The gain adjust control associated with this amplifier controls the level of the VF and the signaling frequencies. The amplifier stage raises the voice frequency level to the required value for application to the four-wire VF circuit. The signaling pulses of 3825 Hz are selected, by a band-pass filter, at the output of the pre-stage channel amplifier and converted back to DC signaling pulses.

As shown on Figure 3.21B hybrid decouplers are used to combine the three channels into a subgroup, and the four subgroups into a basic group, in the transmitting and receiving directions. Because their passbands are extremely close, the neighbouring filters cannot readily be interconnected if frequency distortion is to be kept to a minimum. For this reason, channels 1 and 3 are connected to one side of a hybrid and channel 2 to the other. This allows a 4 kHz separation between the cut-off frequencies of the two filters connected directly in parallel. The design of the band-pass filters is simplified by allowing a 12 kHz separation between the cut-off frequencies of the two filters connected directly in parallel. The separation is sufficient to insure that the reactive components of one filter have negligible effect on the other. This is shown in Figure 3.22.

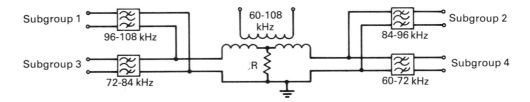

Figure 3.22   Hybrid Decoupler and Band-Pass Filters

3.7.2 LD–G Channel Modem

Figure 3.23 shows a block diagram of the LD–G channel modem. The channel modem provides the first step of transmit modulation (VF to HF) and the final step of receive demodulation (HF to VF) for one LD–G multiplex channel. The unit also performs the transmitting and receiving signaling functions and generates the carrier frequency for the modulation and demodulation processes. End-to-end synchronization is accomplished by optional master/slave strapping. The carrier frequency and the signaling tone are generated by the channel modem. All monitoring jacks are found on the front panel; hence a separate jack-field is not used in this system.

3.7.2.1 Channel Transmit
In 4 W operation, the VF signals enter the unit at the 4 W VF Tx pins, and bridging access is obtained by means of the Tx MON (EQ) (transmit equipment side monitor) front panel jack. To send a test tone into the channel the jack Tx MOD is used. If a dummy plug is placed into the Tx MOD jack, the Tx MON (EQ) jack permits access to the external equipment. The transmit signal is first limited to prevent overloading by the speech signals, then passes to a 0.3 to 3.4 kHz band-pass filter. The output of the filter is applied to the modulator, and mixed with the carrier frequency pertinent to that channel. The modulator output is then amplified and combined with the signaling tone from the channel oscillator. The Tx LEV potentiometer adjusts the resulting signal before leaving the modem at the HF Tx pins.

3.7.2.2 Channel Receive
The HF receive signals enter the unit at the HF RCV pins and are amplified. The Rx LEV (receiver level) potentiometer adjusts the amplifier output before the signals enter into the channel band-pass filter. This filter selects the desired 4 kHz channel frequency band pertinent to that modem, and passes the signals to the demodulator.

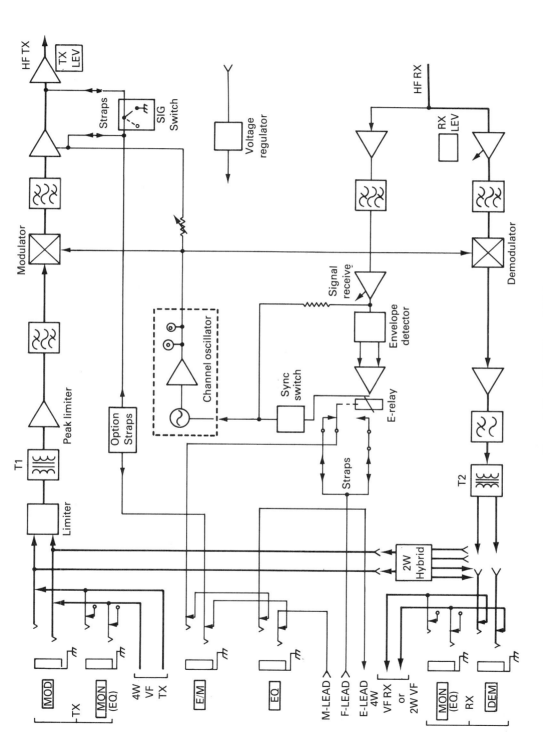

Figure 3.23 LD-G Channel Modem

After demodulation with the carrier frequency, the output is amplified and filtered to the speech band (0.3 – 3.4 kHz). The Rx DEM front panel jack provides a means to perform terminated measurements on the received signals, and the Rx MON (EQ) allows bridged measurements to be made. If a dummy plug is inserted into the Rx DEM jack, the Rx MON (EQ) jack provides direct access to the external equipment.

3.7.2.3 Channel Oscillator Card

Test points are provided on this card, which is a part of the channel modem, to monitor the carrier frequency.

3.7.3 Signaling

3.7.3.1 Transmit and Mute

M-lead conditions enter the channel modem (Figure 3.23) at the M-LEAD pins and pass through the EQ and E/M front panel jacks to the transmit signaling circuits. M-lead options include the muting of the channel transmit output until such time as an M-lead seizure occurs, as well as ground or battery keying.

3.7.3.2 Receive

The receive signaling tone is separated from the audio input path by the means of filters. This tone is then detected and used to provide information for the activation of the relay which opens or closes the E-lead contacts. An F-lead from external equipment enters the channel modem at the F-LEAD pins and feeds the E-lead relay contacts. The E/M front panel jack allows direct access to the channel modem E/M leads while breaking the EQ connection. The EQ jack allows direct access to external equipment while breaking the E/M connection. The E-lead conditions are present at the E-LEAD pins.

The M-lead can be strapped for the following options:

Idle condition	Open, ground, or battery
Seized condition	48 V battery, 24 V battery, ground

The E-lead can be strapped for the following options;
Tone on	E-lead contacts close
Tone off	E-lead contacts close

Strapping also includes whether the modem is to be the master or slave terminal, or master-end synchronization without signaling. End-to-end operation may be permitted by strapping with the M-lead unseized, or with the transmit circuit muted and the M-lead seized only.

3.8   THE GROUP OR SUBGROUP MODEM, AND THE SUPERGROUP MODEM[8,9]

Figure 3.24 shows a block diagram of a typical group modem. The five basic groups

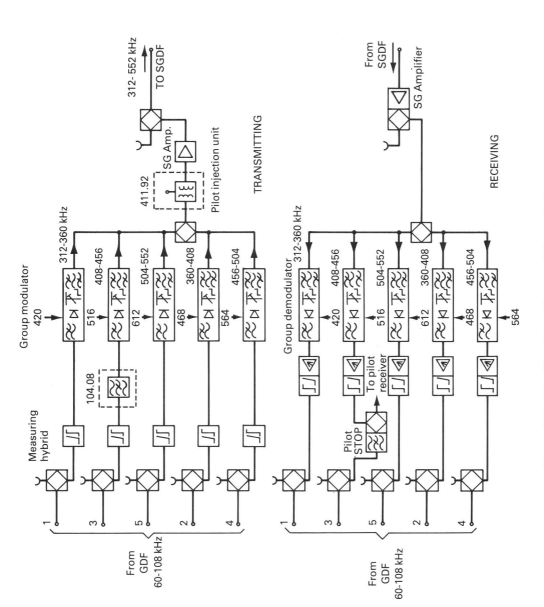

Figure 3.24 Simplified Block Diagram of Group Modem

pass via the Group Distribution Frame (GDF) to the group modulator for conversion by modulation to the basic supergroup (SG) frequency range. In the receive direction a basic supergroup is separated by filters and demodulated into five basic groups and applied via the GDF to the 12-channel modem equipment or to a through group filter for connection to another group modulator.

The circuit arrangement of the various equipment units varies with different manufacturers but the same general principles apply. In the typical group modem equipment, measuring hybrids are used to facilitate testing. The hybrid transformer matches the measuring point, which is situated on a supervisory panel, to the measuring equipment. An associated resistive network ensures that the measuring point is isolated from the transmission path to prevent interference to the channels due to faulty operating techniques.

In the transmitting direction, each basic group passes through a measuring hybrid to an equalizer which can be adjusted to ensure that the loss at 84.08 kHz is midway between the minimum and maximum loss in the 60 – 108 kHz range. This ensures that the group pilot, if used, gives the best regulation over the frequency range. From the equalizer the signal is applied to the group modulator via a low-pass filter which restricts the upper frequency to about 108 kHz. The output level of the group modulator is controlled, with a step attenuator, to obtain the same transmission level for all five groups. The band-pass filter selects the lower sideband and rejects the upper. Groups 1, 3, and 5 are connected in parallel to one side of a hybrid decoupler and groups 2 and 4 are connected in parallel to the other side. The supergroup amplifier following the optional pilot injection equipment raises the transmitted power level to that required for the particular system. The different power levels at the GDF and the supergroup distribution frame (SGDF) can be found, for different countries, in Table 3.2 on page 91. A measuring hybrid is included to allow a check on the transmission level.

In the receiving direction, the supergroup amplifier raises the power level before application to the decoupler and filters, which split the basic supergroup into five groups. A band-pass filter at the input to each group demodulator selects the appropriate group and applies it to the group demodulator via an attenuator. A low-pass filter at the output of the demodulator rejects the upper sideband but passes the 60 – 108 kHz basic group. Equalization is achieved in a similar manner to that of the transmitting direction and the group amplifier raises the power level to the level required as per Table 3.2 on page 91. This power level can be checked via the measuring hybrid circuit.

A supergroup pilot frequency of 411.92 kHz (see CCITT Recommendation G241) can be injected, if required, through a pilot injection unit which precedes the supergroup transmitting amplifier. This frequency accompanies the supergroup through all equipment to the distant station where it can be used to control regulation and/or supervision of the received signals for that supergroup only. A 104.08 kHz band-stop filter is included before the group demodulator to group 3 to suppress any frequencies (signaling, carrier, etc.) which could introduce an error in pilot level. In the receiving direction provision is made for the pilot to be picked off, with a pilot receiver, at the output of the group modulator for group 3. This receiver is tuned to the frequency 104.08 kHz, which is the difference between the group 3 carrier and the pilot frequency (516 – 411.92 kHz). A 104.08 kHz band-stop filter follows the pilot pick-off equipment to prevent the pilot frequency causing interference in any speech or signaling channel.

3.8.1 Supergroup Modem

Figure 3.25 shows the block diagram of the supergroup modem equipment. The frequency range of 60 – 4028 kHz is combined from 16 basic supergroups, one of the basic supergroups, namely supergroup 2, being applied directly to line. In the receiving direction the 60 – 4028 kHz frequency range is separated into 16 supergroups for application to supergroup demodulators. The resulting basic supergroups are applied to group demodulating equipment or other supergroup modulators if in a

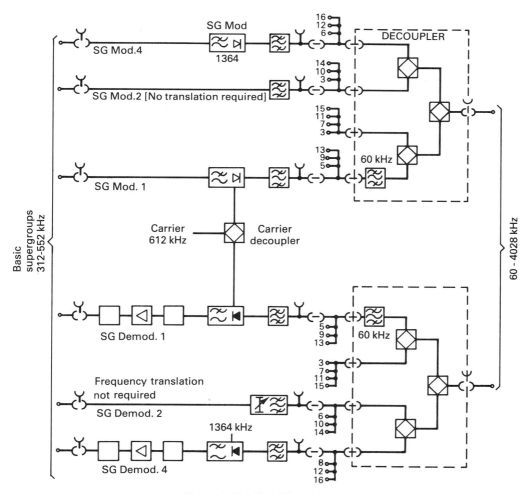

Figure 3.25   Simplified Supergroup Modem

repeater station. Each of the basic supergroups, SG 1, SG 3 to SG 16, is applied via a low-pass filter to the supergroup modulator to which the appropriate carrier frequency is connected through a carrier decoupler circuit. The carrier decoupler is a hybrid used to prevent coupling of the send and receive circuits through the common carrier supply. A band-pass filter at the modulator output selects the lower sideband and rejects the upper. Combining of the 16 supergroups is brought about in the decoupler, a hybrid combination as shown in Figure 3.25. A 60 kHz band-stop filter is included in the combined output of SG 1, 5, 9 and 13 to prevent any unwanted 60 kHz signals falsifying the pilot frequency level when a 60 kHz line pilot frequency is used.

In the receiving direction a combination of hybrids in the decoupler, in conjunction with band-pass filters preceding the supergroup demodulator, is used to split the received line frequencies into 16 supergroups. Fifteen supergroups are applied to individual demodulators, together with the appropriate carrier frequencies which are supplied through the same carrier decouplers used for the transmit circuits. A low-pass filter rejects the unwanted sideband but selects the 312–552 kHz basic supergroup. This is applied through the two-stage supergroup amplifier and the SGDF to the group modem or through a supergroup filter for connection to another system. A 60 kHz band-stop filter is included in the combined input of SG 1, 5, 9, and 13 to prevent the 60 kHz line pilot, if provided, interfering with the signaling circuit of a channel in supergroup 1. Supergroup 2, which is already in its line frequency position, is taken through a band-pass filter to the combining hybrids in the transmitting direction. In the receiving direction supergroup 2 is already in its basic supergroup range and is taken through a band-pass filter and attenuator to the SGDF and group modulator or through a supergroup filter for connection to another system.

3.8.2 LD–G Subgroup Modem

Figure 3.26 shows a block diagram of the LD–G subgroup modem. The subgroup modem provides the final two steps of transmitting modulation and receiving demodulation for up to six low-density voice channels. This unit generates two carrier frequencies for the modulation and demodulation processes.

Transmitting signals from up to six channel modems are applied through two modulation processes to produce a unit output in the 4 – 2788 kHz frequency range. The unit provides similar demodulation of an identical receive signal, which passes to the channel modems. Two crystal-controlled oscillators generate the carrier frequencies for both the modulation and demodulation. The transmitting 12 – 36 kHz signal passes through an amplifier and a 36 kHz low-pass filter to the first modulator. This signal is next mixed with the 10.676 MHz oscillator signal. The modulator output signal then passes through a 10.7 MHz band-pass filter to the second modulator.

The carrier frequency from the second crystal oscillator applied to the second modulator is chosen to ensure that the transmitted spectrum will be upright or inverted, as necessary to permit alignment with CCITT and other authorities' frequency plans. The output from the second modulator is applied through a 2.8 MHz

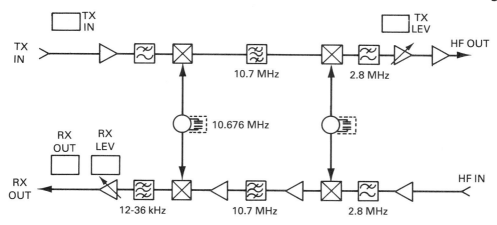

Figure 3.26   LD-G Subgroup Modem

low-pass filter to an amplifier, whose gain can be adjusted by the Tx LEV potentiometer before entering the final amplifier which provides a balanced HF OUT.

Received signals in the 4 – 2788 kHz frequency range enter the unit and pass
into an amplifier before entering the 2.8 MHz low-pass filter. The output of this filter
enters into the second modulator whose carrier frequency is selected according to the
subgroup modem number. The resulting demodulation product is applied through
another amplifier before entering the 10.7 kHz band-pass filter. This output is then
amplified before being applied to the first modulator whose carrier frequency is
10.676 MHz. The resulting demodulation product is applied to a 12 – 36 kHz band-
pass filter before entering an amplifier whose gain can be adjusted by the Rx LEV
potentiometer.

To change the number of the subgroup modem, and hence the position in the
spectrum which the output signal will take, only the crystal in the crystal oscillator
supplying the second demodulator has to be changed. This is convenient in the field,
where corrective maintenance may be required.

3.9   THE GROUP OR SUBGROUP COMBINER

The supergroup modem was treated in Section 3.8. In this section only the simple
LD–G combiner will be discussed (Figure 3.27). The combiner is used in this multiplex system to combine passively the outputs of up to 50 subgroup modems into one
composite multiplex signal for transmission, and to distribute the received composite
signal to the same subgroup modems. In the transmitting direction, balanced outputs
from the subgroup modems enter the combiner via a transformer. The transformer
combines the inputs from the subgroups into one composite multiplex signal. This
combined signal is passed through an attenuator which provides, in 0.5 dB steps,

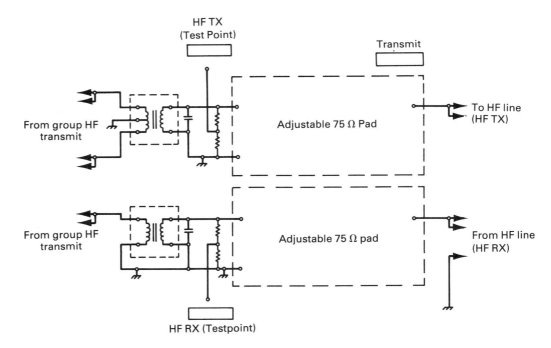

Figure 3.27   LD–G Combiner

attenuation up to 15.5 dB, before being transmitted at the HF Tx as the composite or part of the composite baseband signal. The output of the transformer is also fed to a 30 dB fixed pad which drops the signal for measurement at the HF Tx test point. In the receiving direction, HF signals enter the unit and pass to another attenuator, which again provides up to 15.5 dB attenuation in 0.5 dB steps. The adjusted output of this attenuator is fed through a transformer to the receiving inputs of the subgroup modems. The output of the attenuator is also routed to a 30 dB pad to provide the HF Rx test point. This test point and the HF Tx test point provide a signal which is −30 dBm0.

A technical summary is given below:

HF Transmitting Path

Input level	−21.5 dBm
Output level	−21.5 to −37 dBm adjustable in 0.5 dB steps
Level at HF Tx (−30 dBm0)	−51.5 dBm ±0.25 dB
Input impedance	75 Ω, balanced
Output impedance	75 Ω, unbalanced

HF Receiving Path

Input level	-13 to -30 dBm
Output level	-82.5 dBm ± 0.5 dB
Level at HF Rx (-30 dBm0)	-58.5 dBm ± 0.25 dB
Input impedance	75 Ω, unbalanced
Output impedance	75 Ω, balanced

3.10 ALIGNMENT AND TEST PROCEDURES

These test procedures give an indication of the tests which in general should be performed on multiplex equipment of most makes, and a general indication of the test equipment which should be used to perform these tests. As some equipment will have jack-fields, and others not (as in the LD–G multiplex) common sense must prevail if these procedures are to be used in practice. More often than not the manufacturer will supply detailed test procedures of his own to be used on his equipment. This section is not intended to replace these procedures, but only to enhance the understanding of them. By continually referring to the theory as set down earlier in this chapter the tests made should be understood. When there are no detailed manufacturer's test procedures, those laid down in this section should suffice to provide the means to align or to rectify a fault or faults if one or more has occurred. All that is required is the block and level diagram, and the modulation plan used. The wiring diagrams and the circuit diagrams would also be helpful as would the carrier or multiplex equipment room records – if they exist.

3.10.1 Preliminary Tests

Each of the channel, group and supergroup modems may be slide-out units, or they may require the use of extender cards to gain access to the circuit whilst live.

Channel Modem
On each of the channel modems, check with a multi-meter that the required operating voltage is present at the output of its voltage regulator. If no regulator is present, then check for the correct voltage on the voltage supply rail of the modem. If the requirement is not met, either adjust the regulator output voltage to that required, or for the case where there is no regulator, adjust the power supply voltage until the requirement is met.

Group and Supergroup Modem
On each of the group and supergroup modems, check with a multi-meter that the required operating voltage is present at the output of each modem's voltage regulator. If no regulator is present, check for the correct voltage on each modem's

voltage supply rail. If the requirement is not met, either adjust the regulator output voltage to that required, or where there is no regulator, adjust the power supply voltage output until the requirement is met.

3.10.2 Channel, Group and Supergroup Modem Oscillator

Channel Modem

With the use of appropriate test leads, e.g. BNC to alligator clip, or pin PRODS, connect a frequency counter to the output of the channel oscillator of each channel modem. Record the counter reading with a gate time of 1 s, and an appropriate input attenuator setting. The requirement is that the carrier frequency should read within ± 2 Hz of the following:

CCITT single modulation (Type B): See Figures 3.8(a) and 3.21(a).

Channel	1	2	3	4	5	6	7	8	9	10	11	12
Frequency, kHz	108	104	100	96	92	88	84	80	76	72	68	64

CCITT double modulation: see Figures 3.8(b) and 3.21(b).

Channel	1	2	3	4	5	6	7	8	9	10	11	12
Frequency, kHz	12	16	20	12	16	20	12	16	20	12	16	20

Subgroup modem	1	2	3	4
Frequency, kHz	120	108	96	84

LD–G channel: see Figures 3.16 and 3.22.

Channel	1	2	3	4	5	6
Frequency, kHz	16	20	24	28	32	36

If the requirement is not met, check the frequency of the 4 kHz master frequency generator at the group carrier supply rack and after verification by another technical person that the frequency counter has been calibrated against a reference, adjust the frequency of the 4 kHz master. For smaller systems, each channel modem may have its own oscillator. This oscillator should be adjusted by adjusting its trimmer capacitor to the proper frequency to within ± 2 Hz. When the requirement cannot be met, there may be an inductor which can be varied. If so, place the trimmer capacitor to mid-range, and adjust the inductor for the proper frequency ± 5 Hz, then adjust the trimmer capacitor for the proper frequency ± 2 Hz. If unable to meet the requirement, replace the channel modem.

Note that when the channel modem is returned to its normal position, the frequency may change slightly. This is due to it not being in thermal equilibrium, and possibly the influence of adjacent units on the oscillator circuitry. The change should not be enough to degrade the specified performance of the channel.

Group Modem

With the use of appropriate test leads, connect a frequency counter to the output of the group oscillator of each group modem. Record the counter reading with a gate time of 1 s and an appropriate attenuator setting. The requirements are (a) and (b) below:

CCITT. See Figures 3.9 and 3.24.
(a) Carrier frequency should read within ± 2 Hz of the following:

Group	1	2	3	4	5
Frequency, kHz	420	468	516	564	612

LD–G subgroup modem. See Figures 3.16, 3.17 and 3.26.
(b) Oscillator frequencies should read within ± 5 Hz of the following:

Oscillator 1 : 10 676 kHz

Oscillator 2

Group n	010 to 019	020 to 029	030 to 039	040 to 049
Frequency, kHz	$10868 - 24n$	$10544 + 24n$	$10844 - 24n$	$10836 - 24n$

Group n	050 to 059	060 to 069	070 to 079	080 to 089
Frequency, kHz	$10828 - 24n$	$10820 - 24n$	$10812 - 24n$	$10804 - 24n$

Group n	090 to 099	100 to 109
Frequency, kHz	$10796 - 24n$	$10788 - 24n$

If the requirement is not met, adjust the oscillator trimmer capacitors to the correct frequency ± 1 Hz. If neither oscillator can meet the requirements, replace the sub-group modem.

Note that the CCITT group carrier frequencies will possibly arise, not from individual oscillators, but from the group carrier supply, found on the group carrier supply rack. If the channel frequencies were found to be out, then, as the group carriers are derived from the same 4 kHz master oscillator, the group carriers will be expected to be off-frequency also.

CCITT supergroup carrier oscillators: see Figures 3.10 and 3.24.
Using the same test set up as above, record the frequencies of each of the supergroup carrier frequencies. The requirement is that the carrier frequencies should read within ± 2 Hz of the following:

Supergroup	1	3	4	5	6	7	8	9	10	11	12	13
Frequency, kHz	612	1116	1364	1612	1860	2108	2356	2604	2852	3100	3348	3596

Supergroup	14	15	16
Frequency, kHz	3844	4092	4340

If the requirement cannot be met, check on the group carrier supply rack for the

SG 1 carrier, and for supergroups 3 to 16 on the supergroup carrier supply rack, or the 124 kHz master oscillator.

3.10.3 Transmit Path Alignment

The CCITT does not recommend any specific 4 W levels for input to the channel modem, nor to the group and supergroup distribution frames. Usually, however, the levels at the 4 W channel modem transmitting and receiving points are:

Transmitting	−16 dBr
Receiving	+7 dBr

Table 3.2 shows the relative power levels at group and supergroup distribution frames in different countries. The source of the information is CCITT Recommendation G233, Malaga - Torremolino, 1984.

3.10.3.1 General Procedure
The transmitting alignment procedure uses an 800 Hz test tone for CCITT systems, or 1000 Hz for US systems. The level of this tone into a 2 W circuit is usually 0 dBm (0 dBr), or −16 dBm (−16 dBr) into the 4 W transmit. Some authorities may prefer the transmitter test-tone level to be at a level of −16 dBm0 in order not to produce excessive crosstalk on a system which is carrying traffic while being tested. If this is so then the 2 W transmitter test tone will be −16 dBm, and the 4 W test tone will be −32 dBm. A reference channel modem in a group or subgroup is aligned to the proper power level at the GDF or the HF output level. The remaining channels are then aligned for the same level. The procedure is repeated for the remaining subgroups in the terminal.

3.10.3.2 Test Equipment Required
1. Electronic voltmeter (EVM)
2. Frequency selective voltmeter (SVM), otherwise known as a selective level meter (SLM)
3. Audio oscillator (OSC)
4. Various patch cords to interface the test equipment with the multiplex equipment
5. Various adapters as required.
6. Terminating plug (600 Ω TERM.)

PROCEDURE FOR USING TEST EQUIPMENT
1. Connect OSC to the input of EVM with the 600 Ω TERM also across the EVM's input.
2. Adjust the OSC for either the 800 Hz or 1000 Hz test tone at the appropriate level.

Table 3.2
(*See Reference 9*)

Country	Relative power level at group distribution Frame Transmit (dBr)	Basic Receive (dBr)	* Group at DF	Impedance at GDF, Ω (Bal.)	Relative power level at supergroup distribution frame Transmit (dBr)	Receive (dBr)	Impedance at SGDF, Ω (Unbal.)
Australia/ Denmark[†]							
System 1	−36.5	−30.5	B	150	−35	−30.5	75
System 2	−42	−5	B	135	−35	−30	75
Belgium	−37	−8	B	150	−35	−30	75
France	−33	−15	B	150	−45	−35	75
German DDR (New	−36	−30	B	150	−35	−30	75
equipment)	−36	−23	B	150	−36	−23	75
Germany FDR	−36	−30	B	150	−35	−30	75
India	−36.5	−30.4	B	150	−34.8	−30.4	75
Hungary, Italy, Netherlands	−37	−30	B	150	−35	−30	75
Japan (NTTPC)	−36	−18	B	75	−29	−29	75
Mexico (Tele- fonos d.M)	−47	−10	B	150	−47	−24	75
Poland (New	−36	−30	B	150	−36	−23	75
equipment)	−36	−23	B	150	−36	−23	75
Spain, Norway, Ireland, New Zealand, UK	−37	−8	B	75 (unbalanced)	−35	−30	75
Switzerland (New	−41	−7.8	A/B	75	−35	−26	75
equipment)	−36.5	−30.5	A/B	(unbalanced)	−35	−26	75
USSR	−36	−23	B	150	−36	−23	75
USA (ATT)	−42	−5	B	135	−25	−28	75

* See Section 3.4.2 for an explanation of type-A and type-B groups.
[†] System 1 only

3. Insert the calibrated test-tone into the reference channel, either CH6 for CCITT, or CH3 for LD–G.
4. Connect the SLM to the particular group being tested, ensuring that the impedance of the SLM matches that at the measuring hybrid, or test point.
5. Tune the SLM to the reference channel frequency. Table 3.3 shows the frequencies of each of the translated channels test tone, for either CCITT or LD–G.

Table 3.3   ***Translated 800 Hz (CCITT) and 1 kHz (USA) Test Tones at Group Modem Input Test Points***

CCITT Channel	1	2	3	4	5	6	7	8	9	10	11	12
Frequency, kHz	107.2	103.2	99.2	95.2	91.2	87.2	83.2	79.2	75.2	71.2	67.2	63.2

LD–G Channel	1	2	3	4	5	6
Frequency, kHz	15	19	23	27	31	35

The CCITT level requirements are those given in Table 3.2. The LD–G level requirements are −20 dBr on initial alignment, and −20 dBr ±0.5 dB on routine checks.

If the requirement is not met, adjust Tx LEV control on the channel modem for appropriate level at the group modem input test point. If the requirement still cannot be met, check that the M-lead is not strapped such that in the unseized condition the transmit audio circuit is not muted.

6. Connect the SLM switched to the 75 Ω (unbalanced) terminated position to the SGDF test point, or to the HF Tx test point on the LD–G combiner unit.

7. Tune the SLM to the HF frequency of the reference channel used in step 3, as given in Table 3.4.

A simple expression to find the test tone frequency at HF for the various subgroups and channel numbers is given by the expressions below, for channel number C greater than or equal to 1 and less than or equal to 6.

For subgroups 010 to 019
24 × (Group number) + 4 × (Channel number) − 181 = HF test tone frequency, kHz.

$$\text{or } 24G + 4C - 181 = f \text{ kHz} \tag{3.1}$$

Example Find the frequency f of an HF test tone, if 1 kHz is transmitted on channel 3, group 18.

From equation 3.1
$$f = 24 \times 18 + 4 \times 3 - 181 = 263 \text{ kHz}$$

For *subgroups 020 to 029* the expression becomes

$$f = 24G - 4C - 143 \tag{3.2}$$

For *subgroups 030 to 039*

$$f = 24G + 4C - 157 \tag{3.3}$$

For *subgroups 040 to 049*

$$f = 24G + 4C - 149 \tag{3.4}$$

For *subgroups 050 to 059*

$$f = 24G + 4C - 141 \tag{3.5}$$

Table 3.4

CCITT
Group 1

Channel	1	2	3	4	5	6	7	8	9	10	11	12
Frequency, kHz	312.8	316.8	320.8	324.8	328.8	332.8	336.8	340.8	344.8	348.8	352.8	356.8

Group 2

Channel	1	2	3	4	5	6	7	8	9	10	11	12
Frequency, kHz	360.8	364.8	368.8	372.8	376.8	380.8	384.8	388.8	392.8	396.8	400.8	404.8

Group 3

Channel	1	2	3	4	5	6	7	8	9	10	11	12
Frequency, kHz	408.8	412.8	416.8	420.8	424.8	428.8	432.8	436.8	440.8	444.8	448.8	452.8

Group 4

Channel	1	2	3	4	5	6	7	8	9	10	11	12
Frequency, kHz	456.8	460.8	464.8	468.8	472.8	476.8	480.8	484.8	488.8	492.8	496.8	500.8

Group 5

Channel	1	2	3	4	5	6	7	8	9	10	11	12
Frequency, kHz	504.8	508.8	512.8	516.8	520.8	524.8	528.8	532.8	536.8	540.8	544.8	548.8

LD–G HF Test-Tone frequencies, kHz

LD–G

Channel	1	2	3	4	5	6
Subgroup						
010	63	67	71	75	79	83
011	87	+(4)=91	+(4)=95	+(4)=99	+(4)=103	+(4)=107
	+(24)=					
012	111 etc.					
etc.						

The CCITT level requirements are those given in Table 3.2 at the supergroup distribution frame. The LD–G level requirements are −51.5 dBm on initial alignment, and −51.5 dBm ±0.5 dB on routine checks.

   If the requirement is not met, adjust the Tx LEV on the group or the sub-group modem.

(*Procedure for Using Test Equipment, 8–10. Alignment of Remaining Channels*)

8. Transfer OSC to channel 1 or the lowest channel in the group or subgroup under test.

9. Tune SLM to the HF frequency, as in steps 6 and 7.
   The CCITT level requirements are those given in Table 3.2 at the supergroup distribution frame. The LD–G level requirements are −51.5 dBm on initial alignment and −51.5 dBm ±0.5 dB on routine checks.

   If the requirement is not met, adjust Tx LEV control on the channel modem under test for required level. If the requirement still cannot be met, replace the channel modem.

10. Repeat steps 8 and 9 for the remaining channels in the group or subgroup under test.

(*Procedure for Using Test Equipment, 11–13. Alignment of Remaining Groups or Subgroups*)

11. Repeat entire procedures for the remaining groups or subgroups.

12. If testing supergroups, and master-groups, the same principles will apply as in this procedure, but the channel frequencies at the next higher stage will need to be calculated.

13. Remove test equipment.

3.10.4 Transmit Signaling Level

For this test consider only out-of-band signaling frequencies. CCITT Recommenda-tion 21 gives four recommended out-of-band systems. Three of these systems are recommended for use with carrier systems having 12 channels per group, and the other for those systems having 8 channels per group.

3.10.4.1 Systems for Use with 12-Channel Groups
Type I. Compatible with only those group and supergroup reference pilots having a displacement from the virtual carrier of 140 Hz.

Frequency Virtual carrier (zero frequency)
Level   High (-3 dBm0 approx.)
Signals   Discontinuous

Type II. Compatible with only those group and supergroup reference pilots having a displacement from the virtual carrier of 80 Hz.

Type IIa. For use with discontinuous signals
  Frequency 3825 Hz
  Level   High (-5 dBm0 approx.)

Type IIb. For use with semi-continuous signals
  Frequency 3825 Hz
  Level   Low (-20 dBm0 approx.)

3.10.4.2 Systems for Use with 8-Channel Groups
Frequency 4400 Hz
Level
  For discontinuous signals -6 dBm0
  For semi-continuous signals -17.4 dBm0 to -20 dBm0.

NON-CCITT OUT-OF-BAND SIGNALING SYSTEM
The LD–G channel modems provide the following:

Frequency Channel carrier
Level -16 dBm0

In this section only out-of-band signaling will be considered. Chapter 4 gives details of other signaling systems.

3.10.4.3 General Procedure
The test procedure described in this section allows checking and adjusting the transmitting level of the signaling oscillator of each channel modem where out-of-band signaling is used.

3.10.4.4 Test Equipment Required
 1. 1 frequency selective voltmeter or level meter (SLM)
 2. 1 card extender (if necessary)
 3. Various patch cords and adapters

PROCEDURE FOR USING TEST EQUIPMENT
 1. Check that the channel modem is correctly strapped for the M-lead option required.
 2. Prepare patch cords to the M-input of the channel under test.

3. Connect the SLM (75 Ω unbalanced) to the SGDF test point, or to the HF Tx test point on the LD–G combiner unit.
4. Apply TONE ON M-lead condition to the unterminated end of the patch cord at the M-input, by either battery or ground, according to the option.
5. Refer to Tables 3.5 and 3.6, and tune the SLM to the signaling tone frequency of the group or subgroup option and channel under test.

Table 3.5   CCITT 3825 Out-of-Band Signaling Tones as observed at the SGDF

Group 1

Channel	1	2	3	4	5	6	7	8	9	10	11	12
Frequency, kHz	315.825		323.825		331.825		339.825		347.825		355.825	
		319.825		327.825		335.825		343.825		351.825		359.825

Group 2

Channel	1	2	3	4	5	6	7	8	9	10	11	12
Frequency, kHz	363.825		371.825		379.825		387.825		395.825		403.825	
		367.825		375.825		383.825		391.825		399.825		407.825

Group 3

Channel	1	2	3	4	5	6	7	8	9	10	11	12
Frequency, kHz	411.825		419.825		427.825		435.825		443.825		451.825	
		415.825		423.825		431.825		439.825		447.825		455.825

Group 4

Channel	1	2	3	4	5	6	7	8	9	10	11	12
Frequency, kHz	459.825		467.825		475.825		483.825		491.825		499.825	
		463.825		471.825		479.825		487.825		495.825		503.825

Group 5

Channel	1	2	3	4	5	6	7	8	9	10	11	12
Frequency, kHz	507.825		515.825		523.825		531.825		539.825		547.825	
		511.825		519.825		527.825		535.825		543.825		551.825

For the CCITT signaling tone, the dBr level as measured at the SGDF or the test points of the supergroup modem hybrid is found from Table 3.2 and the dBm0 level as given above. For example, assume that, in Mexico, the dBr transmit level at the SGDF measured with the SLM set for 75 Ω (unbalanced) = −47 dBr. If the signaling tone is 3825 Type IIb, then the signaling level will be −20 dBm0, i.e., on measurement, the value should be −67 dBm. Assume also that channel 5, group 4, is to be measured for this signaling tone at the SGDF. Then, referring to Table 3.5, the SLM is tuned to a frequency of 475.825 kHz, and if there is alignment, the requirement is that the level should read −67 dBm. For the LD–G, the level read should be −67.5 dBm (−16 dB below the −51.5 dBr level). If the requirements cannot be met, adjust if possible the transmitting signaling level on the channel modem, or on the 3825 Hz oscillator.
6. Repeat the above five steps for the remaining channels.
7. Remove test equipment.

Table 3.6 LD–G Out-of-Band Signaling using the Channel Carrier at the Combiner HF Tx

C = Channel number (1 to 6 inclusive)
G = Group number
f = Frequency of the tone in kHz at the baseband or HF Tx on the combiner

Subgroups 010 to 019

$$24G + 4C - 180 = f \tag{3.6}$$

Subgroups 020 to 029

$$24G - 4C - 144 = f \tag{3.7}$$

Subgroups 030 to 039

$$24G + 4C - 156 = f \tag{3.8}$$

Subgroups 040 to 049

$$24G + 4C - 148 = f \tag{3.9}$$

Subgroups 050 to 059

$$24G + 4C - 140 = f \tag{3.10}$$

3.10.5 Receiving Path Alignment

3.10.5.1 General Procedure

The receiving path alignment procedure uses an 800 Hz test tone for CCITT systems, or a 1 kHz test tone for US systems. There are two ways that this test can be done. In the first method the distant terminal inserts the test tone into the reference channel of the group, subgroup, etc., under test, at a level of −16 dBr (4 W) or 0 dBr (2 W), and the corresponding group, subgroup and reference channel at the local terminal are aligned and adjusted as required to the proper VF output level, which is +7 dBr (4 W), or −4 dBr (2 W). The remaining channels in the group are aligned for the same level. The procedure is then repeated for the remaining groups or subgroups in the terminal. The second method makes use of a loop-back procedure. The receiving equipment is aligned out-of-service at the group or supergroup level if an amplifier is placed between the output (transmit) and input (receive) of the group or supergroup under test. The value of the amplifier input and output impedance must match the impedance of the equipment, and the value of the amplifier must be such that the transmitting level becomes that of the receiving level upon entering the group, supergroup or combiner input. In cases where the transmitting level is higher than the receiving level, an attenuator must be used. For example, assume that we are to do a loop test at the supergroup level in Spain. Table 3.2 shows that the transmitting level at the SGDF is −35 dBr, and the receive level is −30 dBr. Hence a 5 dB amplifier is required to increase the transmitting level from −35 dBr to the required −30 dBr. Before proceeding with this alignment, it is necessary to complete the transmit path alignment.

3.10.5.2 Test Equipment Required

1. 2 electronic voltmeters (EVM), one at each terminal if end-end alignment is to be done. (If the alignment is in the back configuration, only 1 EVM is required.)
2. 1 selective level meter (SLM)
3. 1 audio oscillator (OSC)
4. Various patch cords for connecting the test equipment to the multiplex equipment
5. Various adapters as required
6. 2 Ω terminating plugs (600 Ω TERM)

TEST PROCEDURE

1. At the transmitting terminal, insert a test tone at the correct frequency (800 Hz or 1 kHz) into the reference channel of the subgroup, group, etc., under test. Select channel 6 for the CCITT system, or channel 3 for the LD–G, as the reference channel, and set the test tone level to −16 dBm for a 4 W channel, and 0 dBm for a 2 W channel.
2. At the receiving terminal, connect the SLM, set to the correct impedance, to the HF Rx test point at the SGDF, GDF or combiner.
3. Refer to Table 3.4 for the translated test-tone frequency of the reference channel belonging to the particular subgroup, group, etc., under test.
4. Tune the SLM to the frequency found in step 3.

   The requirement is that the level measured at the receive SGDF or GDF should be as in Table 3.2, and for the LD–G combiner there should be a level of −58.5 dBm ±0.75 dB on routine checks. If the requirement is not met, check the receiving level on the baseband from the radio, or the amplifier/attenuator if the loop-back configuration is used. If a problem still exists check the transmitting level, and the test equipment impedance.
5. Connect the SLM to the output of the demodulator or to an Rx LEV monitor at the GDF or to receiver output from the group modem (see Table 3.2 for the correct measuring impedances) or to Rx OUT (600 Ω, balanced, bridged) on the LD–G – of the channel under test.
6. Tune the SLM to the frequency of the reference channel as per Table 3.3.

   The requirement is that the level should be as in Table 3.2, and, for the LD–G, −20 dBm on initial alignment and −20 dBm ±0.5 dB on routine checks. When the requirement cannot be met, adjust the Rx LEV control on the group or sub-group modem for the proper level. If the requirement still cannot be met, replace the group or subgroup modem.

REFERENCE CHANNEL

7. Connect the EVM terminated in 600 Ω to the DEM jack of the channel modem under test.

   The requirement is +7 dBm for a 4 W channel, −4 dBm for a 2 W channel on initial alignment, and ±0.5 dB variation on routine checks. If the requirement is not met, adjust RCV LEV control on the channel modem for the proper level. If the requirement still cannot be met, replace the channel modem.

ALIGNMENT OF REMAINING CHANNELS

8. At the transmit terminal, transfer OSC to channel 1 or the lowest channel in the group or subgroup under test.

9. At the receive terminal, transfer EVM to DEM jack of the channel under test.
   The requirement is +7 dBm for a 4 W channel, −4 dBm for a 2W channel on initial alignment, and ±0.5 dB variation for this on routine checks. If the requirement is not met, adjust RCV LEV control on channel modem for the proper level. If requirement still cannot be met, replace the channel modem.

10. Repeat steps 8 and 9 for the remaining channels in the group or subgroup under test.

ALIGNMENT OF REMAINING GROUPS OR SUBGROUPS

11. Repeat the entire procedure of this section for the remaining groups or subgroups.

12. Remove the test equipment.

3.10.6 Receive Signaling Check

3.10.6.1 General Procedure

The test uses a signaling tone from the distant terminal, or from the local terminal when in the loop-back configuration, to check the relay operation controlling the E-lead contacts in each channel modem.

3.10.6.2 Test Equipment Required

1. 1 multi-meter (MM)

2. A patch cord to interface the MM with the E-lead, and ground. And a patch lead at the distant terminal to key the M-lead.

TEST PROCEDURE

Note that the previous sections must be completed prior to starting this test. The test procedure is:

1. During initial alignment check that the channel modem is correctly strapped for E-lead TONE ON requirement of the local office. Set the MM to the proper voltage scale and range if the E-lead* produces battery upon activation. (This will occur in some cases if the F-lead is connected to the battery.) If E-activation produces ground, (if the F-lead in some cases is connected to ground) set MM to test for continuity by placing it on the ohms reading position.

2. At the transmit terminal key the M-lead of the channel under test for TONE-ON E-lead status.

3. At the receiving terminal connect the E-lead to the MM.
   The requirement is for a proper TONE-ON E-lead status. If the requirement

* See Section 3.7.3.

is not met, verify the test set-up, including the F-lead condition, and complete steps 4 and 5 before taking any corrective action.

4. At the transmit terminal, apply TONE-OFF M-lead condition
5. At the receive terminal, observe the MM indication.
    The requirement is for a proper TONE-OFF E-lead status. If requirements of step 3 and/or step 5 are not met, replace the channel modem.
6. Repeat steps 2 through 5 for remaining channels.
7. Remove test equipment.

3.10.7 Synchronization Check

3.10.7.1 General Procedure
The test describes the procedure to check those channel modems which have the facility to become master or slave terminals, for synchronization.

3.10.7.2 Test Equipment Required
1. 1 Selective level meter (SLM)
2. 1 Frequency counter (COUNTER)
3. 1 Card extender (if required)
4. Various patch cords as required
5. Various adapters as required

TEST PROCEDURE
Note 1 The system must be fully aligned before starting this test.
Note 2 The test can only be performed on equipment which has channel carrier signaling and channel oscillators with injection locking facilities, i.e. is NON-CCITT channel modems.

1. Verify that the channel modem is strapped as a slave.
2. Connect COUNTER (attenuator 100 ×, gate time 1 s) to test points at the channel oscillator output, on the channel modem under test.
3. Apply the TONE-ON M-lead condition at the corresponding far-end channel modem.
4. Record COUNTER indication.
5. Change the channel modem strapping to master. Adjust the channel oscillator frequency to that recorded in step 4, ±50 Hz.
Note Do not adjust the oscillator to a frequency that is not within ±50 Hz of the nominal shown in step 7.
6. Restrap the channel modem back to slave.
    The requirement is that the COUNTER indicates the same frequency as that of step 4. If the requirement is not met, replace the channel modem.
7. Again restrap the channel modem to master.
    The requirement is to be able to reset the channel oscillator using the trimmer capacitor to its proper frequency ±2 Hz according to the following table:

Channel	1	2	3	4	5	6
Frequency, kHz	16	20	24	28	32	36

When the requirement cannot be met, adjust the trimmer capacitor to mid-range, and adjust the oscillator variable inductor to the proper frequency ±5 Hz; then adjust the trimmer capacitor to the proper frequency ±2 Hz.

8. If the terminal is to be used for slave channel modems, leave the channel modem strapped for slave.
9. Repeat entire procedure of this section for remaining slave channels.

3.10.8 Signaling Distortion

3.10.8.1 General Procedure
This procedure describes how to measure signaling pulse distortion over a voice channel.

3.10.8.2 Test Equipment Required
1. 2 Pulse test sets (PULSE SET) – one for each location.

Note 1 The system must be fully aligned before starting this test.

Note 2 Verify that the E/M options of the channel agree with options provided on the test set.

TEST PROCEDURE
1. At the transmitting terminal, connect PULSE SET (12 pulses/s signal with 58 per cent break) to the M-lead of the channel modem under test.
2. At the receiving terminal, connect PULSE SET to the E-lead of the channel under test.
   The requirement is for a 58 ±5 per cent break. If the requirement is not met, recheck transmit signaling level. If it still cannot be met, replace the channel modem.
3. Repeat entire procedure of this section for remaining channels under test.

3.10.9 Idle or Thermal Channel Noise

3.10.9.1 General Procedure
This test describes an end-to-end procedure to measure idle noise in each voice channel.

3.10.9.2 Test Equipment Required
1. Noise measurement test set (NOISE SET)
2. Patch cord to connect noise set to channel.
3. 600 Ω termination for the distant terminal (TERM)

Note 1 Any measurement made with only one channel removed from service, and all remaining channels in service, may contain noise from the radio and associated equipment. Such a measurement may also contain interfering tones generated by an adjacent channel containing traffic.

Note 2 For the most accurate measurement of idle noise, perform the test immediately before the multiplex terminal is placed onto the radio facility.

Note 3 The system must be fully aligned and in normal condition before starting this test.

TEST PROCEDURE
1. At the receiving terminal, set NOISE SET or SLM for measuring in dBrnC0, pWp, with one of the following reference levels:

   +7 dBm for 4 W channels
   −4 dBm for 2 W channels

2. At transmitting terminal, insert TERM into one of the following for the channel under test:

   MOD jack for 4 W channels
   DEM jack for 2 W channels

Noise check
3. At the receiving terminal, connect the NOISE SET or SLM (600 Ω, balanced, terminated) to the DEM jack on the channel modem of the channel under test.
   The requirement is that the noise level should not be greater than the value in dBrnC0 or pWp provided by the system authority, or manufacturer, as computed from the idle noise of the radio baseband plus the multiplex contribution.

Note: Individual channels, or the entire assembly, can be tested in loop-back configuration. The requirement specification in loop-back is 25 dBrnC0 for some American manufacturers, or less, depending on the number of groups equipped. The more channels, the more idle noise there will be.

4. Repeat the entire chart for the remaining channels under test.

3.10.10 Crosstalk Tests

3.10.10.1 General Procedure
Usually after the radio link has been aligned, far-end crosstalk tests are performed. Two types of crosstalk test will be described. These are:

   Far-end
   Near-end

3.10.10.2 Test Equipment Required
1. Noise measurement test set (NOISE SET) or SLM
2. Audio oscillator (OSC) for the far-end or distant terminal
3. 600 Ω terminations (TERM), eleven for each terminal in a 12-channel group. If

an LD–G subgroup is being tested, only five TERMs are required at each terminal
4. Various patch cords to connect the test equipment to the channel modems

TEST PROCEDURE FOR FAR-END CROSSTALK
1. Insert an 800 Hz (CCITT) or a 1 kHz tone at a level of −16 dBm into the MOD jack of the lowest 4 W channel, and the lowest group or subgroup at the distant terminal.
2. Terminate the 4 W MOD jacks of all the other channels in this group or subgroup at both the distant and local terminals with the TERMs.
3. At the local end, using the NOISE SET or SLM, measure and record the receiving level at the DEM jacks for all channels in the group or subgroup under test, except for that channel carrying the tone.
4. Repeat steps 1 through 3 for a tone inserted in each other channel in this group or subgroup.
5. Repeat steps 1 through 5 for transmission of the tone from the local terminal to the distant terminal.
   The requirement is that the highest recorded level should be not more than that specified by the system authority or the manufacturer: the value should be ⩽65 dBm0.

PROCEDURE FOR NEAR-END CROSSTALK
1. Insert an 800 Hz (CCITT) or a 1 kHz tone at a level of −16 dBm into the MOD jack of the lowest 4 W channel, and the lowest group or subgroup at the LOCAL terminal.
2. Terminate the 4 W MOD jacks of all the other channels in this group or subgroup at both the distant and local terminals with the TERMs.
3. At the local end, using the NOISE SET or SLM, measure and record the receive level at the DEM jacks for all channels in the group or subgroup under test.
4. Repeat steps 1 through 3 for a tone inserted in each other channel in this group or subgroup.
5. Repeat steps 1 through 5 at the distant terminal, with all of the TERMs inserted at the local terminal.
   The requirement is that the highest recorded level should be equal to or less than that specified by the system authority or the manufacturer: the value should be ⩽65 dBm0.

3.11 TEST EQUIPMENT DESCRIPTION

The Frequency Selective Level Meter (SLM)

The Cushman CE–24A is suitable for these tests as it has;

   200 Hz to 5 MHz range
   (−120) to (+12) dBm level range
   AFC to eliminate errors due to drift and time-consuming signal peaking
   Light Emitting Diode (LED) read-out with 10 Hz resolution

10^{-5} frequency accuracy
Phase-lock stability
1.74 kHz equivalent, or 3.1 kHz wide bandwidth for channel noise measurements
45 Hz narrow bandwidth for single tone measurements
Five switch-selectable impedances

For good general purpose applications the W & G SPM–3 is suitable also

Multi-meter (MM)

Digital Multi-meter, Fluke 4000A or equivalent, or volt-ohm meter (20 kOhms/volt)
Triplett 630 NA or equivalent.

Electronic Counter (COUNTER)

Hewlett-Packard 5382A with option 001, TCX0 or equivalent.

Oscillator (OSC)

Multi-impedance generator, able to cover entire baseband frequency range.
Hewlett-Packard 204C, or 651B, option 02, or W & G PS–3, or equivalent.

Electronic Voltmeter (EVM)

Hewlett-Packard 400 FL or EL, or equivalent.

Noise Measuring Test Set (NOISE SET)

Northeast Electronics TTS 35 BAQ or equivalent. Note that this equipment is used
only for performance evaluation and not alignment.

Pulse Test Set (PULSE SET)

Northeast Electronics Type 26 or equivalent. Note this equipment is used only for
performance evaluation and not alignment.

3.12 EQUIPMENT TEST RECORDS

Figures 3.28 and 3.29 are typical test record sheets for the procedures of Sections
3.10.3 to 3.10.9 and of Section 3.10.10 respectively.

SITE _____   SYSTEM _____

Record GROUP MODEM TX IN LEVEL, RX OUT LEVEL and SUPERGROUP
or COMBINER HF RX LEVEL only for reference channels.

GROUP or SUBGROUP Number	CHANNEL Number	GROUP or SUBGROUP MODEM TX IN LEVEL	SUPER GROUP or COMBINER HF TX IN LEVEL	TX SIGNALLING TONE LEVEL	SUPERGROUP or COMBINER HF RX OUT LEVEL	GROUP or SUBGROUP MODEM RX OUT LEVEL	CHANNEL DEMOD OUT LEVEL	TONE-ON E-Lead CHECK	TONE-OFF E-LEAD CHECK

Figure 3.28   Multiplex Test Data Sheet

SYSTEM _____   SITE _____

TRANSMIT LEVEL _____ dBm   RECEIVE LEVEL _____ dBm

NEAR or FAR END? _____

GROUP	TX Channel	1	2	3	4	5	6	7	8	9	10	11	12
	CROSSTALK / IDLE												
	1												
	2												
	3												
	4												
	5												
	6												
	7												
	8												
	9												
	10												
	11												
	12												

Figure 3.29   Crosstalk Data Sheet

REFERENCES

1. Application Information for Thin Film Cascadable Amplifiers, Watkins-Johnson Company (Relcom) Applications Engineering. 3333 Hillview Avenue, Palo Alto, California, 94304, Catalogue 1984, p. 220.
2. Mixer Application Information, Watkins-Johnson Company Catalogue, 1984, (see Reference 1), pp. 433–473 (several articles).
3. Tant, M. J., *The White Noise Book* (Marconi Instruments, 1974).
4. ITT, *Reference Data for Radio Engineers*, 6th Ed. (Howard W. Sams. 1981).
5. CCIR Recommendation 401–2.

6. CCITT Recommendations G232, G241, Orange Book, Vol. III–1, *Line Transmission*, VI Plenary Assembly, 1976.

7. Farinon LD–G Multiplex, Farinon Electric, 1691 Bayport Avenue, San Carlos, California, 94070.

8. Broadband Terminal Equipment, Australian Post Office Training Publication CP225, 1966, Telcom Australia, Engineering Training, 210 Kingsway, South Melbourne, Australia 3205.

9. CCITT Recommendations G231, G232, G233, Orange Book, Col. III–1 *Line Transmission*, VI Plenary Assembly, 1976.

4 SIGNALING – AND THE DIRECT LINE AND DIRECTORY NUMBER TEST LINE

4.1 SIGNALING FOR TELEPHONE NETWORKS[1,2,3,4]

When a telephone call connection is set up through a switched telephone network, signaling conveys the intelligence needed to interconnect one subscriber to any other in that network. Signaling tells the switch that a particular subscriber requires service and then provides the local exchange or switch with the information necessary to identify the required or called subscriber to route the call properly. Signaling also provides the supervision of the call during its set-up, duration and clear-down, as well as certain status information such as dial tone, busy tone and ringing. Metering pulses for call charging during the conversation or on-line period is also another form of signaling. Below is a tree diagram of the different classification and types of signaling commonly encountered.

CLASSIFICATION OF SIGNALING

I

I _____I_____ I

Supervisory	Address	Audio-visual
I	I	I

I _____ I	I _____ I	I _____ I

Forward	*Backward*	*Station*	*Routing*	*Alerting*	*Progress*
Seize	Idle	Rotary dial	Channel	Ringing	Dial tone
Hold	Busy	Pushbutton	Trunk	Paging	Busy tone
Release	Disconnect	Digital		Off-hook warning	Ring-back tone

Signaling information can be transmitted from one subscriber to an exchange, or between exchanges by such means as:

Duration of pulses
Combination of pulses
Frequency of signal
Presence or absence of a signal
Binary code
For DC systems, the direction or level of transmitted current

4.2 LINE OR SUPERVISORY SIGNALS

These signals are associated with inter-exchange links, where, in a fully automatic service, exchanges must interchange information amongst themselves, and also include the link between the customer's telephone and the exchange. They are used to control the interface equipment at either end of the individual links. To summarize, the setting up of a call through an automatic exchange system or network requires the passing of signals:

Between the calling subscriber and the originating exchange
Between items of equipment within the originating exchange
Along trunk and junction circuits connecting exchanges
Between the terminating exchange and the called subscriber

Fundamentally, supervisory signaling is two-state signaling – that is the calling party for example may be 'on-hook' or 'off-hook'. It provides information on the line or on the circuit condition and indicates whether a circuit is in use. It informs the exchange and interconnecting trunk circuits of line conditions such as whether the calling party is off- or on-hook, the called party on- or off-hook, etc.

Some examples of the types of line signal which may be passed over two-wire or four-wire physical or carrier-derived circuits when the period between the two-state condition creates a current pulse of varying duration are: seizure, answer, meter, clear forward, release guard, forced release, clear back, blocking, trunk-code dialled, call-code receiver and forward transfer. Most of these are discussed below (with some additional signals).

Seize forward. An indication that a call is to be established over the link; the signal initiates the transition of the circuit at the distant exchange from idle to busy, and prepares it to receive further signals.

Seizure acknowledge, (proceed to send). A signal sent in the backward direction from the incoming line relay set upon receipt of the seize forward signal. It indicates that the equipment is ready to receive information signals.

Answer. An indication to the outgoing exchange that the B-party (called party) has answered. The signal has a supervisory function and initiates the call charging process.

Answer acknowledge. A signal sent in the forward direction from the outgoing line relay set upon receipt of an Answer signal.

Metering (charging) signal. A signal sent in the backward direction from the line relay set performing the charging function (the charge point). Signals are sent at regular intervals at a rate determined by the administration's charging plan.

Clear-back. (An alternative name in common use is *cleardown*). A signal sent in the backward direction from the incoming line relay set to indicate that the B-party has cleared. The signal has a supervisory function and initiates time supervision (see *Forced release*). It is not repeated beyond the charge point.

Re-answer. A signal identical to *Answer* which indicates that the B-party, having cleared back, has re-answered the call. The re-answer signal cancels the time supervision initiated by the clear-back signal.

Clear-forward. A signal sent in the forward direction from the outgoing relay set in order to initiate the release of the connection.

Release guard. A signal sent in the backward direction in response to a *Clear-forward* signal. Either the beginning or end of the signal (depending on the signaling system) indicates that disconnect operations at the incoming end are complete. The signal thus protects the circuit from re-seizure during the clearing period. Non-receipt of the signal (due to an inter-exchange link fault) causes the outgoing inter-face relay set to be busied, and a *Clear-forward* signal is sent until the *Release guard* signal is received.

Forced release. A signal sent in the backward direction from the line relay set associated with the charge determining equipment (charge point). The receipt of the *Forced release* signal by the outgoing line relay set of the originating exchange initiates a *Clear-forward* signal to release the connection. A *Forced release* signal is only sent at the expiry of a time supervision period such as that initiated by a *Clear-back* signal.

Blocking (Back busy). A signal sent in the backward direction from the incoming line relay set, which causes a busy (engaged) condition to be applied to the outgoing line relay set.

Fleeting test reversal. A short duration (125 ms) reversal of the line current during the setting up of the call. The resulting signal tests for the presence of a high-resistance condition (generally caused by a diode) in the subscriber's line, which indicates that the service has a restricted trunk access category. The signal is usually sent from an originating exchange over the subscriber's line, but will also be sent over an inter-exchange link from a parent exchange which performs trunk-access analysis for a satellite terminal.

Forward transfer (ring forward, or recall). This signal may be initiated only by operators. It is used to attract or recall the attention of an operator on a national call, or to obtain operator assistance in the destination country on an international call.

Usually in exchanges there exists a 'line signaling' relay set whose function is to:

Recognize and act on signals received over it.
Transmit and repeat signal received for action at the exchange, or telephone connected at the other end.
Interchange signals between parts of the exchange.

Figure 4.1 shows a basic circuit of a telephone and an exchange line signaling relay set.

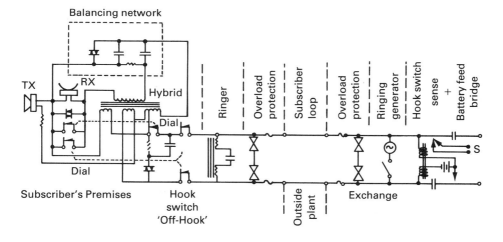

Figure 4.1   Signaling with a Conventional Telephone

4.2.1 Line Signaling over Carrier Circuits

The systems used for line signaling over carrier circuits can be broadly classified into two main groups: out-of-band; and in-band.

Out-of-band signaling systems are those in which the signals are passed over the link on a path other than the speech path. This separation can be either a frequency difference, such as signaling on a carrier channel using a frequency outside the speech bandwidth but within the channel bandwidth, or where the signaling and speech are carried out on two different circuits.

4.3   E & M SIGNALING

Figure 4.2 is a diagram incorporating the use of the E & M leads. At both ends of a carrier system, the connection between the line signaling relay set and the channel equipment of the carrier system comprise the speech wires plus the E & M signaling leads. For a signal to be sent from one end of the link to the other end, a ground is applied to the M-lead (connecting the channel modulator or 'mouth' for 'M'), and the carrier system causes the ground to reappear on the E-lead at the far end (connecting the switching equipment or 'ear' for 'E'). Usually the M-lead is activated by battery and not ground as described above. Battery on the M-lead gives rise to ground on the far-end E-lead. It is true E and M signaling only where the trunk interfaces with the exchange.

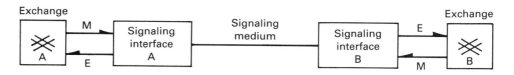

Figure 4.2   E & M Signaling

4.4   REVERSE BATTERY SIGNALING

This form of signaling was used on metallic pair trunks. The on- and off-hook conditions are given by the polarity across the loop (i.e. direction of current flow), i.e. battery or ground.

4.5   AC SIGNALING

There are three categories: low frequency; in-band, and out-of-band.

4.5.1 Low Frequency

This form of signaling operates at a single frequency less than 300 Hz, namely 25 Hz (Germany), 50 Hz (Italy, France, Germany, Spain), 80 Hz (Spain), and 135 or 200 Hz in other countries. These systems cannot be operated over carrier-derived channels due to excessive distortion and band limitations. Their use is limited to metallic-pair transmission systems, and even on these systems cumulative distortion limits the circuit length.

4.5.2 In-Band Signaling

This form of signaling uses tones placed inside the speech channel (0.3 – 3.4 kHz). The advantage of this form of signaling is that on patching audio circuits from one system to another, etc., only four wires are required to be patched instead of six as in 4 W E & M out-of-band signaling. The disadvantage is in the additional complexity of the circuits, in order to prevent talk-down, where the voice energy falsely triggers the signaling circuits to disconnect the call. The additional circuitry permits signaling to occur only on the setting up of the call, and during the disconnect phase. There are three categories of in-band signaling: single frequency (SF) or 1 VF (one voice frequency); two frequency (2 VF), and Multi-frequency (MF).

The tones are usually at frequencies above 2000 Hz, because in normal conversation most of the speech energy is concentrated at frequencies below 2000 Hz. Thus there is a reduction in the probability of talk-down occurrence.

4.5.2.1 Single Frequency (SF)

This tone is used almost exclusively for supervision. The frequency most commonly used is 2600 Hz (as in the D260 systems and Tellabs equipment). Figure 4.3 gives a circuit using this form of signaling.

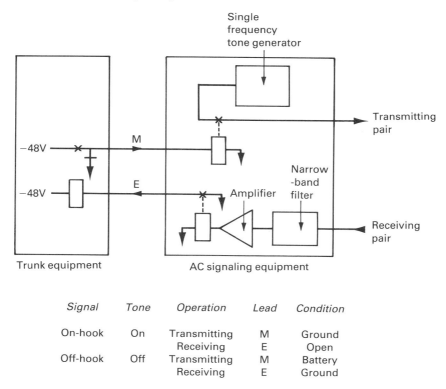

Signal	Tone	Operation	Lead	Condition
On-hook	On	Transmitting	M	Ground
		Receiving	E	Open
Off-hook	Off	Transmitting	M	Battery
		Receiving	E	Ground

Figure 4.3   *Single-Frequency (SF) Signaling*

4.5.2.2 Two-Frequency Signaling (2 VF)

This form of signaling is used for both supervisory or line signaling and for information or address signaling. Often SF and 2 VF supervisory signaling systems are associated with carrier (FDM) operation. In line signaling, 'idle' refers to the 'on-hook' condition, and 'busy' to the 'off-hook' condition. Thus for such types of line signaling that are governed by audio tones the conditions of 'tone-on when idle' and 'tone-on when busy' exist. Similarly for out-of-band signaling methods.

As mentioned previously a major problem with in-band signaling is the possibility of talk-down, which refers to the premature activation or deactivation of supervisory equipment by an inadvertent sequence of VF tones through normal usage of the channel. Such tones could simulate the SF tone. Chances of simulating a 2 VF tone set are much less likely. To avoid the possibility of talk-down on SF circuits, a time-delay circuit or slot filters to bypass signaling tones may be used. Such filters do offer some degradation to speech unless they are switched out during conversation. Thus it becomes apparent why some administrations and telephone companies have turned to the use of 2 VF for supervision. For example, a typical 2 VF line signaling arrangement is the CCITT No. 5 Code (CCITT Recommendation Q144), where f_1 is 2400 Hz, and f_2 is 2600 Hz. Another form is CCITT No. 4 Code, which uses f_1 as 2040 Hz and f_2 as 2400 Hz in both directions. North America uses the R–1 system, in which the SF is 2600 Hz continuous in both directions. The UK uses the SS–AC11 system of 2280 Hz. 2 VF signaling is also widely used for information or address signaling (see Section 4.6).

4.5.3 Out-of-Band Signaling

With this signaling, supervisory information is transmitted out-of-band (i.e. greater than 3400 Hz). In all cases it is a single-frequency system. Some out-of-band systems use 'tone-on idle', indicating the on-hook condition. Tone-on idle keeps a constant load on the channel when not in use. When the condition changes from tone-on to tone-off, a line seizure takes place. Thus any interruption such as a deep radio fade would cause a massive seizure, appearing to the exchange as if all trunks on that particular circuit group required service at once. A precaution used to overcome this problem is to sense the group pilot tone, which should always be present. If the pilot is lost, appropriate action to avoid massive seizure is taken. Some out-of-band systems use 'tone-off idle'. This provides a solid indication of a seized trunk and does not suffer the drop-out problem that can cause massive seizure. The tone being on when the trunk is busy does not affect the speech path at all, for the supervisory signaling channel is separate from the speech channel. The disadvantage of out-of-band signaling is that, when patching of the channel is required, not only the four-wire voice leads, but also an additional pair of leads for the E & M, must be patched.

The preferred CCITT out-of-band frequency is 3825 Hz, whereas 3700 Hz is commonly used in North America (see Section 3.10.4).

4.6 INFORMATION OR ADDRESS SIGNALING

The signals are directly associated with that part of the switching equipment which requires information to perform routing, control and charging functions during the setting up of the call. The essential information signals are those which convey the number being called, but in more comprehensive information signaling schemes it is also possible to forward information about the calling party, and in the backward direction to transmit signals which are used either to predetermine what forward signals are to be sent, or to give information concerning the party being called.

Decadic pulses have a relationship with both line and information signaling. The relationship with address or information signaling is that the decadic pulses constitute all or part of the called party's, or B-party's, address. However, as they are transmitted through the network by means of line signals, they must be included as line signals in any line signaling scheme. The length of the pulse distinguishes information signaling from supervisory signaling. For example, forward and backward supervisory signals may have a pulse length of 150 ms or 600 ms, while the length of decadic pulses may be 65 ms. The exchange is sensitized to the pulse duration, and is thus able to differentiate between supervisory and decadic dialling information. Decadic dialling pulses are formed by the interruption of the subscriber's loop by the on- or off-hook conditions. The off-hook, or make, permits loop current to flow, whereas the on-hook, or break, prevents loop current from flowing. Figure 4.1 is a diagram of a typical telephone subset circuit which generates decadic pulses by means of a make and break relay operated by the rotary dial finger plate.

If more than one exchange is involved in the call set-up, information and line signaling is required between exchanges. Address signaling between exchanges in conventional systems is called inter-register signaling. Some types of inter-register signaling for providing the exchange with routing and charging information are:

By AC pulses of the 1 VF or 2 VF type
By multi-frequency coded (MFC) pulses
By multi-frequency coded (MFC) compelled sequence
By common-channel signaling

By definition, inter-register signaling is 'machine to machine'. There is no intervention of man. Hence it is worked exclusively with machine intelligence, which consists of digits identifying the subscriber's telephone, and at the same time, providing routing information and often charging information. Pulse sending may comprise simple pulse trains, binary codes, or there may also be a meaning as shown above in line signaling, in the duration of the pulse or where the pulse fits in the sequence of sending.

As the single frequency (1 VF) has already been dealt with in Section 4.5 and Figure 4.3, attention will be paid to MF signaling methods and the SF discussion left until it is taken up again when dealing with the direct-line equipment.

In most cases inter-register signaling is bi-directional. Information is sent from the initiating exchange to a receiving exchange, and information is sent from that receiving exchange back to the initiating exchange. This may be achieved by two-way 'junctors' or trunks, or by two separate trunks, one outgoing from the initiating exchange, and the other incoming. The two main forms of inter-register signaling are *compelled sequence* and *non-compelled sequence* signaling.

4.6.1 Non-Compelled-Sequence Signaling

This form of signaling occurs when signals from an initiating exchange are sent independently of signals returning from the far-end exchange. Pulse length is important for the operation of non-compelled signaling systems. An example of non-compelled signaling frequencies is the CCITT No. 5 MF Code, shown in Table 4.1.

Table 4.1   The CCITT No. 5 MF Code

Signal	Frequencies, Hz	Remarks
KP1	1100 + 1700	Terminal traffic (start of pulsing)
KP2	1300 + 1700	Transit traffic (start of pulsing)
1	700 + 900	
2	700 + 1100	
3	900 + 1100	
4	700 + 1300	
5	900 + 1300	
6	1100 + 1300	
7	700 + 1500	
8	900 + 1500	
9	1100 + 1500	
0	1300 + 1500	
ST	1500 + 1700	Forward direction (end of pulsing)
Code 11	700 + 1700	Code 11 Operator
Code 12	900 + 1700	Code 12 Operator

4.6.2 Compelled-Sequence Signaling

A continuous forward signal (combination of two frequencies) is transmitted from the MFC sender. When both frequencies are received in the appropriate MFC receiver, the signal is identified, and a continuous backward (or *revertive* or *controlling*) signal, requesting another forward signal, is returned. When the backward signal is recognized by the MFC sender, the transmission of the forward signal is stopped. After the MFC receiver recognizes that the forward signal has ceased, the transmission of the backward signal is stopped. On the MFC sender recognizing that the backward signal has ceased, the new forward signal, requested by that backward signal, is sent.

4.6.2.1 Forward and Backward Signals

Various administrations use pairs of frequencies selected from a set of frequencies for both the forward and backward direction. Commonly used pairs out of a set are:

Forward. 2 out of 6, and 2 out of 5
Backward. 2 out of 5, and 2 out of 4

For example, take the 2 out of 6 in the forward direction and the 2 out of 5 in the backward direction, as discussed in the next two paragraphs.

FORWARD SIGNALS

The forward signals all comprise two frequencies chosen out of a set of six, giving 15 possible combinations. These signals are numbered 1 to 15. Since the 15 signals are insufficient for all forward information signaling requirements the same numbered signal is used to mean different things. Thus the signals are arranged in groups, so that the 15 signals in group 1 (for example) convey the B-party's or called party's address digits, or special routing requirements, while the 15 group 2 signals convey the service category of the caller (whether subscriber, coin telephone, operator, etc.). Which group a forward signal belongs to is determined by the particular backward signal which requests it, and the stage that the setting up of the call has reached. The terminology for forward signals is, for example, that signal 5 in group 2 is written 'signal 2/5'.

Table 4.2 The CCITT MF Code R–2

Index number for groups I/II and series A/B	1380	1500	1620	1740	1860	1980	Forward direction I/II
	1140	1020	900	780	660	540	Backward direction A/B
1	×	×					
2	×		×				
3		×	×				
4	×			×			
5		×		×			
6			×	×			
7	×				×		
8		×			×		
9			×		×		
10				×	×		
11	×					×	
12		×				×	
13			×			×	
14				×		×	
15					×	×	

Frequencies, Hz

BACKWARD SIGNALS
The backward signals all comprise two frequencies chosen out of a set of 5, giving 10 possible combinations. In a similar manner to the forward signal groups, the backward signals are assembled in a number of series, so that the 10 signals in the 3A Series (for example) request information required for switching, charging or barring the call; while the ten B Series signals convey information to the originating exchange about the condition of the B-party's line. Which series a backward signal belongs to is determined by the group, and possibly the number, of the forward signal which precedes it. The terminology for backward signals is, for example, that signal 3 in Series B is written 'Signal B3'.

The above is a form of the CCITT R–2 system (CCITT Recommendation Q361). Table 4.2 shows the code. Other special signals that may be used in signaling systems as accepted for international use may be found in CCITT Recommendation Q140.

4.6.3 Common-Channel Signaling

The concept of signaling, both line and information, has been in the past such that the individual channel which carries the speech during the conversation time is also the medium over which the signals for that particular call are conveyed. This may be termed *channel-associated signaling*. Modern switching and transmission technology has made possible a new technique whereby all signaling requirements for a number of speech circuits are carried separately by a single high-speed high-capacity data link. This technique, known as *common-channel signaling*, was initially developed because of problems in the international network, particularly when satellite links were involved. These problems centred around signaling delays brought about by long propagation times, and also by the long recognition times (100 ms. or more) of individual line signals in the signaling systems being used.

The following are some of the features of the common channel signaling method:

The inherently large capacity of these systems means that a greatly increased range of information signals is possible compared with the present MFC system.
There is no need to distinguish between line and information signals.
Double seizure problems on both-way circuits are avoided.
The system may be used to facilitate provision of sophisticated network facilities in the future where communication is required with a central point.

The CCITT has developed two systems, namely No. 6 and No. 7. The CCITT system No. 6 is capable of providing all signaling requirements for 2048 telephone speech channels. The North American system CCIS (common channel inter-office signaling) can handle 8192 telephone channels. The new system, No. 7, is better suited to smaller circuit groups and to application in national networks.

4.7   SINGLE FREQUENCY SIGNALING

Outstations connected by carrier systems, but which do not have an exchange, may require telephone facilities. To provide the outstation with a telephone or telephones, what is required are circuits which will individually behave like a short-circuit between the far-end telephone and the exchange at the central office. Figure 4.4 shows this requirement.

4.7.1 Operation – Subscriber to Exchange

With a telephone connected to an exchange, if the handset goes off-hook, the subscriber's line to the exchange is looped, and the exchange line relay set is activated, etc. To perform the same operation at an outstation using the equipment in Figure 4.4, when the telephone handset goes off-hook and loops the line, the foreign exchange subscriber (FXS) is activated, generating an M-lead condition. The activation of the M-lead in turn activates the single frequency convertor (SFC), which stops the in-band signaling tone (2600 Hz) – for a tone-on-idle condition – going into the appropriate multiplex channel. At the central office end, this loss of tone is detected by the SFC, which generates on the E-lead an open-circuit condition. The foreign exchange office (FXO) responds by looping the two-wire line to the exchange. Decadic dialling information is transmitted in the same manner as described above, whereas MFC information will pass directly to the exchange.

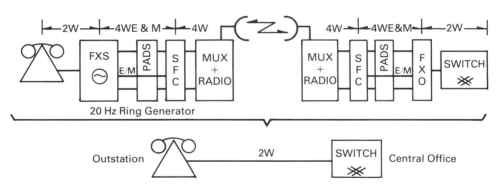

Figure 4.4   SF Signaling

4.7.2 Operation – Exchange to Subscriber

The exchange when calling the subscriber sends ringing current to the telephone. When the subscriber goes off-hook and loops the line, that informs the exchange to switch out the ring-current generator, and switch in the audio circuit.

Where the subscriber is at an outstation, the ringing current sent by the exchange is received by the central office FXO, which converts it to an M-lead signal to the central office SFC. This in turn breaks the 2600 Hz tone for the same duration as the 'ring'. This tone break (and make), after entering the central office multiplex and carrier system, is received at the appropriate far-end MUX channel and the information is passed to the outstation SFC. The subscriber SFC detects this loss (or restoration) of tone and provides an E-lead open-circuit (or ground) condition. This permits the subscriber's FXS to switch in (or out) the 20 Hz ring-current generator to the subscriber's telephone. The telephone thus rings at the same rate and duration as that of the exchange. When the subscriber goes off-hook, the same process informs the central office exchange that the line has been looped, preventing any further use of the exchange's ring-current generator. Also as the FXS detects the subscriber looping the line, it cuts out the subscriber's ring-current generator, and switches in the audio path.

4.8   DIRECT-LINE EQUIPMENT OPERATION

Direct-line telephone sets are usually located in buildings or offices at microwave outstations. In effect the direct line is an extension of the central office exchange, or of the public exchange. The only difference between this telephone, and one hard-wired into the central office or public exchange, is that it is connected via the microwave backbone and ancillary equipment. The purpose of the direct line may be only to gain independence from the outstation exchange. This may be required if communication must be maintained to the outstation personnel if the exchange goes 'down', thus providing a stand-by telephone link to the rest of the system. Figure 4.5 shows a block and level diagram of such a subsystem.

4.9   DIRECTORY NUMBER TEST LINE OPERATION

A system which incorporates the directory number test line (DNTL) feature permits the rudimentary testing of outstations from any other outstation or Central Office. By placing a call to any site containing the equipment for this facility, a 1000 Hz tone at a fixed level, or a quiet termination, can be obtained. In addition, a call can be looped through that location from either one, or two, other locations, enabling tests to be performed through a particular outstation without actually having any personnel present at that outstation.

4.9.1 Test tone

From an outstation or central office, a call made to the DNTL number of another outstation equipped with the DNTL equipment will via the distant exchange access a 1000 Hz tone. Receipt of this tone, which is being generated at the distant DNTL equipment at a preset level, will indicate that the distant exchange, multiplex and

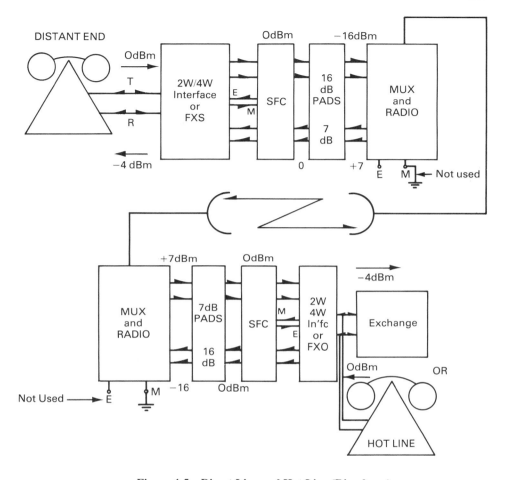

Figure 4.5   Direct Line and Hot Line (Ringdown)

microwave radio equipment are operating correctly or not, according to the received level of the tone.

4.9.2 Quiet Termination

From an outstation or central office, a call made to the DNTL number of another outstation equipped with the DNTL equipment will via the distant exchange access a quiet termination, which is below −50 dBm0. This will permit assessment of the quality of the line and link to be observed for crosstalk, or other forms of noise.

4.9.3 Loop-back

The loop-back connection uses two telephones at any location. The call is looped through the selected outstation exchange by the following procedure: the selected exchange is first called using the 1000 Hz test-tone number from the first telephone; then with the first telephone still connected to the 1000 Hz tone, the selected exchange is called from the second telephone, using the quiet termination number. This stops the 1000 Hz tone, and completes the connection between the two telephones.

Between these two telephones, or open lines, various tests as required can be made.

4.10 RINGDOWN CIRCUITS

Ringdown signals are spurts of ringing current (16 – 25 Hz) applied usually through the ringing key of an operator and intended to operate a bell, ringer or drop at the called end. The current may be generated by a manually operated magneto or by a ringing machine with or without automatically inserted silent periods. Ringing to telephone subscribers in automatic central offices is stopped or 'tripped' automatically by relay action resulting from the subscriber's off-hook condition. Ringing signals may be converted to 500 Hz or 1000 Hz, usually interrupted at a 20 Hz rate, to pass through voice channels of carrier equipment. A ringing signal to a manual switchboard usually lights a switchboard lamp, which can be extinguished only by local action and not by stopping or repeating the ringing signal. This characteristic makes ringdown operation unsuitable for fully automatic operation. Ringdown signaling over carrier circuits has the advantages of simplicity and of not requiring the distinct signaling channels of E- and M-systems. A single ringdown circuit may be used as a hot line, which does not require the dialling of any number to reach the called party. Figure 4.5 shows this option when commercial interfacing equipment is used over a multiplex voice channel.

REFERENCES

1. Schroeder, R., *Switching Systems – Signaling* (Telecom Australia Engineer Development Programme, 1978), ED 0017.
2. Freeman, R. L., *Telecommunications Transmission Handbook* (Wiley, 1975). (Appendix C,pp. 547–558).
3. CCITT Recommendations Q140, Q151, Q361, Q364, Q365 (Orange book).
4. Welch, S., *Signaling in Telecommunications Networks* (Peter Peregrinus, 1981) IEE Telecommunications Series, Vol. 6.

5 NOISE[1]

5.1 INTRODUCTION[2]

The drift or flow of electrons in a conductor, semiconductor, etc., constitute current. In a sufficiently short period of time this current can be considered as being made up of closely, but not evenly, spaced pulses. From $q = it$ (or charge = current × time), if the current is small, then the charge, or the number of electrons making up the current will be small; the current may not be large enough to appear as a constant or continuous flow, but only as random pulses. This then sets the lower limit on the current that an amplifier can amplify.

According to kinetic theory, if a conductor, etc., is above 0K ($-273°C$) there will always be thermal agitation of bonds holding the material together, and hence a random motion of the free electrons contained in the material. If a signal current passing through a conductor is not greater than this random current, the signal will be masked by the random current or – as otherwise known – *noise current*. If the signal current is amplified, then so will be the noise current, and the signal will remain masked. In addition to the thermal noise in the conductor, noise currents generated by the transistors or active devices will also be produced. These noise powers will add and result in a drop in the signal-to-noise ratio at the output of the amplifier, when compared to that at the input to the amplifier.

The noise which will be dealt with in this chapter is that noise which is produced by the random notion of the free electrons in a conductor and which is caused by thermal agitation of the molecules or atoms of that conductor.

5.2 THERMAL OR JOHNSON NOISE

If a length of conductor is taken at some temperature above absolute zero, the random notion of the free electrons caused by thermal agitation produces a voltage at the open ends of the conductor. This voltage is an AC voltage, and the frequency components cover the complete range of the RF spectrum uniformly. Due to this wide range of frequencies, an analogy to the visible part of the electromagnetic spectrum can be made, where the total of all colors produces 'white' light; hence broad-band noise is termed *white noise*. The value of the voltage *e* occurring at the conductor ends can be calculated as:

$$e^2 = 4KTBR \tag{5.1}$$

where B = the noise bandwidth in Hz.
K = Boltzmann's constant = 1.38×10^{-23} J/K.
T = absolute temperature in kelvins.
R = the resistance of the conductor in ohms.

5.3 AVAILABLE POWER

Consider a Thévenin generator with a series resistance of R Ω. The maximum amount of power which can be transferred by this generator is when the generator and series resistance R feeds into a load of value R Ω (maximum power transfer theorem). The value of this power delivered to the load is:

$$(e/2R)^2 \times R = e^2/4R \tag{5.2}$$

If the Thévenin generator of voltage *e* is replaced by a noise generator of voltage *En* then equation 5.2 becomes with the aid of equation 5.1:

$$En^2/4R = 4KTBR/4R = KTB \tag{5.3}$$

Hence the available power P_n from a noise generator is:

$$P_n = KTB \tag{5.4}$$

where B = the bandwidth over which the noise is measured in Hz.
K = Boltzmann's constant = 1.38×10^{-23} J/K.
T = the absolute temperature in kelvins.

Equation 5.4 shows that the available noise power produced by a resistor is independent of the size of the resistor producing the noise.

5.4 AVAILABLE POWER GAIN

Consider a Thévenin generator with a series resistance R Ω, supplying an amplifier or a network, which has an output resistance R_o. Let the unloaded or open-circuit

Figure 5.1   A Two-Part Network and Thévenin Generator

output voltage be E_o, as shown in Figure 5.1: then the available power gain G in the ideal case is given by:

$$G = \frac{\text{Available signal power from the amplifier or network output}}{\text{Available signal power from the Thévenin generator}}$$

The available signal output power S_o from the network is given by

$$S_o = E_o^2/4R_o \tag{5.5}$$

Notice that the *available* signal output power from the amplifier or network is independent of the load resistance R_L, because it is the maximum available power, and not the actual power delivered into the load R_L, which of course would be less. The available signal power S_i from the input source is given by

$$S_i = E_g^2/4R \tag{5.6}$$

Thus the available power gain $G = (E_o/E_g)^2.R/R_o$ (5.7)

The available power gain of a network or an amplifier is thus dependent on the impedance match between the signal source and the network or amplifier:

$$G = A^2.R/R_o \tag{5.8}$$

where A is the voltage gain of the network or amplifier.

Examples

1. Consider Figure 5.2. We have:

$$R_o = (R_1.R + R_1.R_2 + R_1.R_3 + R.R_3 + R_2.R_3)/(R + R_1 + R_2)$$

$$E_o = E_g.R_1/(R+R_1+R_2)$$

Hence $S_o = E_o^2/4R_o$

$$= R_1^2.E_g^2/4(R+R_1+R_2)(R_1.R+R_1.R_2+R_1.R_3+R.R_3+R_3.R_2)$$

$$S_i \qquad = E_g^2/4R$$

$$\text{Hence} \quad A \qquad = R_1/(R + R_1 + R_2)$$

$$G \qquad = \frac{R.R_1^2}{(R + R_1 + R_2).(R_1.R + R_1.R_2 + R_1.R_3 + R.R_3 + R_3.R_2)}$$

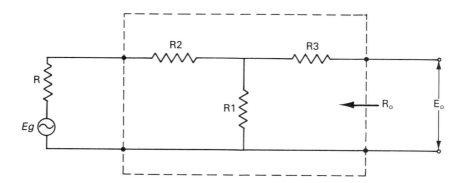

Figure 5.2   For Example 1

2. Consider Figure 5.3. The Thévenin generator is now replaced with a noise generator and the network is active, i.e. it amplifies. As the available output signal power $S_o = G \times$ available input signal power S_i; i.e.

$$S_o = G.S_i \tag{5.9}$$

where, as in equation 5.8,

$$G = A^2.R/R_o$$

The available output power is a noise power given by:

Available output noise power $P_{on} = E_o^2/4R_o = A^2E_n^2/4R_o$
$= A^2.4KTBR/4R_o = A^2.RKTB/R_o$

where $E_o = AE_n$ from Figure 5.3

From the expression for G, we have

$$P_{on} = G(KTB) = G \times \text{(available input noise power)}$$

Since every resistor has a noise voltage across its terminals, the internal resistance of any signal source will also produce a noise voltage. Hence an antenna connected to the input of a radio receiver will produce a noise voltage as well as the received signal voltage. If the received signal voltage is small and hence comparable to the noise voltage, E_n of the internal resistance of the source, both voltages must be considered at the input of an amplifier.

If the input available noise power $P_n = KTB$ and the input available signal power $= S_i$, the available output power is

$$(S_o + P_{on}) = GS_i + G(KTB) = G(S_i + KTB)$$

The ratio of the signal-to-noise at the input to the signal-to-noise at the output is given by:

$$\frac{S_i/KTB}{S_o/P_{on}} = \frac{S_i/KTB}{GS_i/GKTB} = 1$$

Thus for an ideal noise-free amplifier this ratio is unity, and there is no degradation in the signal-to-noise ratio.

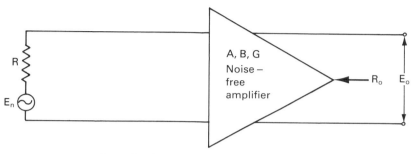

Open circuit voltage gain A, Bandwidth B,
Available power gain G

Figure 5.3   Resistor at Input of Ideal Amplifier

5.5   NOISE BANDWIDTH

The B in equation 5.1 is the noise bandwidth of a square-topped passband as shown in Figure 5.4. In practice, however, to construct an amplifier whose passband would approximate closely to that of a square-topped passband, would be difficult and expensive. The passband of an amplifier does not have vertical sides and a flat top, but rolls off at a rate according to its design. Because of this, an equivalent bandwidth or an equivalent noise bandwidth can be defined for an amplifier or any other device. This equivalent noise bandwidth can be obtained by finding the area under the actual passband curve on the graph of available power gain plotted against frequency, and converting it to an equivalent flat-top passband.

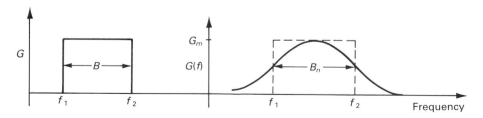

Figure 5.4   Definition of Noise Bandwidth

Consider the selectivity curve shown in Figure 5.4. The vertical dashed lines can be moved closer or further apart until the area above the curve inside the dashed box is equal to the area under the curve outside the dashed box. Once these areas are equal, then we have effectively found f_1 and f_2 which would produce an equivalent flat-topped, vertical side, passband. On a more formal basis, the noise bandwidth B_n can be defined as:

$$B_n = \frac{\int_0^\infty [H(f)]^2 df}{[H(f_o)]^2} \quad (5.10a)$$

where $H(f)$ = transfer function of the impedance, admittance, voltage or current.
f_o = frequency at which the response is measured. It is usually when the response is at a maximum.

Due to the squared term in both the numerator and denominator in equation 5.10a, the constants which would convert either into power cancel out. Hence equation 5.10a can be written as:

$$B_n = \frac{\int_0^\infty [H(f)]^2 df}{G_m} \quad (5.10b)$$

$$= 1/G_m \times \text{area under the selectivity curve}$$

where G_m = the maximum available power gain.

If the equation for the selectivity curve is known, the integration may be performed directly to obtain the area under the selectivity curve; otherwise approximation methods may be used. B_n for the multistage amplifier is very nearly equal to the ordinary 3 dB bandwidth, i.e. the bandwidth at which the available gain G has fallen to one-half of its value at the maximum response level of the selectivity curve. For convenience, most calculations concerning the noise bandwidth define it as being that band of frequencies lying between the 3 dB or half-power points. This is not the correct definition although it does provide a quick and easy method of roughly determining the noise bandwidth. The more exact method is that as given by equation 5.10b.

5.6 NOISE TEMPERATURE

In actual life there is no such thing as an amplifier which does not add its own noise to the input signal or input noise. Thus when the input noise changes, only one of the two parts of the output noise signal of the amplifier will change, i.e. the amplified input noise. The noise added by the amplifier will remain the same. The noise which is added by the amplifier is described in terms of 'effective noise temperature'. If Figure 5.3 is considered, where the amplifier is no longer noise-free, then for an amplifier with a matched resistor at its input:

$$\text{Noise out} = (KTB_n + N_a)G \tag{5.11}$$

where N_a = the measure of the noise added by the amplifier.

As this noise may be added at different stages in the amplifier and in different ways, the noise contribution by the amplifier N_a is first apparent at the output as given by equation 5.11. N_a is multiplied by the available power gain G in equation 5.11; thus the amplifier noise contribution is referred back to the input of the amplifier. Hence

$$N_a = KT_e B_n \tag{5.12}$$

This means that the added noise N_a, in terms of K, T, and B_n, can be expressed as an 'effective temperature' (symbol T_e).

T_e is a completely fictional temperature because N_a is not necessarily thermal noise. For example, it may be the power supply ripple, or harmonics of it which occur within the bandwidth of the amplifier. The 'standard noise temperature' T_0 for noise measurement work is 290°C.

5.6.1 Noise Temperature of Resistors in Series

Consider a number of resistors, $R_1, R_2, \ldots R_n$ placed in series. The total resistance R_t can be expressed as

$$R_t = \sum_{i=1}^{i=n} R_i$$

If they were all at the same temperature T, the total mean square noise voltage E_n^2 would be given by

$$E_n^2 = 4KT\,R_t B = 4KT\sum^{n} R_i B \tag{5.13}$$

Thus the total mean square noise voltage is equal to the *sums* of the squares of the individual noise voltages. If the resistors had been at different temperatures Ti the total mean square noise voltage E_n^2 would be

$$E_n^2 = 4K \sum_{i=1}^{i=n} T_i R_i B \tag{5.14}$$

The available noise power P_n would be

$$P_n = \frac{(E_n/2)^2}{\displaystyle\sum_{i=1}^{i=n} R_i} = \frac{4KB\displaystyle\sum_{i=1}^{i=n} T_i R_i}{4\displaystyle\sum_{i=1}^{i=n} R_i} = KTB \tag{5.15}$$

From this equation it can be seen that the noise temperature T of the series resistors is

$$T = \frac{\displaystyle\sum_{i=1}^{i=n} T_i R_i}{\displaystyle\sum_{i=1}^{i=n} R_i} \tag{5.16}$$

Considering the individual contributions of each resistance to the total temperature, we have:

$$T = \sum_{i=1}^{i=n} a_i T_i \tag{5.17}$$

where $a_i = \dfrac{R_i}{\displaystyle\sum_{i=1}^{i=n} R_i}$ $\tag{5.18}$

If a matched signal generator of power P_n is connected to the set of resistances the total signal power available from that generator would be absorbed by the resistances and each resistance would absorb a fraction (P_i) of the available power:

$$\frac{P_i}{P_n} = \frac{R_i}{\displaystyle\sum_{i=1}^{i=n} R_i} = a_i \tag{5.19}$$

Any real power losses in a passive microwave system can generally be treated as though they were due to an attenuator. This is true for the absorption losses which occur in a propagation path between antennas or between a microwave source and a receiving antenna. Thus the noise temperature of a radiating antenna is the summation of all of the losses and absorbing bodies temperature which the antenna sees, as given by equation 5.17. The background sky temperature is likely to be of the order of 4 K, whilst the atmosphere which the signal must traverse may be of the order of 290 K.

5.6.2 Noise Temperature of a Pad[4]

Consider Figure 5.2 where R is at a temperature T_1, and the pad, which comprises R_1, R_2, and R_3, has a loss L, and is at a temperature T_L. To determine the effective temperature (T_o) of the attenuator being fed from a matched resistor, the output noise power from the attenuator is divided into that which is contributed by the attenuated input noise, and that which results from the attenuator itself. Since the input resistor is at a temperature T_1, the attenuated input noise is KT_1B/L. The total available noise power at the output of the attenuator is the sum of that noise contribution from the noise resistor R and the pad itself:

$$T_o = a_1 T_1 + a_2 T_L$$

and, from equation 5.18: $a_1 + a_2 = 1$

If $a_1 = 1/L$, then $a_2 = (1 - 1/L)$

Hence: $T_o = (1/L)T_1 + (1 - 1/L)T_L$ (5.20)

Equation 5.20 is thus the effective temperature of an attenuator being fed from a matched resistor. The practical implication of this example is where a receiving antenna feeds into the front end of a receiver via a duplexer. The duplexer will have loss, since it is only a filter which prevents the transmitted signal from a transceiver from being fed back into the receiver, so that both the transmitter and the receiver can operate using one antenna.

Example Assume that an antenna is receiving a satellite carrier signal via a duplexer which has 5 dB of loss, and is in a hut at a temperature of 290 K. The sky temperature is assumed to be 4 K. Converting a 5 dB loss figure into a noise loss of 3.16, we find from equation 5.20:

$$T_o = 0.316 \times 4 + 0.684 \times 290$$
$$= 199.558 \text{ kelvins}$$

This indicates that as the maximum contribution of noise comes from the duplexer itself due to its high temperature, there is a need to cool the duplexer if there is to be a reasonable signal to noise ratio for received signals which are very weak. Equation

5.20 also shows that if both input and pad temperatures are the same, the greatest contribution of noise comes from the pad if the loss factor of the pad is greater than 2, or the loss figure is greater than 3 dB.

5.7 EFFECTIVE INPUT NOISE TEMPERATURE

As mentioned in Section 5.6 in a practical amplifier, noise is generated by the active elements, such as transistors and diodes. The noise which is generated in the early stages of an amplifier is considerably amplified before reaching the output. The ways in which this additional noise can be quantified so that comparisons between amplifiers can be made are by means of:

1. The effective input noise temperature, T_e.

2. The noise figure or noise factor F.

Noise factor shall be considered in Section 5.8. Previously it has been seen how noise powers may be characterized by noise temperatures. Taking this further, it is possible for an output noise power from a system to be represented in terms of an operating temperature T_{op}. The operating temperature is attributable to noise contributions outside the amplifier or, say, the receiver, and from the device or system itself. The total output noise power N_{T_o} as given by equations 5.11 and 5.12 can be expressed as:

$$N_{T_o} = GKB(T_i + T_e) \tag{5.21}$$

where T_i is the noise temperature of the input termination which delivers the incident noise power.

T_e is the effective input noise temperature of the receiver.

If a radio receiver has two responses, such as a superheterodyne receiver which has one response at the signal frequency, and due to its local oscillator, another at its image frequency, the total output noise power is given by:

$$N_{T_o} = G_1 K B_1 (T_{i_1} + T_e) + G_2 K B_2 (T_{i_2} + T_e)$$

This equation can be extended to n responses if so required.

5.7.1 Amplifiers in Cascade

In the design of a radio receiver the first amplification stage is considered the most important for noise considerations. For it is the first stage which will end up having its noise being the largest contributor to the total system or receiver noise appearing at the output. In order to obtain the overall effective input noise temperature of a number of stages or amplifiers in cascade in terms of the effective input noise temperatures of the individual units, the noise contributed by each unit must be known. Consider Figure 5.5, in which amplifiers in cascade are shown. The total output noise power from the first cascaded pair is:

$$N_{T_o} = G_1 G_2 KB(T_i + T_{e_{12}})$$ (5.22)

where $T_{e_{12}}$ is the overall effective input noise temperature of the two amplifiers in cascade.

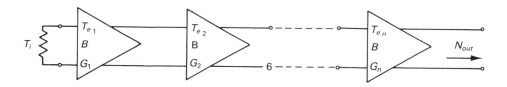

Figure 5.5 Cascade of n Amplifiers

The total output noise power of the second amplifier can be divided into two components: the noise contributed by itself, and the noise incident upon its input and amplified by its gain. Thus

$$N_{T_o} = G_2 KB\, T_{e_2} + G_2.(G_1 KB(T_i + T_{e_1}))$$ (5.23)

where $G_1 KB(T_i + T_{e_1})$ is the output noise power of the first amplifier and hence the input noise to the second amplifier.

Combining equations 5.22 and 5.23 and solving for $T_{e_{12}}$ gives:

$$T_{e_{12}} = T_{e_1} + T_{e_2}/G_1$$ (5.24)

For n amplifiers in cascade:

$$T_{e_{1.2...n}} = T_{e_1} + T_{e_2}/G_1 + T_{e_3}/G_1.G_2 + ... + T_{e_n}/G_1.G_2...G_{n-1}$$ (5.25)

Since the effective input noise temperature of an amplifier is dependent upon the impedance of the network connected to its input port, the T_e of each amplifier in the cascade must be that T_e which is associated with the output impedance of the preceding amplifier.

5.7.2 Effective Input Noise Temperature of a Matched Attenuator Pad

Consider Figure 5.2, where the input resistor is at a temperature T_i, followed by a matched attenuator having a loss L at a temperature T_L. Applying equation 5.21 which defines T_e, we have:

$$N_{T_o} = GKB(T_i + T_e) = 1/L \; KB(T_i + T_e)$$

The output noise in turn is, by equations 5.20 and 5.4, given as

$$N_{T_o} = (1/L).KB(T_i + (L-1)T_L)$$

Comparing the above two equations it is evident that the effective input noise temperature T_e of the attenuator is given by

$$T_e = (L - 1)T_L \tag{5.26}$$

5.7.3 Matched Attenuator in Cascade with an Amplifier

Consider a microwave link using standard equipment, i.e. equipment which has a fixed transmitter power output, etc. Suppose that the distance between the transmitter and receiver is less than for other links; hence the received signal strength is too high for the receiver to operate optimally. If the input to the receiver is padded down, what will be the effect upon the signal-to-noise ratio of the receiver output signal?

Let the attenuator have a loss L, at a temperature T_L, in cascade at the input of an amplifier having an effective input noise temperature T_e. Combining equation 5.25 with 5.26, the overall effective input noise temperature $T_{e_{12}}$ is, if $T_{e_1} = (L-1)T_L$, and $G = 1$, given by

$$T_{e_{12}} = (L-1)T_L + LT_e$$

Thus, the attenuator ahead of the amplifier multiplies the effective temperature of the amplifier by the loss factor and also adds $(L-1)T_L$ to the effective input noise temperature. This may look as if one should not ever place an attenuator ahead of an amplifier due to the increase in the overall effective temperature of the system by the attenuator loss factor. As will be shown when dealing with noise figure, this is not as bad as may first be assumed, since it reflects only a decrease in the value of the received signal level according to the value of the attenuator, and nothing more.

5.8   NOISE FIGURE

First some simple definitions which will be used in this section.

The *noise factor* is a numerical ratio value only. The *noise figure* is equal to $10 \log_{10}$ (noise factor), and is expressed in decibels (as it is formed from a ratio). The noise factor F is defined as the ratio of the available signal-to-noise power at the signal generator terminals when the temperature T of the input termination is 290 K, to the available signal-to-noise power at the output of the network. Thus

$$F = \frac{\dfrac{S_i}{290\,KB}}{S_o/N_{T_o}} \qquad (5.27)$$

For an amplifier where $S_o = G.S_i$

$$F = \frac{N_{T_o}}{290\,GKB} \qquad (5.28)$$

From equation 5.27

$$S_o/N_{T_o} = \frac{S_i/(290\,KB)}{F}$$

This shows that the output signal-to-noise ratio of an amplifier is decreased if the amplifier adds noise, because F is always greater than or equal to unity or 0 dB. *F is a measure of the degradation of the input signal-to-noise ratio that has occurred on passing through a network if the temperature of the signal source is 290 K*

5.8.1 Amplifiers in Cascade

Figure 5.6 shows a block diagram of amplifiers in cascade. Assume that the bandwidths of both amplifiers are the same, i.e. $B_1 = B_2 = B$. The total available noise power output N_{To} for two stages in cascade is:

$$N_{T_o} = 290\,F_{12}.G_1.G_2KB \qquad (5.29)$$

where F_{12} is the overall noise factor for the cascade.

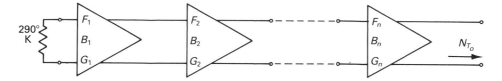

Figure 5.6   Cascade of n Amplifiers

The available output power of the first stage becomes the input noise power for the second stage. In addition to this, noise will be added by the second stage which can be represented by the noise factor of this stage F_1. Thus the input noise power N_1 entering the second stage is amplified by the available power gain G_2 of the second stage and appears at the output together with the second-stage amplifier noise. The total available output noise power equals the noise power at the output of the first stage plus the noise power added by the second stage, thus:

$$N_{T_o} = N_1 G_2 + \text{second-stage amplifier noise}$$

or

$$F_{12}.G_1.G_2.K.B.290 = (F_1.G_1.K.B.290).G_2 + (F_2 - 1).G_2.K.B.290$$

which gives

$$F_{12} = F_1 + (F_2 - 1)/G_1 \tag{5.30}$$

Similarly for n stages, the overall noise factor becomes:

$$F_{1,n} = F_1 + \frac{(F_2 - 1)}{G_1} + \frac{(F_3 - 1)}{G_1 G_2} + \ldots + \frac{(F_n - 1)}{G_{1,n-1}} \tag{5.31}$$

where
$$
\begin{aligned}
n &= \text{the number of stages.}\\
F_1, F_2, \ldots F_n &= \text{the noise factors of the individual stages.}\\
G_1 &= \text{the gain of the first stage.}\\
G_{1,n-1} &= \text{the gain of the first } (n-1) \text{ stages} = G_1.G_2 \ldots G_{n-1}.
\end{aligned}
$$

Note that the 290 K in equations 5.30 and 5.31 has been cancelled out. This indicates that these equations are independent of the input termination temperature. Also, and more practically important, if the gain of the first section is high, the overall noise factor is practically equal to the noise factor of the first stage. Good design practice follows two criteria:

1. High gain for the first and if possible second amplifier stages

2. Low noise factor for the first amplifier stage

It was assumed in the derivation of equations 5.30 and 5.31 that the amplifier stages were of equal bandwidth. The error will be small if the available power gain per stage is high and the noise factor per stage is small.

5.8.2 Superheterodyne Receiver Application

Equation 5.30 can be applied to the mixer stage of a superheterodyne receiver. Thus:

$$F_{12} = F_m + (F_{if} - 1)/G_m \qquad (5.32)$$

where F_{12} = the overall noise factor of the receiver.
 F_m = the noise factor of the mixer.
 F_{if} = the noise factor of the IF amplifier.
 G_m = the conversion gain of the mixer, which in some cases may be a loss.

If the conversion gain of the mixer is actually a conversion loss, the value of F_{if} becomes important to the overall noise factor of the receiver, and should, within the constraints of insuring a minimum of delay distortion, be made as small as possible in the design of the IF filter. The value of F_m for a crystal mixer decreases as the IF is increased whereas F_{if} increases with an increase in the *IF*. The IF that will give the lowest overall noise factor can be determined with the aid of equation 5.32.

5.8.3 Noise Figure of a Matched Attenuator Pad

Refer again to Fig. 5.2 (and, out of interest, Section 5.7). If the input termination R is at a temperature of 290 K, followed by a matched attenuator with a loss L and at a temperature T_L, then from equation 5.28:

$$N_{T_o}(T_i{=}290) = F.G.K.B.290 \qquad (5.33)$$

As $G = 1/L$

$$N_{T_o}(T_i{=}290) = (F/L).K.B.290$$

From equations 5.20 and 5.4, the following equation is derived:

$$N_{T_o} = (1/L).K.B.(290 + (L{-}1)T_L) \qquad (5.34)$$

Combining equations 5.33 and 5.34, we have:

$$F = 1 + (L{-}1)T_L/290 \qquad (5.35)$$

If T = 290 K

$$F = L \qquad (5.36)$$

5.8.4 Matched Attenuator in Cascade with an Amplifier

If a matched attenuator having a loss L at a temperature T is in cascade with amplifier having a noise factor F, then, combining equation 5.30 with equation 5.35, we have:

$$F_{12} = 1 + (L-1)T_L/290 + (F-1)L \qquad (5.37)$$

or when $T = 290$ K

$$F_{12} = FL$$

Thus when an attenuator is introduced ahead of an amplifier, the amplifier noise factor is degraded by the loss L. This is important when a microwave radio receiver must be padded down in order to prevent overload of the receiver. If equation 5.37 is considered in the form where the noise factor is the ratio of the signal-to-noise ratio out of the receiver to the signal-to-noise ratio at the input to the receiver, then:

$$F_{12} = \frac{(S/N)_i}{(S/N)_o} = FL \qquad (5.38a)$$

or

$$(S/N)_i = FL(S/N)_o$$

or

$$\frac{S_i/L}{N_i} = F(S/N)_o$$

$$\text{or } F = \frac{(S_i/L)N_i}{(S/N)_o} \qquad (5.38b)$$

Hence the only difference between the noise factor of a matched amplifier in cascade with an amplifier, and the noise factor of an amplifier, is that the input signal has been reduced by the loss factor L. Intuitively, this is what would be expected. Usually the attenuator is not put at the transmitter output to pad down the transmitter power, because of the heat that may be dissipated in the pad.

5.8.5 Relationship between *F* and *T_e*

From the definition of noise factor given by equation 5.28, we find:

$$\begin{aligned}
N_{T_o} &= (F)\,GKB.290 \\
&= GKB.290 + (F)GKB.290 - GKB.290 \\
&= GKB.290 + (F-1)\,GKB.290 \\
&= \text{Input noise due to resistor at 290 K} + \text{amplifier noise}
\end{aligned}$$

If the input temperature is other than 290 K, say T_i, then

$$N_{T_o} = GKBT_i + (F-1)GKB.290 \tag{5.39}$$

The 290 K in the second term of equation 5.39 remains, due to the definition of the noise factor, F.

Comparing equation 5.39 with equation 5.21, we get:

$$T_e = (F-1).290$$

and hence

$$F = 1 + T_e/290 \tag{5.40}$$

The reference temperature should always be taken at 290 K, but if it is not, then

$$F = 1 + T_e/T_o \tag{5.41}$$

If T_o is used as the reference temperature, the system or network appears noisier when it is used with a cold generator than when used with one which is hot. A manufacturer could make his equipment appear superior by quoting a noise figure based on a reference temperature which is far higher than 290 K. Care must be exercised when comparing equipment of different quoted noise figures, to ensure that each is a noise figure referenced to 290 K.

EXERCISES

1. Consider a receiver with a noise figure of 10 dB. Its output signal-to-noise ratio is 50 dB. What is its input equivalent signal-to-noise ratio? (*Answer* 60 dB)

2. Consider a receiver with an equivalent noise temperature of 500 K. What is its noise figure in dB? (*Answer* 4.35 dB)

3. A receiver has a noise figure of 10 dB. What is its equivalent noise temperature in Kelvins? (*Answer* 2610 K)

4. A matched attenuator pad has a temperature of 300 K and a loss of 19.5 dB. What is its effective input temperature in K? (*Answer* 26437 K)

5. A matched attenuator having a loss of 19.5 dB and an input temperature of 300 K is in cascade with the input of an amplifier having an effective noise temperature of 400 K. Determine the overall effective noise temperature. (*Answer* 62087 K)

6. What is the noise figure of the attenuator and amplifier in cascade? (*Answer* 23.32 dB)

5.9 CALCULATION OF THERMAL NOISE IN THE SIGNAL CHANNEL[5,6]

In transmitting a carrier using angle modulation (frequency or phase), the information signal produces a required phase deviation of the RF carrier. Noise, however, produces an unwanted, yet determinable, phase deviation of this same carrier. The signal and noise powers after demodulation are directly proportional to these phase deviations squared.

The signal-to-noise ratio (S/N) for the thermal noise in the resultant telephone channel after reception, demodulation, etc., can be expressed in its non-logarithmic form as:

$$S/N = \phi_s^2/\phi_n^2 \tag{5.42}$$

where ϕ_s = the RMS carrier phase deviation in radians due to a sinusoidal test-tone.

ϕ_n = the RMS noise phase deviation in radians appearing due to an upper and lower sideband of frequencies $(\omega_c + \omega_n)$, and $(\omega_c - \omega_n)$ respectively.

By definition, from FM theory, an RMS frequency deviation Δf produced by a dBr signal at the transmitter input gives:

$$\phi_s = \frac{\Delta f}{f_m} = \frac{\text{Frequency deviation}}{\text{Modulating frequency}} \tag{5.43}$$

As ϕ_s is a RMS quantity, so then must be the frequency deviation Δf.

What is now required is to express ϕ_n in terms of the carrier and the noise power. Referring to Figure 5.7, we can see that the noise voltages v_{n_1} and v_{n_2} are contra-rotating on the carrier. At some time these voltages will produce a resultant voltage which is in quadrature with the carrier, and hence will produce the maximum phase deviation possible. As the noise voltages are rotating at different frequencies, the combined noise power is not formed from the summation of the two voltages, but from the summation of the individual powers which these voltages represent. Hence if $v_{n_1} = v_{n_2}$, the resultant voltage magnitude v_R is:

$$v_R = (v_n^2 + v_n^2)^{1/2} = \sqrt{2}\, v_n$$

The angle subtended between the carrier v_c and the resultant noise voltage v_R is the phase deviation produced by the noise. In Figure 5.7, for the purpose of explanation, the noise voltages have been exaggerated, and for narrow-band systems are much smaller. Thus for small angles, as the length of the arc = radius × angle, we have approximately:

$$\phi_n = v_R/v_c = \sqrt{2}\, v_n/v_c \tag{5.44}$$

and hence, since both the noise and the signal voltage work into the same impedance,

$$\phi_n^2 = 2P_n/P_s \tag{5.45}$$

where $2P_n$ = the double sideband noise in the telephone channel.

P_s = the carrier power.

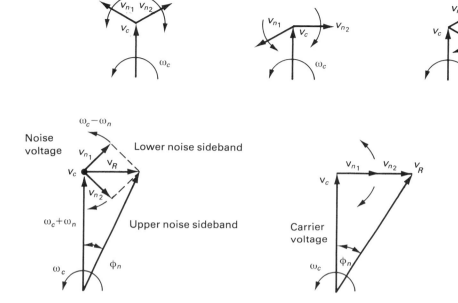

Figure 5.7 *Thermal Noise Phasor Diagrams*

By substituting equations 5.43 and 5.45 into equation 5.42, we get:

$$S/N = \frac{P_s\,(\Delta f\,/f_m)^2}{2P_n} \tag{5.46}$$

The signal and noise entering the receiver will meet with noise contributed from the receiver itself; hence the thermal noise coming out of a telephone channel at the receiving site will have a noise power equal to:

$$2P_n = F.KT.2B \tag{5.47}$$

where F = the receiver noise figure
     B = the bandwidth of a telephone channel (3.1 kHz)
     K = Boltzmann's constant (1.38×10^{-23} J/K)
     T = absolute temperature in kelvins

Substituting equation 5.47 in equation 5.46 gives

$$S/N = \frac{P_r\,(\Delta f\,/f_m)^2}{FKT.2B} \tag{5.48}$$

Usually in a radio receiver there is de-emphasis to reduce the distortion in the high-frequency region of the baseband. If the noise out of the telephone channel is weighted to CCITT recommendations then 2.5 dB is added to improve the channel signal-to-noise ratio. Thus the final equation for the 'triangular' Thermal noise out of a receiver is

$$S/N = 10 \log \frac{P_r(\Delta f / f_m)^2}{FKT2B} + D + 2.5 \text{ dB} \tag{5.49}$$

where S/N = the signal-to-noise-power ratio in the telephone channel in dB.
  P_r = the received power in watts.
  F = the receiver noise factor.
  K = Boltzmann's constant (1.38×10^{-23} J/K).
  T = the absolute temperature in kelvins.
  B = the telephone channel bandwidth in Hz (equal to 3100 Hz).
  Δf = the RMS frequency deviation of the telephone channel due to a test tone.
  f_m = the highest baseband frequency in the baseband frequency spectrum.
  D = the de-emphasis factor in dB.
  2.5 = the weighting factor as per CCITT recommendations.

5.10 HYPOTHETICAL REFERENCE CIRCUIT

The CCITT/CCIR definition of the *hypothetical reference circuit* (HRC) is: 'This is a complete telephone circuit (between audio frequency terminals at each end) established over a hypothetical international carrier system of definite length. It comprises a definite number of modulations and demodulations of the groups, supergroups, and master-groups, the number of these processes being reasonably large, but not the greatest number possible.'

The purpose of the recommendation is to apply to radio relay links, as far as possible, signal-to-noise-ratio requirements found valid for the proper operation of long-distance cable systems. This is because most of the circuits may be routed via radio relay links as well as via cables. If interchangeability of the transmission facilities is necessary without there being a change in the quality of transmission at the end of the system or circuit, the noise power contributions by each type of equipment should be approximately in proportion to the length of the path. In an attempt to approach this objective as closely as possible hypothetical reference circuits were defined for both cable and radio relay transmission systems.

All CCITT and CCIR hypothetical reference circuits are 2500 km (1550 miles) in length. Other administrations also have defined their own hypothetical reference circuits. North American Bell use a route length of 4000 miles and DCA* 1000 and

* US Defense Communications Agency

840 km
840 km
840 km
2500 km
420 km
420 km

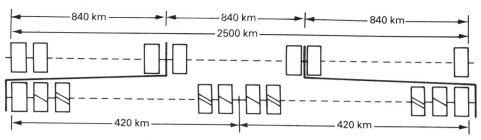

Figure 5.8   CCIR, 12-60 Channels, FDM

840 km
840 km
840 km
2500 km
280 km
280 km
280 km

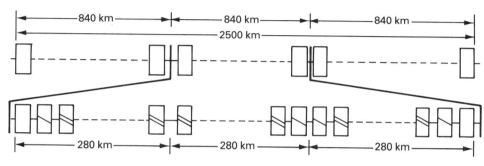

Figure 5.9   CCIR, > 60 Channels, FDM

840 km
840 km
840 km
2500 km
280 km
280 km
280 km

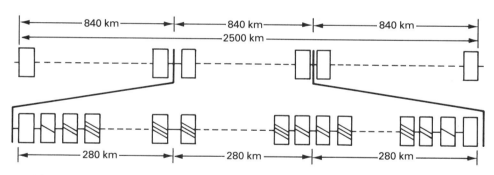

Figure 5.10   CCITT 12 MHz Hypothetical Reference Circuit, (Recommendation G332)

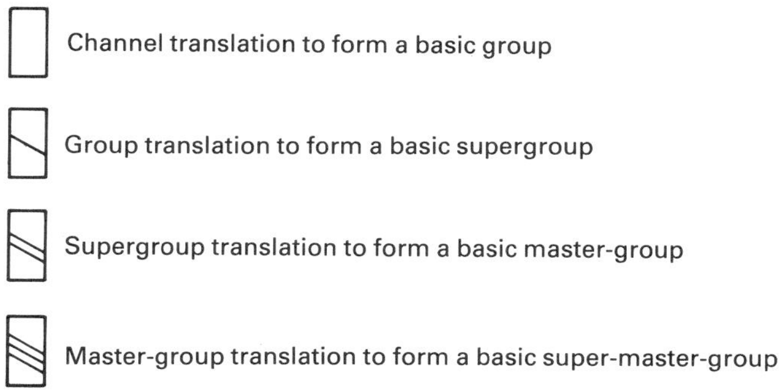

Channel translation to form a basic group
Group translation to form a basic supergroup
Supergroup translation to form a basic master-group
Master-group translation to form a basic super-master-group

Figure 5.11   CCITT/CCIR Translation Symbols

6000 nautical miles. If the circuits are short and cannot become a part of a long-distance circuit, it is sufficient that with all types of transmission the same quality as that defined for the longer circuits should be attained. If higher noise powers per kilometre are permitted by an administration, it is important to have a feel for the effect that the resulting noise from the telephone receiver will have on the subscriber. The following ratings are not by any means definite, and are cited only to give a feel for the effect of noise at different levels.

Noise Power at dBr point		Subjective Response at the Subscriber's Handset
10 000 pW	= 50 dBm0	Just perceptible
100 000 pW	= 40 dBm0	Good intelligibility for soft speakers
1 000 000 pW	= 30 dBm0	Noise is quite audible and intelligibility poor

5.10.1 Configuration of the CCIR Reference Circuits

Figures 5.8–5.11 show three configurations recommended by the CCIR and CCITT, with associated symbols. No recommendations exist for the length of individual hops. The hop length, as will be seen in Chapter 8, is determined by the propagation conditions, etc. Usually the hop is no greater than 80 km long. If an average of 50 km is used, then it can be seen from Figure 5.9 that there are six hops containing six pairs of transceivers for one section of the hypothetical reference circuit.

5.10.2 Permissible Noise Power in the Hypothetical Reference Circuit[9]

Table 5.1 is a summary of noise objectives specified by the CCITT and the CCIR for telephone circuits.

Table 5.1

Types of system	General Objectives		Noise allocation of a national system
	Cable[2] or radio-relay link		
Telephone circuits considered [1]	National four-wire extension and international circuits from 250 to 2500 km	Circuits from 2500 to about 25 000 km	

Recommendations
  Of the
  CCITT
	G152
	G212[3]
	G222
	G226
		G153	G123

Of the CCIR	391, 392, 393, 394, 395, 396-1, 397-1		

Table 5.1 *continued*

Hypothetical reference circuit (HRC) or typical circuit considered			General Objectives		
			HRC of 2500 km[4] or similar real circuit	Circuit or 7500 km[4]	Total length L in km of the long line FDM carrier systems in the national chain
Recommended objectives	Psophometric power	Hourly mean — Total power	10 000 pW		$(4000+4L)$ pW or $(7000+2L)$ pW
		Hourly mean — Terminal equipment	2500 pW		
		Hourly mean — Line	7500 pW i.e. 3pW/km	15 000 pW (provisional) 2 pW/km or better[5]	
		For one minute exceeded during 20% of the month — Line	7500 pW		
		Percentage of a month during which the psophometric power for one minute due to the line indicated can be exceeded	47 500 pW 50 000 pW	0.1 0.3 (provisional)	
	Unweighted power	Percentage of the month during which 10^5 pW (5 ms) can be exceeded		0.01 0.03 (provisional)	

Special Objectives

Radio-relay links					For very short distances
Circuits not very different from HRC 280 $<L<2500$ km[7]	Composition of links very different from HRC				Two circuits at most in each of the two terminal national networks
	50$<L<$280 km[7]	280$<L<$840 km	840$<L<$1670 km	1670$<L<$2500 km	
					G125
395	395	395	395	395	

Table 5.1 *continued*

Special Objectives					
Radio-relay links					For very short distances
Circuits not very different from HRC 280 $<L<2500$ km[7]	Composition of links very different from HRC				Two circuits at most in each of the two terminal national networks
	$50<L<280$ km[7]	$280<L<840$ km	$840<L<1670$ km	$1670<L<2500$ km	
					1000 pW[6] at most
$3L$ pW	$(3L+200)$ pW		$(3L+400)$ pW	$(3L+600)$ pW	
$3L$ pW	$(3L+200)$ pW		$(3L+400)$ pW	$(3L+600)$ pW	
$\dfrac{L}{2500}\times 0.1$	$(280/2500)\times0.1$ $\dfrac{280\times0.1}{2500}$	$(L/2500)\times0.1$ $\dfrac{L\times0.1}{2500}$	$(L/2500)\times0.1$ $\dfrac{L\times0.1}{2500}$	$(L/2500)\times0.1$ $\dfrac{L\times0.1}{2500}$	

Notes
1. Special objectives for telegraphy are indicated in Recommendations G143, G153, G222 and G442.
2. For these systems, it is sufficient to check that the objective for the hourly mean is attained.
3. See, in this recommendation, the details of the hypothetical reference circuits to be considered.
4. The objectives for line noise, in the same column, are proportional to the length in the case of the shorter lengths.
5. Objective 3 pW/km for the worst circuits: if a real circuit has more than 40 000 pW, it should be equipped with a compandor.
6. Desirable value: 500 pW. Highest value for one circuit: 2000 pW.
7. L is the radio hop length.

5.10.3 Noise Objectives of Modulation and Demodulation Equipment

The hourly mean noise objectives for terminal equipment is shown below.

Equipment	CCITT noise per pair pW0p	Number of modular pairs	Total noise pW0p
Channel	200	1	200
Group	80	3	240
Supergroup	60	6	360
Master-group	120	12	1440
Super-master-group	120	–	–
15-supergroups assembly	120	–	–
Through connecting equipment	–	–	260

It can be seen that the most numerous items of equipment, namely the channel modems, are allocated the most noise. This permits cutting of costs in providing a large channel number system.

CCITT Recommendation G221, (Red Book), 1984.

EXERCISES

1. What is the available power in dBm, from a noise generator, whose bandwidth is limited to 1 kHz, at 17°C? (*Answer* −144 dBm)

2. What is the noise power in dBm at the output of an amplifier of power gain 10 dB, if the input noise to the amplifier is 100 pW? (*Answer* −60 dBm)

3. An amplifier contributes 100 pW of noise. If the noise measured at the output of the amplifier is 10 nW, and the amplifier bandwidth is 10 Hz at 17°C, what is the power gain of the amplifier in dB? (*Answer* 18.54 dB)

4. What is the effective noise temperature of the amplifier of question 3, in °C? (*Answer* 724°C)

5. A LOS antenna is receiving a carrier signal at 17°C, which is then passed into a duplexer which has a 5 dB loss and which has been cooled to −173°C. What is the noise power expected out of the duplexer in dBm, if the duplexer has a 25 KHz bandwidth? (*Answer* −102.6 dBm)

6. A receiver consists of a 10 dB RF amplifier with a noise figure of 3 dB. This is followed by a mixer of 25 dB conversion gain and noise figure of 6 dB. Finally, the IF amplifier has a 30 dB gain and a noise figure of 8 dB. What is the system's overall noise figure? (*Answer* 3.62)

7. What will be the output noise power from the receiver of question 6 if the IF bandwidth is 10 MHz? (*Answer* −35.38 dBm)

8. What is the effective temperature of the receiver of question 6? (*Answer* 104.58°C)

REFERENCES

1. Mumford, W. W. and Schelbe, E. G., *Noise Performance Factors in Communication Systems* (Horizon House, 1968).
2. Johnson, J. B., 'Thermal Agitation of Electricity in Conductors', *Physical Review*, 1928, Vol. 32, pp. 97–109.

3. Nyquist, H., 'Thermal Agitation of Electric Charges in Conductors', *Physical Review*, 1928, Vol. 32, pp. 110–113.
4. Raff, S. J. *Microwave System Engineering Principles* (Pergamon, 1977).
5. Brodhage, H. and Hormuth, W., *Planning and Engineering of Radio Relay Links* (Siemens and Heyden, 1977).
6. Robins, W. P., *Phase Noise in Signal Sources* (Peter Peregrinus, 1982) IEE Telecommunications Series 9.

For Further Reading
7. Bell, D. A., *Electrical Noise* (Van Nostrand, 1960).
8. *Reference Data for Microwave System Engineers* (Telettra, Telefonia Elettronica e Radio SPA, Milano, 1976).
9. Tant, M. J., *The White Noise Book* (Marconi Instruments, 1974).
10. Freeman, R. L., *Telecommunication Transmission Handbook* (Wiley, 1975).

6 MULTICHANNEL TRAFFIC LOADING

6.1 INTRODUCTION

Most FDM carrier equipment is used to carry speech information, although there is an increasing amount of digital information being transmitted on voice channels these days. This chapter is concerned with the problem of how much power is produced at the baseband by people talking on the voice channels of the multiplex equipment. The baseband power which varies with time, type of talker and the number of talkers at any one time, will of course produce a deviation of the transmitted carrier whose variation is also dependent upon the same variables. The design of the transmitter is also likely to be based on the maximum permissible deviation which can be produced by a given input power. Hence, if the carrier system is loaded too heavily and an input level to the transmitter is produced which is higher than it is designed to take, the intermodulation and crosstalk will become intolerable, and would thus eventually lead to the state where the system would be unusable. If the system is not loaded sufficiently, the signal-to-noise ratio will suffer, due to the ever present thermal noise, against a smaller input signal. When the input signal is very small in power, the thermal noise being constant, the signal-to-noise ratio is small. As the signal power increases, so does the signal-to-noise ratio. In between these two extremes in signal level, there is an optimum loading level for the transmitter. Figure 6.1a shows a typical curve of the effect of signal loading level on the signal/noise ratio out of the far-end receiver, and a typical optimum loading level. As the loading level at the transmitter produces a deviation, the same argument applies as above, and an optimum deviation can be found for a particular transmitter, which will produce an output signal-to-noise ratio in 3.1 kHz slot which is at a maximum. Figure 6.1b shows the curve of N/S (inverted S/N) in a voice-channel slot plotted against transmitter deviation, and the optimum deviation, where N/S is a minimum, or S/N is a maximum. This type of curve is commonly known as a *bucket curve*.

One would not think that S/N in a voice channel could in anyway be related to the carrier-to-noise ratio of a transmitter, but it is. It is easier to think of it the other way around, and that is that the signal-to-noise ratio of a voice channel out of a far-end receiver is dependent on the signal level being received by that receiver. Intuitively, if a carrier level falls below a certain threshold, the receiver will not function correctly, and no signal will be received by the receiver, nor will one emerge

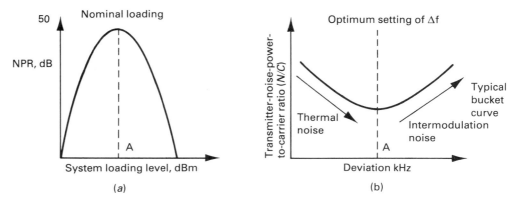

Figure 6.1 Loading Effect on Noise

from the multiplex voice channel. This threshold is called the *noise threshold*. From the last chapter, it was found that, for an ideal amplifier, or in this case, a receiver, the level at which the signal power cannot be discerned from the noise power occurs when the signal power level is equal to KTB, or -174 dBm/Hz. For the case where the receiver is not ideal, but has a noise figure F, the minimum level out of the receiver at which the signal cannot be discerned from the noise is given by $KTBF$. Hence, at this level, the received signal power or carrier-to-noise ratio C/N is unity, or 0 dB. It is at this level that the 'noise threshold' is defined.

$$
\begin{aligned}
\text{Noise threshold} \quad &= 10 \log KTBF \, \text{dBm} \\
&= -174 \, \text{dBm} + 10 \log B + F \, \text{dB} \\
\text{Carrier-to-noise ratio} &= 0 \, \text{dB}
\end{aligned}
\tag{6.1}
$$

As the received signal increases in level from this noise threshold and thus C/N increases, there is a decibel to decibel improvement in S/N. Once the received carrier level has risen to a level which is 10–12 dB above the noise threshold, the 'FM capture' effect takes place, and S/N out of the voice channel suddenly increases by another 20 dB or so. This point is known as the *FM improvement threshold*, and can be defined as:

or

$$
\left.
\begin{aligned}
\text{FM improvement threshold} &= -174 \, \text{dBm} + 10 \log B + F + 10 \\
\text{Carrier-to-noise ratio} \quad &= 10 \, \text{dB}
\end{aligned}
\right\}
\tag{6.2}
$$

where B = the receiver IF bandwidth in Hz.
  F = the receiver noise figure.
  K = the Boltzman constant (1.38×10^{-23} J/K).
  T = the absolute temperature in kelvins.
  KT = -174 dBm at 15.34°C.

After the 20 dB jump, giving an S/N from the voice channel of 30 dB (or 32 dB if C/N = 12 dB for the FM capture effect), there is again a 1 dB improvement in S/N for each 1 dB increase in C/N. This trend continues until the receiver saturation point is reached. If then, S/N increases as C/N increases, it can be said that C/N has increased for an increase in S/N. Thus, as the deviation of the transmitter increases, and the RF bandwidth to accommodate this increase in deviation increases also, there is an increase in C/N. Thus there is, up to the receiver saturation point, a trade-off of bandwidth for thermal noise improvement. Figure 6.1b may, therefore, equally be a curve of transmitter noise-power-to-carrier-power versus deviation.

6.2 SOURCES OF NOISE IN A TELEPHONY SYSTEM

Section 5.10 discussed, in brief, the noise objectives for multichannel telephone carrier systems. As discussed in Chapter 5, noise is, in the main, either thermal, or caused by intermodulation or crosstalk. Thermal noise, of course, will be introduced in every stage of the system where there is a resistor or semiconductor, but may also arise from, and become quite severe under, conditions of fading due to the fluctuations of the receiver automatic gain control (AGC). The level of thermal noise due to this mechanism is not a function of the traffic loading level on the baseband, and cannot be related to the bucket curve of Figures 6.1a or 6.1b.

Baseband crosstalk may arise from the imperfect filtering in the multiplex equipment, and can be minimized by correct cable screening and by the use of sharp cut-off filters. Intermodulation is caused by non-linearities in the amplifiers, modulators, etc., of the multiplex or radio, or corrosion or mismatch in the antenna system. The formation of the intermodulation products which may fall into any voice channels are usually not related to the signals in that channel, and thus appear as unintelligible random interference or noise.

6.3 LOADING OF A FDM–FM COMMUNICATIONS SYSTEM

In an FM transmitter the frequency deviation from the carrier frequency is proportional to the instantaneous voltage of the modulating signal. If the modulating signal is due to somebody talking, the signal voltage is quite erratic and peaks many decibels above the RMS value may occur. In a FDM–FM system, the modulating signal applied to the transmitter is usually the sum of many people talking on their individual telephone voice channels. The characteristics of this random composite signal depend on the channel activity, individual talker levels, and the number of talkers, etc. If the number of talkers is large enough, this composite signal is similar to that produced by a white noise generator, and can be specified in part by its probability distribution in terms of levels exceeded for particular percentages of time. The knowledge of the value of the peak levels produced below a certain level for a particular percentage of time is important in the design of the RF bandwidth of the transmitter, and consequently the IF bandwidth of the receiver, so as to accommodate the number of telephone channels to be carried over that radio link.

6.3.1 Activity Factor

Speech is characterized by large variations in amplitude, which may range from 30 to 50 dB about some mean value. The volume unit (VU) is often used to measure speech levels, as mentioned in Section 1.7.9.2. The VU can be equated to the dBm for a simple sine wave in the VF range, across 600 Ω. For a complex signal, such as speech, the average power measured in dBm of a typical single talker is given by equation 1.26:

$$P = (VU - 1.4) \text{ dBm} \qquad \text{(based on 1.26)}$$

or that a 0–VU talker has an average power of -1.4 dBm. Empirically, the peak power is about 18.6 dB higher than the average power for a typical talker. The peakiness of speech level means that carrier equipment must be operated at low average power to withstand voice peaks and avoid overloading and distortion. The average power level is not constant, but differs from talker to talker, and hence this must be taken into account. The VU level and the average power can be related to an activity factor K, which is defined as that proportion of the time that the rectified speech envelope exceeds some threshold. If the threshold is about 20 dB below the average power, the activity dependence on threshold is fairly weak. Equation 1.26 for average talker power in dBm can now be rewritten in relation to the activity factor as follows:

$$P = VU + 10 \log K \qquad (6.3)$$

If $K = 0.725$, the results will be the same as that of equation 1.26.

If a second talker is added at a different frequency band to the baseband signal, the system average power will increase by 3 dB. If N talkers are taken, and each is translated and added to the baseband signal, the average power in dBm developed will be given by:

$$P = -1.4 + VU + 10 \log N \qquad (6.4)$$

where P is the power developed across the frequency band occupied by all talkers in dBm.

Empirically, it has been found that the peakiness or peak factor of many talkers over a multichannel analog system reaches the characteristics of random white noise peaks when the number of talkers N exceeds 64.

Figure 6.2 shows the probability of the ratio of peak-to-RMS voltage being exceeded for a given value of peak-to-RMS voltage, for 1, 4, 16, and 64 active tele-

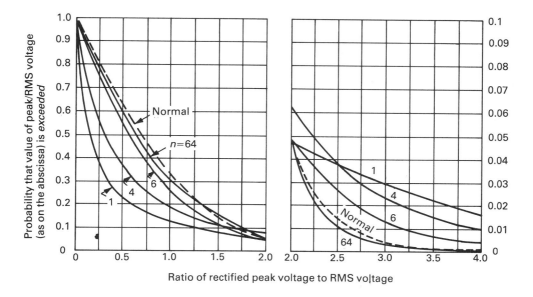

Figure 6.2 *Peak Voltage Distributions for n Active Channels at the Baseband (Courtesy of Marconi Instruments Ltd. Ref. 7)*

phone channels. The graphs show that there is a greater probability, for large ratios of peak to RMS voltage, that the peak will exceed that RMS voltage for a lower number of talker channels, than for a larger number of talker channels. This is to be expected since the lower number of channels produce a more peaky baseband signal than do the larger number of channels. An activity factor of unity means that, on any and every channel, somebody is talking all the time. This is assumed in the derivation of equation 6.4, and the discussion following it. The assumption is quite unrealistic.

To dimension correctly a system, the activity factor of 1 is dropped to 0.7. This in effect permits more channels to be available than those which are actually in use; the additional channels prevent congestion, or unsuccessful calls being high in number, owing to a call not being made, exactly at the time a talking party hangs up. The activity factor is again reduced from 0.7 to 0.5, to lower the expected power of the baseband signal due to the inactivity of a channel during the setting up of a call, as well as the pauses which occur due to thinking by the talkers during conversation. During a conversation however, on a full duplex channel, normally only one person is talking. The other person being in the listening mode causes his or her channel to be idle during this period. If it is assumed that each party listens for 50 per cent and talks for 50 per cent of the time, the activity factor can be halved again, producing the traditional figure for the activity factor of 0.25. This value of 0.25 is accepted by the CCITT and North America for a single channel.

In large channel groups, the activity factor of 0.27 is rarely exceeded even during the busiest hours. For channels in small circuit groups, the activity may be appreciably lower than 0.25. In the design of multichannel equipment, the maximum number of channels which are likely to be active or continually in operation within a baseband is a deciding factor. If one channel is expected on the average to be active for 25 per cent of the time it is in use, then, in a multichannel system, the percentage of the time which all the channels will be active is expected to be quite low. This is because some of those channels will not be used whilst others are, and so will bring down the percentage use per channel, when averaged over all channels. Figure 6.3 provides the relationship between the nominal system channel capacity N, and the number of active channels n.

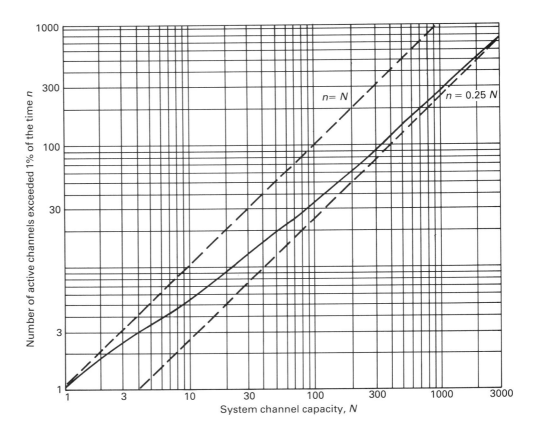

Figure 6.3   Number of Active Channels n as a Function of System Channel Capacity N
Activity factor of a single telephone channel = 0.25
(Courtesy of Marconi Instruments Ltd. Ref. 7)

Activity coefficients are plotted in Figure 6.4 and for common channel capacities are tabulated in Table 6.1[3]. From these figures, the higher channel capacity systems tend to have an activity factor tending towards 0.27. The CCITT however, take 0.25 as their activity factor, since it is not expected to be exceeded for a little over 1 per cent of the time.

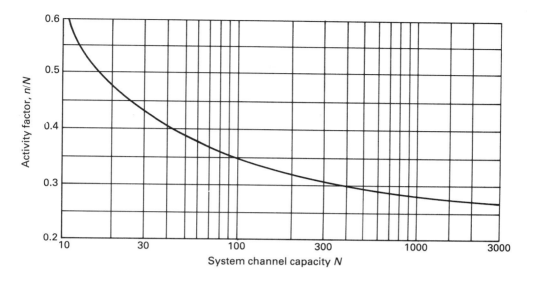

*Figure 6.4   Activity Factor for Different N
(Courtesy of Marconi Instruments Ltd. Ref. 7)*

Table 6.1

Nominal system capacity N	Number of active channels n	Activity coefficient n/N
12	7	0.583
24	11	0.458
36	15	0.417
60	23	0.383
120	41	0.342
240	76	0.317
300	93	0.310
600	175	0.292
900	256	0.284
960	272	0.283
1200	335	0.279
1260	351	0.279
1800	493	0.274
2700	728	0.270

Figure 6.5 has been derived from Figure 6.2 for the two probability levels equivalent to 99 per and 99.9 per cent of the time. These curves show, for example, that for 3 channels, an 11.2 dB peak factor will be exceeded for 1 per cent of the time and 16 dB for 0.1 per cent of the time. It should also be noted that, for about 80 channels or more, the peak factors are within 0.5 dB of the Gaussian distribution.

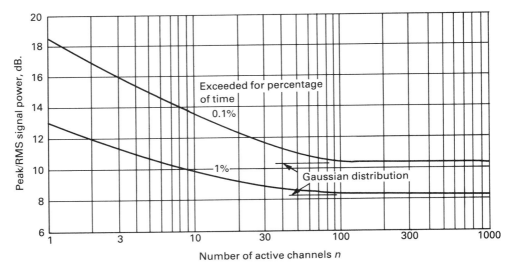

Figure 6.5 ***Variation of Peak Factor with n***
(Courtesy of Marconi Instruments Ltd. Ref. 7)

Example

This example is given to show how Figure 6.5 and the activity factor may be applied to the design of a transmitter.

Assume that the transmittter is to transmit a baseband signal containing 24 channels without going into saturation and producing intermodulation distortion for more than 0.1 per cent of the time.

If all the 24 channels were active, the linear range of the transmitter would be: 24 × the single-channel RMS power + 11.9 dB from Figure 6.5 for 24 channels. Thus:

The dynamic range = 13.8 + 11.9 = 25.7 dB above the single-channel RMS power

It is known, however, that a 24-channel system is not always used to full capacity, and that, referring to Table 6.1 or Figure 6.4, the number of active channels is only 11.

Reworking the above calculation for 11 channels shows that the transmitter's dynamic range is:

11 (10.4 dB) × the single-channel RMS power + 13.5 dB (from Figure 6.5, for 11 channels)

Thus

The dynamic range = 10.4 + 13.5 = 23.9 dB above the single-channel RMS power

The second calculation shows that the dynamic range of the transmitter is not as high as that calculation where 24 channels were considered active (actually it is closer to that of a 60-channel system).

If the case had been taken where the peaks of a 24-channel system had been assumed to be the same as that of a single channel instead of being smoothed out a little, as above, then the dynamic range would be:

11 (10.4 dB) × the single-channel RMS power + 18.5 dB from Figure 6.5 for a single channel

Thus

The dynamic range = 10.4 + 18.5 = 28.9 dB above the single-channel RMS power

As can be seen, this is 5 dB higher than the transmitter would in fact require. For larger-channel systems the discrepancy becomes more apparent.

6.3.1.1 Summary of the Calculation of the Transmitter Dynamic Range

1. Given a system channel capacity N, from Figure 6.4, or Table 6.1, determine the expected number of active channels n.

2. Calculate from this the RMS level of n active channels above the single-channel RMS level, using the expression $10 \log n$ dB.

3. From Figure 6.5, determine the peak figure (peak/RMS in dB) for n active channels and the percentage of time that this peak figure is permitted to be exceeded (0.1 or 1 per cent). Then:

Peak figure for n active channels in dB
 = Peak power for n channels in dBm − RMS power for n channels in dBm.

4. Add together the results from steps 2 and 3 to obtain the dynamic range:

The dynamic range for n channels in dB = $10 \log n$ + peak figure for n channels (6.5)

EXERCISES

1. Determine the linear dynamic range of a transmitter which is to have a 60-channel capacity, and which will produce intermodulation distortion for only 0.1 per cent of the time (*Answer* 25.6 dB above single-channel RMS power)

2. An audio amplifier has a rated output power of 100 W. If distortion is not to occur for more than 1 per cent of the time, what is its maximum power gain, if the mean level of speech entering into it is 88 μW? (*Answer* 48 dB approx.)

6.3.2 Loading

6.3.2.1 Active Channel

So far the peak value of baseband power due to a voice channel or a number of voice channels has been determined relative to the mean power or RMS power of a single channel. To obtain an absolute power, so that measurements may be made to check that the system is operating correctly, we need to determine the RMS power of a single channel. The mean speech power P_m of the 'reference' English talker on an active voice channel has been measured by Holbrook and Dixon[1], and found to be:

$$P_m = -10.1 \text{ dBm0}$$

This is the mean value of a Gaussian distribution, obtained by measuring the actual speech power of a great number of persons. The underlying measurements imply the use of the same type of circuits, and the same level of calibration and gain adjustments, preceding the point of zero reference level for each talker. The distribution is such that, practically the whole area lies below the +7 dBm0 power level; hence the probability of ever exceeding this level is negligible[3]. For this reason the CCITT recommends that a trunk telephone circuit should be capable of accepting the peak voltage of a sine wave with a power of 5 mW at a point of zero level without distortion (+7 dBm0).

6.3.2.2 System Channel (for N ⩾ 240)

The commonly accepted mean power of a single telephone channel in a system which has a channel capacity of greater than 240 channels is −15 dBm0. This is the mean power level during the busy hour and for speech on a large number of circuits in a system. The CCITT have adopted this value as the 'conventional' value, and it has been derived from the following variables;

Mean power level of signaling and tones	$10 \mu W$
Mean power level of speech (88 μW) multiplied by the mean activity factor of 0.25	$22 \mu W$
Total mean power per channel	$32 \mu W$ or -15 dBm0

6.3.2.3 Mean Power Level of the Multiplex Voice Channels at the Baseband

Since the individual channels carry entirely uncorrelated messages, the overall mean power of the multiplex signal is the sum of the mean powers developed over the n active channels. Thus, the total mean power of the multiplex signal at the point of zero relative level is given[4] as:

$$P_{m(total)} = 10 \log n + P_m \qquad (6.6)$$

This value can be obtained under either of the following two conditions:

1. Each channel level is adjusted according to the actual speech power of the talker, such that each channel yields the same power P_m.
2. The number of active channels n is sufficiently large to approximate a Gaussian distribution, which may be represented by its mean value P_m.

In any practical transmission system, the actual mean power is higher by X dB than is shown by equation 6.6, for the following reasons:

1. For the proper operation of the FDM equipment, continuous pilot tones are transmitted in addition to the speech channel spectra.
2. Telephone channels are temporarily overloaded when, as a preliminary step for the establishment of a call, they transmit high-level ringing signals.
3. Some telephone channels, used to pass several telegraph channels in the form of tone telegraphy, or for data signals, may exhibit a much higher mean power level than that given by equation 6.6.

Equation 6.6 is modified, therefore, to:

$$P_{m(total)} = 10 \log n + P_m + X \text{ dBm0} \tag{6.7}$$

Replace n, the number of active channels, by $n = KN$, where K is the activity factor, and N is the total number of voice channels in the system. Then:

$$P_{m(total)} = 10 \log N + 10 \log K + P_m + X \text{ dBm0}$$

For microwave links carrying a large number of channels – i.e. N is equal to or greater than 240 – the CCIR recommends use of a value of -15 dBm0 for the last three factors in equation 6.7. Hence, according to the CCIR[5]:

$$P_{m(total)} = 10 \log N - 15 \text{ dBm0 for } N \geqslant 240 \text{ and 99 per cent of the busy hour} \tag{6.8}$$

When the system channel capacity N is less than 240, the CCIR recommendation is given[6] as:

$$P_{m(total)} = 4 \log N - 1 \text{ dBm0 for } 12 \leqslant N < 240 \tag{6.9}$$
$$\text{(or for } 60 \leqslant N < 240 \text{ see Section 7.1.3)}$$

6.3.2.4 Mean Power Level of the Multiplex Voice Channels at the Baseband: Non-CCITT and $12 \leqslant N < 240$

Referring to Figure 6.4, if a straight line is drawn between the points with co-ordinates (12, 0.583) and (240, 0.317), and another parallel straight line tangent to the curve drawn, then a parallel straight line can be found between these two lines, which will approximate to the values of the curve. The equation of this third line which is given in reference 7, pp. 48, is:

$$10 \log n = 8 \log N - 0.4 \text{ (for } 12 \leqslant N < 240) \tag{6.10}$$

It is worth mentioning that equation 6.10 is based on the activity factors which are 99 per cent probable during the busy hour. If the relationship between the mean power in a channel and the number of active channels n is required, the speech power in an active channel (-10.1 dBm0) must be included in the relationship. Hence, the mean speech power of n active channels is given by

$$P_{mtotal}(active) = -10.1 + 10 \log n \text{ dBm0} \tag{6.11}$$

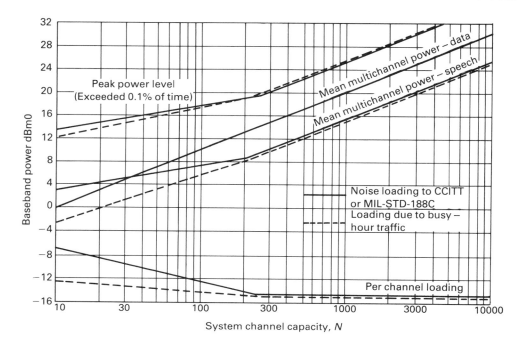

Figure 6.6   Baseband Power Levels for different N
(Courtesy of Marconi Instruments Ltd. Ref. 7)

Combining equations 6.10 and 6.11, we get the total power due to active channels during the busy hour with a 99 per cent probability, for channel capacity between 12 and 240, as:

$$P_{m(total)} = -10.5 + 8 \log N \text{ dBm0} \tag{6.12}$$

Figure 6.6 summarizes equations 6.8 and 6.12. The dashed line is the load due to non-CCITT busy-hour traffic. This shows how the mean power per channel increases from a constant -15 dBm0 above 240 channels to -12.5 dBm0 at 12 channels. At $N = 1$, this value would increase to -10.1 dBm (not shown). The peak power level exceeded only for 0.1 per cent of the time has been taken from Figure 6.5 and converted to nominal system capacity using Figure 6.3. This results in the peak power curve shown in Figure 6.6.

Example

With the same figures as given in the example at the end of Section 6.3.1, use Figure 6.6 to verify this result.

The example showed that the dynamic range of a transmitter for a 24-channel system was 23.9 dB above the single-channel RMS power. Figure 6.6 shows that for 24 channels the peak power level for loading due to busy-hour traffic is approximately 13.5 dBm0. As the mean power level of a single channel in a system which has less than 240 channels is -10.1 dBm0, the peak power level is 23.6 dB above this value. This 23.6 dB dynamic range compares favourably with the calculated value of 23.9 dB.

6.3.2.5 Peak Power Levels[7]

FOR $12 \leqslant N < 240$
Referring to Figure 6.5, the straight-line equation approximating the 0.1 per cent curve is:

$$\text{Peak-power/RMS-power} = 18 - 4.2 \log n \text{ dB for } 12 \leqslant N < 240 \qquad (6.13)$$

Substituting equation 6.10 in equation 6.13 results in:

$$\text{Peak-power/RMS-power} = 18.2 - 3.36 \log N \text{ dB for } 12 \leqslant N < 240 \qquad (6.14)$$

As the RMS power is given by equation 6.12, substituting this into equation 6.14 produces:

$$\text{Peak power level} = 7.7 + 4.64 \log N \text{ dB}$$

This can be approximated without significant error to:

$$\text{Peak power} = 8 + 4.5 \log N \text{ dBm for } 12 \leqslant N < 240 \qquad (6.15)$$
$$\text{(exceeded for 0.1 per cent of time)}$$

FOR $N \geqslant 240$
Again from Figure 6.5, for the 0.1 per cent curve, when $n \geqslant 240$ the peak/RMS signal power is constant at 10.5 dB. From Figure 6.4 the equation relating the system channel capacity N to the number of active channels n is:

$$10 \log n = 9.23 \log N - 3.35 \qquad (6.16)$$

As the peak power $= 10.5 + $ RMS power, by using equation 6.8, the peak power may be expressed as:

$$\text{Peak power} = 10.5 - 15 + 10 \log N \text{ dBm}$$

or

$$\text{Peak power} = -4.5 + 10 \log N \text{ dBm for } N \geqslant 240 \qquad (6.17)$$
$$\text{(exceeded for 0.1 per cent of time)}$$

*6.3.2.6 Peak Power Levels of CCITT**

FOR $12 \leqslant N < 240$
The expression for the peak power level as represented by the CCITT is

$$P_{pk(CCITT)} = 9.4 + 4 \log N \text{ dBm0} \tag{7.1}$$

FOR $N \geqslant 240$

$$P_{pk(CCITT)} = -4.5 + 10 \log N \text{ dBm0} \tag{7.2}$$

6.3.3 Mean Power Level of the Multichannel Signal with Data Loading

6.3.3.1 Frequency-Modulated Voice-Frequency Telegraph Transmission
The resulting loading effect of a number of frequency-shift keyed telegraph sub-channels has been investigated[9] using the percentage time of clipping as a parameter. As an example of application, assume that 18-channel frequency-shift-keying (FSK) telegraphy is carried on a voice channel with 1 per cent clipping and with an activity factor of 0.7. In this case, the channel gain and levels preceding the transmission system input are left unchanged when switching over from calibration to normal operation. The resulting mean loading of the voice channel can be expressed as P_m + teletype loading factor + $10 \log K = -10.1 + 19.6 - 1.5 = +8$ dBm0. When half the total number of voice channels N of a multiplex system each carry the same teletype loading, the equivalent system loading can be written as:

$$P_T = 10 \log N + 5 \text{ dBm0} \tag{6.18}$$

The CCITT recommends another method of computing the loading due to FSK telegraphy carried on multiplexed voice channels[10]. Recommendation R35 states[11] that, for a modulation rate of 50 bauds, the total average power transmitted to telephone lines by all the subchannels of an FSK telegraph system is normally limited to 135 μW at a point of zero relative level, or -8.8 dBm0. These power limits may be raised by agreement between the interested operating authorities; however, they should never exceed those which have been fixed for the power (at 0 dBm0) corresponding to the transmission of a continuous tone for amplitude-modulated telegraphy. All the relevant levels are given in Table 6.2.

These recommendations imply the readjustment of the channel gains and levels preceding the transmission system input (0 dBm0) after 800 Hz calibration.

Denoting the number of voice channels carrying telegraph traffic by N_T, the activity factor by K_1 and the allowable power per voice channel by P_{FSK}, we can write the equivalent total loading of the transmission system as:

$$P_T = 10 \log N_T + 10 \log K + P_{FSK} \text{ dBm0} \tag{6.19}$$

* From Chapter 7.

Table 6.2 CCITT Power Limits in Frequency-Modulated Telegraph Systems

Telegraph system per voice channel	Allowable power per telegraph subchannel, dBm0		Allowable power per voice channel, PFSK, dBm0	
	Normal	Maximum	Normal	Maximum
12 TTY* subchannels or less	−19.5	−14.5	−8.8	−3.7
18 TTY subchannels	−21.25	−18.25	−8.8	−5.8
24 TTY subchannels	−22.5	−20.9	−8.8	−7.1

For the 18-channel FSK teletype operating with the maximum allowable power and with an activity factor of 0.6, equation 6.19 yields:

$$P_T = 10 \log N_T - 2.2 - 5.8 = 10 \log N_T - 8 \text{ dBm0} \qquad (6.20)$$

6.3.3.2 High-Speed Telegraph Systems

For FSK telegraph systems transmitting with modulation rates greater than 50 bauds, the CCITT recommends[12] that the mean power transmitted to line at a point of zero relative level remains normally limited to 135 μW, or −8.8 dBm0, for the aggregate of the channels. Thus equation 6.19 also applies to this case, with P_{FSK} limited to −8 dBm0.

6.3.3.3 Data Transmission on Telephone Channels

For data transmission operating with continuous tones, over leased telephone circuits using frequency-modulation systems, the CCITT recommends that the maximum power level at the zero relative level point should be limited to $P_D = -10$ dBm0. When transmission of data is discontinued for any appreciable time, the power level should preferably be reduced to −20 dBm0 or lower[13]. The corresponding total loading of a transmission system with N_D voice channels, each of them carrying FM data at $P_D = -10$ dBm0 with an activity coefficient of K, can be expressed as:

$$P_{D(total)} = 10 \log N_D + 10 \log K + P_D \text{ dBm0} \qquad (6.21)$$

If $K = 0.75$, we can write:

$$P_{D(total)} = 10 \log N_D - 1.2 - 10 \text{ dBm0}$$

$$= 10 \log N_D - 11.2 \text{ dBm0} \qquad (6.22)$$

For data systems not transmitting tones continuously, as in amplitude-modulated systems, higher peak levels, up to −6 dBm0, may be used, provided that the mean power during the busy hour does not exceed $P_D = -15$ dBm0 on each direction of transmission simultaneously. Thus equation 6.21 still applies in this case, with the value for P_D replaced by −15 dBm0.

* Teletype:(writer)

When data transmission is handled over a switched telephone network the CCITT recommends[13] that, for all types of systems, i.e. FM, AM, etc., the maximum power level should not exceed -6 dBm0. By substituting this value for P_D and assuming a somewhat smaller activity factor, equation 6.21 can again be used for the equivalent loading of the transmission system. Once again, to realize the loading effects exactly as recommended, the circuit gain and levels between the subscriber's data-generating equipment and the transmission system input may have to be readjusted after calibration.

If an activity factor of unity is assumed for systems loaded primarily with data or telegraph signals then, taking the MIL–STD–188C specification[15] of a mean level of 100 μW per channel for P_D, equation 6.21 gives:

$$\text{Mean power level (data)} = -10 + 10 \log N_D \qquad (6.23)$$

Equation 6.23 is shown plotted in Figure 6.6.

6.3.4 Summary of Important Equations

$$\text{Noise threshold} = -174 \text{ dBm} + 10 \log B + F \text{ dB} = 0 \text{ dB carrier/noise} \qquad (6.1)$$

$$\text{FM Improvement threshold} = \text{Noise threshold} + 10 \text{ dB (or thereabouts)}$$

or

$$\text{Carrier/noise} \qquad\qquad = 10 \text{ dB (or thereabouts)} \qquad (6.2)$$

Transmitter dynamic range for n (active) channels in dB =
$$10 \log n + \text{peak figure for } n \text{ active channels} \qquad (6.5)$$

Mean speech power of an English talker on an active channel P_m $\quad = -10.1$ dBm0
Mean power level of a single telephone channel in a system of more than 240
$$\text{channels} \quad = -15 \text{ dBm0}$$

CCITT
The total mean power on a system of $(12 \leqslant N < 240)$ channels is given by:

$$P_{m(total)} = 4 \log N - 1 \text{ dBm0} \qquad (6.9)$$

The total mean power on a system of $(N \geqslant 240)$ channels is given by:

$$P_{m(total)} = 10 \log N - 15 \text{ dBm0} \qquad (6.8)$$

Non-CCITT Loading due to Busy Hour Traffic
The total mean power on a system of $(12 \leqslant N < 240)$ channels is given by:

$$P_{m(total)} = 8 \log N - 10.5 \text{ dBm0} \qquad (6.12)$$

The total mean power on a system of N channels, where $N \geqslant 240$, is given by

$$P_{m(total)} = 10 \log N - 15 \text{ dBm0} \tag{6.8}$$

The relationship between the number n of active channels and the number N of system channels is given by:

For $12 \leqslant N < 240$
$$10 \log n = 8 \log N - 0.4 \tag{6.10}$$

For $N \geqslant 240$
$$10 \log n = 9.23 \log N - 3.35 \tag{6.16}$$

But without significant loss in accuracy the following may be used:

$$10 \log n = 10 \log N - 5$$

Peak Power Levels for CCITT Busy-Hour Traffic
For $12 \leqslant N < 240$
$$P_{pk(CCITT)} = 9.4 + 4 \log N \text{ dBm0} \tag{7.1}$$

For $N \geqslant 240$
$$P_{pk(CCITT)} = -4.5 + 10 \log N \text{ dBm0} \tag{7.2}$$

Peak Power Levels for non-CCITT Busy-Hour Traffic
For $12 \leqslant N < 240$
Peak power level exceeded for 0.1 per cent of the time $= 8 + 4.5 \log N \text{ dBm0}$ (6.15)

For $N \geqslant 240$
Peak power level exceeded for 0.1 per cent of the time $= -4.5 + 10 \log N \text{ dBm0}$ (6.17)

CCITT Loading of Frequency-Modulated Telegraph Systems

$$P_T = 10 \log N_T + 10 \log K + P_{FSK} \text{ dBm0} \tag{6.19}$$

CCITT Loading of Voice Channels with Data

$$P_{D(total)} = 10 \log N_D + 10 \log K + P_D \text{ dBm0} \tag{6.21}$$

6.3.5 The Use of a Sinusoidal Test Tone to Represent the Composite Baseband Signal

Up to this point the composite baseband signal has been considered only as a complex waveform, and equations relating to its mean power level and its peak power level under varying conditions have been discussed. This is realistic if attempts are made to simulate the baseband using noise generator sets, and to perform tests which are noise-related. A problem however exists if attempts are made to replace the peak and mean power levels by a pure sinusoid, for the above equations no longer hold. The reason why it is important to obtain some meaningful relationship between a sinusoidal test tone and a complex composite baseband signal is because in some cases the setting of or lining-up of the transmitter deviation is accomplished using a single tone to obtain the Bessel zero (see Chapter 10). In this case the test procedure in setting the deviation depends on the level of a pure sinusoid, of known frequency and level, and not a complex time-varying waveform. The following discussion attempts to relate in a meaningful way the level and frequency of a test tone to the true baseband composite signal for varying system channel numbers.

The procedure for lining up the transmitter deviation uses a standard or reference sinusoidal test tone of 800 Hz or 1000 Hz. The RMS level of this tone is in this case taken as 0 dBm0, and a particular frequency deviation is assigned to this tone level. What is required to be known in order that the IF bandwidth should be great enough to accommodate all of the system voice channels, is the peak levels which the composite signal reaches with respect to that level of the RMS test tone.

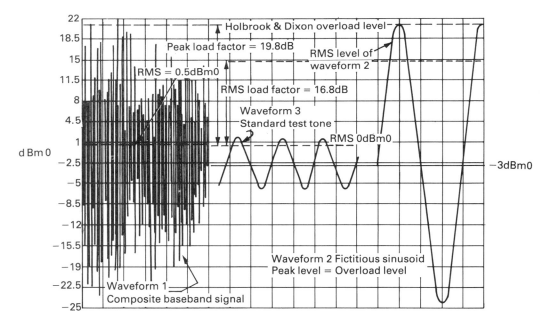

Figure 6.7   Holbrook & Dixon load factors for 24-channel System

Figure 6.7 shows a fictitious sinusoidal tone (waveform 2) whose peak value is the same as the Holbrook and Dixon overload level, and whose RMS level is 3 dB below this peak value (10 log 0.707), and the 0 dBm RMS test tone which is used for lining up the transmitter (waveform 3). The difference between these two RMS levels is known as the *RMS load factor* and may be expressed as: 'The ratio of the RMS level of sinusoid, whose peak value is at the Holbrook and Dixon overload level, to the RMS level of a test tone, is the RMS load factor'. Figure 6.8 shows the Holbrook and Dixon RMS load factor curves for varying number of system channels (N). It follows from this, that the peak load factor is the ratio of the peak value of a fictitious tone representing the Holbrook and Dixon overload level to the level of the RMS test tone, i.e.

$$\text{Peak load factor} = \text{RMS load factor} + 3\,\text{dB} \qquad (6.24)$$

Since the frequency deviations are directly proportional to the voltage of the modulating waveform, the peak load factor can be expressed as:

$$\text{Peak load factor} = 20\log\frac{\text{Peak frequency deviation of composite signal}}{\text{RMS channel test tone deviation}}\,\text{dB}$$
$$\qquad (6.25)$$
$$= 20\log\frac{\text{Peak voltage of composite signal}}{\text{RMS voltage channel test tone}}\,\text{dB}$$

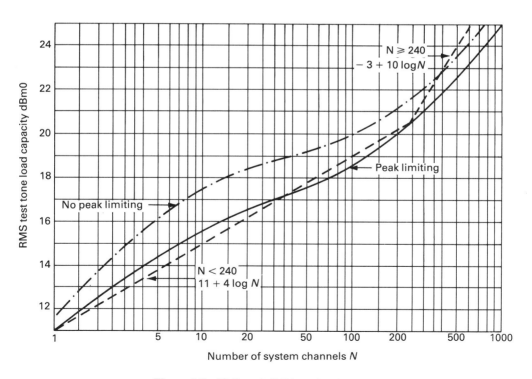

Figure 6.8   Holbrook & Dixon RMS Load Factors
(Courtesy of IEEE, Ref. 14)

Figure 6.8 shows three curves for the Holbrook and Dixon RMS load factors These are: the 'peak limiting'; the 'no peak limiting'; and the dashed line representing the linear approximation to the peak limiting curve. The ordinate is the required RMS load capacity of the system, referred to a point of zero transmission level, and measured by a single sinusoidal tone representing the composite baseband signal. The peak value of the sinusoidal tone representing the composite signal (for which the RMS level is obtained from Figure 6.8) is the Holbrook and Dixon, system overload level. For Holbrook and Dixon, in listening tests, determined a system overload level suitable to multichannel telephone systems. This overload level takes into account the long-term fluctuations due to the number of active channels, pauses in speech, talker volume variations, type of talker, etc., and it also accounts for the rapid fluctuations of a speech wave. A unique combination of the long-term signal peaks and short-term signal peaks is required in order for the overload level to be exceeded. The long-term signal level is such that during only 1 per cent of the time can the short-term fluctuations cause the overload level to be exceeded, and the level of the short-term fluctuations required to exceed the overload point during the 1 per cent interval, is reached only during 0.1 to 0.0003 per cent of the time, depending on the number of channels involved.

To illustrate the above discussion, an example will be given.

Example

Assume that for a 24-channel system, the peak deviation is 488 kHz, and that the transmitter has peak limiting before its modulator section. What is the maximum permissible peak deviation of the test tone, if the test tone level is 0 dBm0?

From Figure 6.8, using the peak limiting curve at 24 channels, the RMS load factor is found to be 16.8 dB. Thus, from equation 6.24, the peak load factor (PLF) is 19.8 dB. From equation 6.25, this 19.8 dB represents a voltage ratio, or a deviation ratio of approximately 9.8. Therefore, if the peak frequency deviation of the composite signal is 488 kHz, the RMS channel test-tone frequency deviation is 488 kHz/9.8 = 49.8 kHz. Thus the peak channel test tone frequency deviation $\sqrt{2} \times 49.8 = 70.4$ kHz. (*Answer*)

6.3.6 Test-tone and Multichannel Deviations

As mentioned above, the reference signal 0 dBm0 can be considered as the sinusoidal signal representing one channel and is referred to as the test tone. As FM microwave systems for the transmission of these signals are being dealt with, this reference signal will result in a test-tone deviation of the transmitter FM carrier. The preferred test-tone deviations are given in CCIR Recommendations 275 and 404–2. Table 6.3 shows some typical peak multichannel deviations for various system channel capacities N.

Table 6.3

Capacity	Multichannel loading factor, dBm0	RMS test tone deviation per channel, kHz	Multichannel deviation MHz	
			RMS	Peak
12	–	35	–	–
24	4.5	35	–	–
60	6.1	200	0.400	1.8
120	7.3	200	0.464	2.1
300	9.8	200	0.618	2.8
600	12.8	200	0.872	3.9
960	14.8	140	0.770	3.4
		200	1.100	4.9
1200	15.8	140	0.860	3.8
		200	1.230	5.5

The peak deviations given in Table 6.3 can be calculated from the RMS test tone deviation per channel using the two expressions:

$$\Delta f_{pk} = 4.47 \, \Delta f_{RMSt.t.} \, (\log^{-1}(-15 + 10 \log N)/20) \text{ for } N \geq 240 \qquad (6.26)$$

and

$$\Delta f_{pk} = 4.47 \, \Delta f_{RMSt.t.} \, (\log^{-1}(-1 + 4 \log N)/20) \text{ for } 12 \leq N < 240 \qquad (6.27)$$

where Δf_{pk} = the multichannel peak deviation.
$\Delta f_{RMSt.t.}$ = the RMS test tone deviation per channel.
N = the number of system channels.

The factor 4.47 is related to the peak load factor as described in Section 6.3.5. The relationship is:

Peak load factor $= 20 \log 4.47 + (-15 + 10 \log N) = -2 + 10 \log N$ for $N \geq 240$
or Peak load factor $= 20 \log 4.47 + (-1 + 4 \log N) = 12 + 4 \log N$ for $60 \leq N < 240$

The two expressions for the peak load factor are derived from the 4.47 factor. The equations representing the straight line approximation to the peak limiting curve of Figure 6.8 differ from the above equations by one decibel. The one decibel difference may be due to their representation of the straight-line approximations of the *no-peak limiting* curve. When the factor 3.76 is used[8], the two expressions for the peak load factor become:

Peak load factor $= 20 \log 3.76 + (-15 + 10 \log N) = -3.5 + 10 \log N$ for $N \geq 240$
or Peak load factor $= 20 \log 3.76 + (-1 + 4 \log N) = 10.5 + 4 \log N$ for $60 \leq N < 240$

These two expressions closely approximate the *peak limiting* straight-line approximations of Figure 6.8. These last two expressions for the peak load factor are used when determining the value of the RF bandwidth of a radio, as per CCIR Report 418–1.

Figure 6.9 shows a curve of the multichannel peak factor plotted against the number of *active* channels n. Equations 6.26 and 6.27 relate to this curve. Equation 6.26 is a fair representation with the 4.47 or 13 dB factor, whereas equation 6.25 is only an approximation (although generally accepted) to the curve of Figure 6.9. This is because, for 12 channels, the form factor is 16 dB, falling to approximately 13 dB around the 100-active-channels mark. The peak factor, as used in Figure 6.9, is defined to be the ratio between the peak power exceeded during 0.1 per cent of the time, and the RMS power of the composite multiplex signal, and is a function of the number of active channels. For $3 \leqslant n < 60$, the peak factor F_{pk} is approximately:

$$F_{pk} = 1/3(59 - 10 \log n)$$

Using equation 6.10 to relate the active channels n to the system channels N gives:

$$F_{pk} = 20 - 2.7 \log N \text{ for } 12 \leqslant N < 190$$

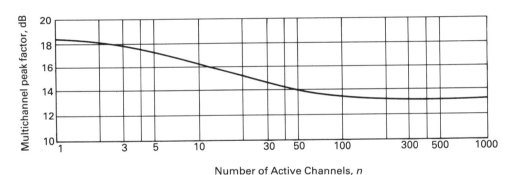

Figure 6.9   Multichannel Peak Factor for n Active Channels

Example

Considering a typical case for a 960-channel system, draw up the levels and deviations, starting at the mean level of one speech channel.

The mean power level of one speech channel is found by dividing the mean power of a multichannel system by the number of channels:

For $12 \leqslant N < 240$
From equation 6.9 $(P_{m(total)}/N) = 4 \log N - 1 - (10 \log N) = -1 - 6 \log N$ dBm0
Thus mean power level of one speech channel in a system of less than 240 channels is:

$$P_m = -1 - 6 \log N \text{ for } 12 \leqslant N < 240 \tag{6.28}$$

Similarly, for $N \geqslant 240$

$$P_m = -15 \text{ dBm0 for } N \geqslant 240 \tag{6.29}$$

The deviation produced by this mean power level of -15 dBm0 (in this example) is 35 kHz. Using equation 6.25, the deviation of the reference signal or test tone is 35 kHz × antilog (15 dB/20) = 196.8 kHz. As 200 kHz is specified in Table 6.3 for the RMS Test tone, 200 kHz will be used. The peak power of the reference signal is 3 dB above the reference RMS value, or 18 dB above the mean power level of the single channel. Hence, using equation 6.25, the deviation is 200 antilog (3/20) = 282.5 kHz. The mean power of the multichannel signal is found from equation 6.8, and is $(-15 + 10 \log N) = (-15 + 10 \log 960) = 14.8$ dBm0. Again using equation 6.25, the deviation is 200 antilog (14.8/20) = 1099 kHz or, as specified in Table 6.3, 1.1 MHz. Finally, using equation 6.26, the peak deviation for the multichannel signal can be found, or knowing that the peak power of the multichannel signal is 13 dB above the mean power of the multichannel signal, the peak deviation can be calculated using equation 6.25. Hence peak deviation = 1100 antilog (13/20) = 4.913 MHz or 4.9 MHz as specified in Table 6.3.

The above results are drawn on a level diagram in Figure 6.10.

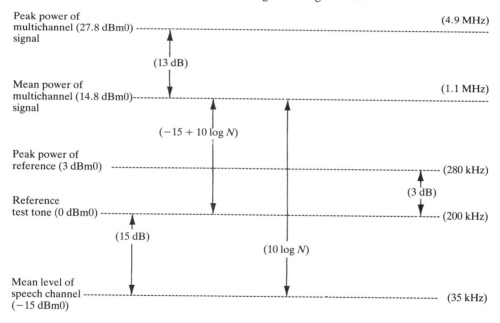

Figure 6.10 Level and Deviation for a 960-Channel System

Note that the CCIR Recommendation 404–2 gives the RMS deviation per channel for a test tone as 35 kHz for 24- and 12-channel systems. The mean level of a speech channel for a large system is -15 dBm0, and the deviation is taken as 35 kHz.

EXERCISES

1. What is the FM improvement threshold expected for a receiver designed to carry 1200 channels (see Table 7.6), if the receiver noise figure is 10 dB? (*Answer* -78 dBm)

2. How many active channels would be expected on a system with a channel capacity of 160? (*Answer* 53)

3. For a system carrying 20 active channels, by what level would the peak signal power exceed the RMS signal power for 1 per cent of the time? (*Answer* 9.3 dB)

4. Determine, from curves in this chapter, the linear dynamic range of a transmitter which is to carry 300 channels, and which will produce intermodulation distortion from this cause, only for less than 0.1 per cent of the time. (*Answer* 30.18 dB above single channel RMS power)

5. How does the answer to question 4 compare if equations are used? (*Answer* Favourably; 30.37 dB above single-channel RMS power)

6. Consider a 300-channel system. Draw up the levels and deviations, starting at the mean level of one speech channel. (*Samples of the answer* RMS test-tone deviation, 200 kHz, 0 dBm0. Peak deviation, 2.75 MHz)

REFERENCES

1. Holbrook, B. D. and Dixon, J. T., 'Load Rating Theory for Multichannel Amplifiers', *Bell System Telephone Journal*, 1939, **18**, pp. 628–629.
2. *ibid*, Figure 1, p. 629.
3. *ibid*, Figure 5, Curve marked 'Exceeded 1 percent of Time', p. 637.
4. Fagot & Magne, *Frequency Modulation Theory* (Pergamon Press, 1961), pp. 158–159.
5. CCIR, Warsaw, 1956, Green Book .
6. CCIR, Los Angeles, 1959, Green Book.
7. Tant, M. J., *The White Noise Book* (Marconi Instruments, 1974), Section 5.
8. CCIR Report 418–1, 1970–74.
9. Lyons, W., 'Considerations in SSB and ISB systems for Long-distance Radiotelegraph Communications', *Trans. AIEE, Pt.1, Communication & Electronics*, 1960, **78**, p. 922, Fig. 2.
10. CCITT, New Delhi, 1960, Red Book, Vol. III, Recommendation H12, p. 243.
11. *ibid*, Vol. VII, p. 33.
12. *ibid*, Recommendation R36, pp. 35–36.
13. *ibid*, Vol. III, Recommendation H41, Vol. VII, Recommendation V2, p. 349.
14. Smith, H. L., 'The Use of Holbrook and Dixon Loading Factors in Setting Receiver Parameters for an FDM–FM Radio Telephone Communications System', *IEEE Trans. on Comm. Tech.*, 1964, p. 155.
15. US Department of Defense, MIL–STD–188C, November 1969, Military Communication System Technical Standards.

7 NOISE LOADING TECHNIQUES AND MEASURING UNITS

7.1 THE COMPOSITE BASEBAND SIGNAL SIMULATED BY WHITE NOISE

In this chapter we are more concerned with the testing of a radio system rather than with the performance of individual items of equipment. As mentioned in Chapter 6, the larger the channel capacity ($\geqslant 240$ channels), the more the composite baseband signal amplitude becomes Gaussian, and its power/frequency spectrum uniform. As white noise has a Gaussian amplitude distribution, and a uniform power/frequency spectrum, it can for large channel capacities represent the composite baseband signal for the purposes of testing. These white noise tests permit a judgement to be made on the performance of the link, as it would behave under live traffic conditions, by comparing the results with the noise objectives outlined in Section 5.10. The results of the tests give an indication of how much intermodulation occurs, and how much thermal noise exists. These tests, together with other tests, permit the item or items of equipment causing unsatisfactory results to be isolated and thus located. In summary, the white noise testing of a radio link, or links, permits:

1. Comparison of the system's noise performance under simulated real traffic conditions, with the noise objectives as specified by some authority (e.g. CCITT, FIA, or a manufacturer)
2. In conjunction with other tests (quieting, etc.), the isolation and location of faulty equipment where applicable.

7.1.1 Simulated Loading where $N \geqslant 240$

The CCITT Recommendation G.223 states that, for the calculation of intermodulation below the overload point, the multiplex signal during the busy hour can be represented by a uniform spectrum of random noise signal, the mean absolute power level of which, at a zero relative level point is given by equation 6.8. For convenience this equation is given again here:

$$P_{m(total)} = -15 + 10 \log N \tag{6.8}$$

where $P_{m(total)}$ is the mean white noise power, and N is the number of channels in the system.

Refer to Figure 6.6. The 'Mean Multichannel Power – Speech' (noise loading to CCITT) line shows equation 6.8 in graphical form, for $N \geqslant 240$.

7.1.2 Simulated Loading where $12 \leqslant N < 240$

The CCITT Recommendation G.233 continues on from that in 7.1.1, to provide equation 6.9, which again is produced below:

$$P_{m(total)} = -1 + 4 \log N \tag{6.9}$$

Again referring to Figure 6.6, the section of the line mentioned above, for $N < 240$ shows equation 6.9 in its graphical form.

For less than 240 channels the composite baseband signal amplitude departs from a Gaussian distribution, and the signal becomes more 'peaky' as the system channel number decreases. To represent this, a higher mean power than that given by equation 6.8 is required. The peak power level expression, which was formulated so that no peaks in the composite baseband signal would exceed for more than 0.1 per cent of the time the value determined by this expression, is given by equation 6.15, i.e.:

$$P_{pk}(0.1\%) = 8 + 4.5 \log N \text{ for } 12 \leqslant N < 240 \tag{6.15}$$

As this equation represents the composite baseband signal itself, and not white noise, it is reasonable to assume that, if intermodulation occurs because the equipment cannot handle the peak levels, the CCITT equation for simulating this composite baseband signal with white noise, should be similar in form. That is so, for the CCITT equation for peak power levels is given by:

$$P_{pk(CCITT)} = 9.4 + 4 \log N \text{ for } 12 \leqslant N < 240 \tag{7.1}$$

Comparison between the composite baseband peak power levels as given by equation 6.15 and the CCITT equation 7.1 shows a marked similarity.

It is assumed that, in order for the CCITT white noise peak power level equation (equation 7.1), for $N < 240$, to exist, there must be a corresponding white noise mean power level equation which continues down from the CCITT equation for $N \geqslant 240$ (equation 6.8). The corresponding CCITT mean power level equation for $N < 240$ is that of equation 6.9.

Regardless of the above discussion, the internationally recognized mean power loading level for white noise when the number of system channels is less than 240, but 12 or greater, is that given by equation 6.9.

For the sake of completeness the companion equation to 7.1, i.e. the CCITT equation for peak power levels where $N \geqslant 240$, is given here:

$$P_{pk(CCITT)} = -4.5 + 10 \log N \text{ for } N \geqslant 240 \tag{7.2}$$

It can be seen that this equation is the same equation representing the composite baseband peak power which is not exceeded for 0.1 per cent of the time (equation 6.17).

7.1.3 Simulated Loading where $12 \leqslant N < 60$

In the CCIR Report 418–1, on 'Examples of Necessary Bandwidth Calculations', the recommended noise loading level for the determination of the peak deviation is taken as:

$$P_m = 2.6 + 2 \log N \text{ for } 12 \leqslant N < 60 \tag{7.3}$$

If this expression is used for noise loading, then equation 6.9 should be taken for the system channel limits of $60 \leqslant N < 240$, and not $12 < N \leqslant 240$.

7.1.4 Data and Telegraph Loading

Equations 6.19 and 6.21 give the required loading levels for telegraph and data respectively. These equations may be used to provide the mean power level of the white noise, if white noise loading is to simulate the real-life channel. For data loading of greater than 12 voice channels, we may use the following equation:

$$P_{m(data)} = -10 + 10 \log N_D \text{ dBm0} \tag{6.23}$$

7.2 NOISE BANDWIDTH

The composite baseband signal from the multiplex equipment is band-limited. The bandwidth depends upon the number of voice channels the multiplex equipment is designed or equipped for. Similarly, if the composite baseband signal is to be simulated by using a noise generator, set to the mean power level as per equations 6.8, 6.9, 6.19, 6.21, 6.23, or 7.3, the white noise must also be band-limited. The noise generator usually has a set of low-pass and high-pass filters. By switching in a high-pass filter whose cut-off frequency is below the cut-off frequency of the low-pass filter, a frequency band can be selected in accordance with the width of the baseband. Table 7.1' provides the details of the limits of the band occupied by telephone channels for different system channel numbers.

Table 7.1 CCIR/CCITT Noise Testing Relevant Frequencies

System Capacity (channels)	Limits of band occupied by telephone channels, kHz	Effective cut-off frequencies of band-limiting filters, kHz		Frequencies of available measuring channels, kHz						
		High pass	Low pass							
60	60–300	60±1	300±2	70	270					
120	60–552	60±1	552±4	70	270	534				
300	60–1300 64–1296	60±1	1296±8	70	270	534	1248			
600	60–2540 60–2660	60±1	2600±20	70	270	534	1248	2438		
900	316–4188	316±5	4100±30			534	1248	2438	3886	
960	60–4028 60–4024	60±1	4100±30	70	270	534	1248	2438	3886	
1200	316–5564	316±5	5600±50			534	1248	2438	3886	5340
1260	60–5636 60–5564	60±1	5600±50	70	270	534	1248	2438	3886	5340
1800	312–8120 312–8204 316–8204	316±5	8160±75			534 7600	1248	2438	3886	5340
2700	312–12336 316–12388 312–12388	316±5	12360±100			534 7600	1248	2438 11700	3886	5340

CCITT Recommendation G228 (Red Book), 1984.

7.3   MEASUREMENT OF NOISE IN A 3.1 kHz SLOT

The idea behind the measurement of the noise in a speech channel or a 3.1 kHz slot is to load the system with white noise at the recommended level and bandwidth, and then have the far end of the link measure the level of the arriving noise to set a reference. Once this has been done a notch filter 3.1 kHz wide and 80 dB deep is switched into the noise generator, creating a 'noiseless' slot. The far end of the link then measures the noise in this slot. Ideally, there should be no noise. Under real life conditions though, the far end will measure noise. This resulting noise is the sum of internal thermal noise, and intermodulation noise or 'total' noise arising from the noise in the rest of the band. The level of this measured noise in the 'noiseless' channel is recorded. In order to differentiate between the thermal noise in the channel or 'noiseless' slot, and the intermodulation noise, the noise generator is turned off. As the baseband now contains no noise loading, there will be no intermodulation occurring, but there will be thermal noise. The far end records the noise from the same 3.1 kHz slot knowing that this is only thermal or 'idle' channel noise. By subtracting the 'idle' noise from the 'total' noise, the result gives an indication of the amount of intermodulation noise present. There will of course be some thermal noise in the subtracted result due to the increase in thermal noise of the unloaded system, but still the results are indicative of problems which are related to high inter-modulation.

The frequencies of the noise measuring channel and the band-stop filter characteristics are subject to CCIR recommendation 399–3 of 1978[4]. This recommendation states that:

> Due to the fact that it is desirable to measure the performance of radio-relay systems for FDM telephony under conditions closely approaching those of actual operation, and that due to the already widespread use of white noise (as it is similar to the multiplex signal if the number of channels is not too small) in the measurement of this performance, some standardizations be made. These standardizations are to be made to the measuring channel centre frequencies and bandwidth, and also to the minimum attenuation and bandwidth of the stop filters used in a white noise generator.

Table 7.1 gives the filter cut-off frequencies, baseband frequency range, and frequencies of available measuring channels. The mean value of speech power in a telephone channel to be taken into consideration during the busy hour (CCITT Recommendation G233, Orange Book, Vol III–1[1]) is found from equations 6.8 and 6.9. Table 7.2 provides the recommended levels for different system capacities, as taken from CCITT Recommendation G228, Red Book Vol. III-2, 1984, and CCIR Recommendation 380–3 (column 2)[2].

Table 7.2 CCIR levels at Baseband for Noise Testing

(1) Number of system telephone channels, N	(2) Relative power level at a point T', dBr	(3) Level of the conventional load, dBm0	(4) Nominal power level of the test signal at point T', dBm
60	−36	6.1	−29.9
120	−36	7.3	−28.7
300	−36	9.8	−26.2
600	−36	12.8	−23.2
	−33		−20.2
960	−36	14.8	−21.2
	−33		−18.2
1260	−33	16.0	−17.0
1800	−33	17.5	−15.5
2700	−33	19.3	−13.7

7.4 NOISE POWER RATIO

In order to evaluate a microwave radio system's noise performance without having to employ a fully expanded multiplex system, a method has been evolved which uses white noise. If such a white noise signal originating from a white noise generator set is transmitted over the system as mentioned previously, intermodulation and thermal noise appear in varying amounts at different places in the receiver baseband. If a translated speech channel in the baseband is arranged such that the noise generator transmits its noise power over the complete band except for that one translated speech channel (or 3.1 kHz slot), then, together with the internal thermal noise of that channel, intermodulation products formed in the transmitting and receiving processes will also appear in that speech channel. The noise power ratio (NPR) is the ratio of the noise measured in the speech channel with the system fully loaded, to the noise power measured in that same speech channel, with the system fully loaded except for that speech channel. Hence

$$NPR_{total} = 10\log \frac{\text{Baseband noise loading power}}{\text{noise power in unloaded channel (Residual noise)}} \text{ dB} \quad (7.4)$$

If noise again is transmitted over the complete baseband, the level at the receiver baseband is measured, and then the noise is turned off, the resulting noise measured at the receiver baseband will consist of the inherent thermal noise of the single channel; and additional noise due to the radio system not being loaded. The resulting NPR is the 'idle' NPR, and gives a reasonable, but not entirely accurate[7], measure of the thermal noise in the translated speech channel. It is not entirely accurate due to the higher thermal noise of an unloaded system. Hence:

$$NPR_{idle} = 10\log \frac{\text{Baseband noise loading power}}{\substack{\text{Noise power in unloaded channel} \\ \text{with system without noise loading}}} \text{ dB} \quad (7.5)$$

This equation may also be called the 'baseband intrinsic noise ratio' (BINR). It is important to note that the baseband noise loading power in both equations 7.4 and 7.5 is the noise that is measured in the 3.1 kHz speech channel or slot, with the baseband completely loaded.

The difference between the values of *NPR* in equation 7.4 and in equation 7.5 gives a measure of the intermodulation noise:

$NPR_{idle} - NPR_{total}$ = Total noise power (thermal + intermodulation)
− thermal noise power dBm.

Hence

$$NPR_{idle} - NPR_{total} = 10 \log \frac{\text{Intermodulation} + \text{loaded thermal noise}}{\text{Unloaded thermal noise power}} \text{ dB} \quad (7.6)$$

As *NPR* is usually measured in decibels, if the intermodulation noise is high, a positive number will result in equation 7.6. But if it is low a negative number will result due to the higher value of the unloaded thermal noise when compared with that of the loaded thermal noise. Positive values less than 3 dB for equation 7.6 are not a problem. At about 5 dB to 7 dB the intermodulation may be becoming a bit much, and above 7 dB there is a problem somewhere in the system (see Section 11.3.2). The less noisy the system is, the higher the value of *NPR*. Typical values of *NPR*$_{total}$ for various working systems range from 45 to 55 dB, with *NPR*$_{idle}$ being 47–57 dB. The higher the system channel number, the worse the value of *NPR* should become (but not worse than the range given above).

7.5   THE MEASUREMENT OF NOISE POWER RATIO

Mention has been made above of the noise generator which generates the noise to be inserted into the baseband input of the radio transmitter. The noise band which is generated is limited to the baseband frequency spectrum as given in Table 7.1, and the 3.1 kHz slots which are to be 'noiseless', are produced by band-stop filters within the noise generator set. The recommended frequencies of these slots are also given in Table 7.1.

At the receiving end of the link a noise measuring set is attached to the receiver baseband output. This noise receiver is also equipped to measure the noise in the band associated with the baseband frequency range, – specifically, *the loaded and the unloaded noise* in the translated speech channel slots according to Table 7.1. This is done with the aid of slot band-pass filters located in the noise receiver unit. As the level of the white noise is constant over the baseband frequency range, when the baseband is fully loaded, the noise measured in the 3.1 kHz slot will be the same over all sections of the baseband if it is assumed that no pre-emphasis or de-emphasis is present and that the baseband frequency response is perfectly flat. As this is not generally so in practice, the received noise power level with the baseband fully loaded is always recorded (or the equipment recalibrated) for each different translated speech channel or slot measurement. Figure 7.1 shows the test set-up.

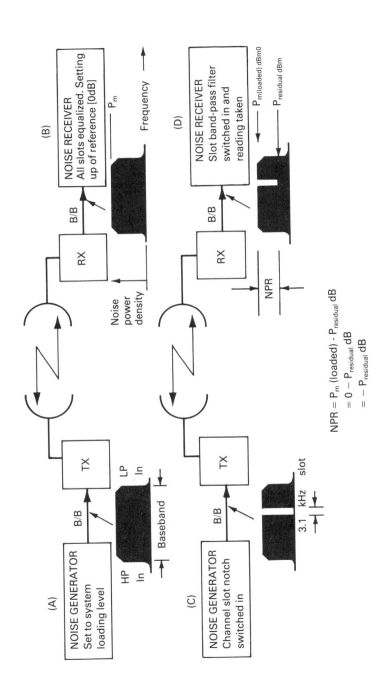

Figure 7.1 Measurement of NPR

7.6 SIGNAL-TO-NOISE RATIO

Imagine that a reading has just been taken of absolute noise coming out of a translated channel in the baseband, and that this noise reads -55 dBm; at first sight it may be thought that the result is very good. With a bit more thought, it may be asked what the level of a standard test tone of input level 0 dBm0 should be, when coming out of this translated channel. In other words, what is the dBr level of the test tone at this point? If the answer were -35 dBm, the signal-to-noise ratio would be calculated as $-35 - (-55) = 20$ dB, and this result would not be good at all. Alternatively, what could have been done, if the dBr point was known in the first place, was to convert the noise to a dBm0 level, or refer the noise back to a zero test level point. Hence from equation 1.12:

$$(\text{Absolute noise level in dBm}) - \text{dBr} = \text{noise in dBm0} \tag{7.7}$$

Hence if the standard test tone is at 0 dBm0:

$$(S/N) \text{ ratio} = 0 \text{ dBm0} - (\text{Noise}) \text{ dBm0}$$
or
$$(S/N) \text{ ratio} = -(\text{Noise}) \text{ dBm0} \tag{7.8}$$

where (Noise)   = the measured noise in dBm0 as determined from equation 7.7
(S/N) ratio = the signal-to-noise ratio.

7.7 CONVERSION OF OTHER UNITS TO (S/N) RATIO(unweighted)

In Section 1.7.6 it was seen that noise in a 3.1 kHz noise bandwidth was given by q dBa $= (q - 82)$ dBm, or, referring to a zero-test-level point,

$$
\begin{aligned}
&q\,\text{dBa0} &&= q\,\text{dBm0} - 82\,\text{dBm0} + 0\,\text{dBm0} \\
\text{or}\quad &q\,\text{dBa0} - (-82\,\text{dBm0}) &&= q\,\text{dBm0} + 0\,\text{dBm0} \\
\text{or}\quad &q\,\text{dBa0} - (0\,\text{dBa0}) &&= -S/N + 82
\end{aligned}
$$

Hence noise power in dBa0 $= 82 - (S/N) \underset{\text{(unweighted)}}{\text{ratio}}$ (7.9)

where $-82\,\text{dBm0}$ $= 0\,\text{dBa}$
    $0\,\text{dBm0}$ $= 82\,\text{dBa0}$
    $q\,\text{dBm0}$ $= -(S/N)$ ratio from equation 7.8

An easier method would be using the signal-to-noise ratio directly, that is:

$$(S/N) \text{ ratio} \qquad\qquad = 0 \text{ dBm0} - N \text{ dBrnC0}$$

or

$$(S/N) \text{ ratio} \qquad\qquad = 88.5 - N \text{ dBrnC0}$$

Hence noise power in dBrnC0 $= 88.5 - (S/N) \underset{\text{(unweighted)}}{\text{rato}}$ (7.10)

Again, this time for pWp,

$$S/N \underset{\text{(unweighted)}}{\text{ratio}} \qquad = 0 \text{ dBm0} - (\text{noise} + 2.5) \text{ dBmp}$$

or

$$(S/N) \underset{\text{(unweighted)}}{\text{ratio}} \qquad = [90 - (\text{noise} + 2.5] \text{ dBmp}$$

Thus

$$\text{antilog } (S/N) \text{ ratio} \qquad = 10^9/[\text{noise power in pWp} \times \text{antilog } (2.5/10)]$$

or

$$\text{Noise power in pWp} = \frac{0.56 \times 10^9}{\text{antilog } (S/N) \underset{\text{(unweighted)}}{\text{ratio}}} \qquad (7.11)$$

The factor of 2.5 converts the flat speech channel to psophometric weighting. If a flat channel noise is required equation 7.11 would read:

$$\text{Noise power in pW} = \frac{10^9}{\text{antilog } (S/N) \underset{\text{(unweighted)}}{\text{ratio}}} \qquad (7.12)$$

For the conversion factors between the various noise units, reference is made to equations 1.19 to 1.22 inclusive.

7.8 CONVERSION OF *NPR* TO (*S/N*) RATIO

If the noise power ratio results from a system, link or RF back-to-back test are obtained, the unweighted signal-to-noise ratio in the speech channel, due to a 0 dBm0 test tone, can be computed. This is achieved as follows.

The noise received by an ideal receiver from a conventional load given by equation 6.8 or 6.9 transmitted by the companion transmitter is $P_{m(total)}$. As this noise power has a bandwidth equal to that of the baseband, it will be reduced in value, if only a bandwidth of 3.1 kHz out of the total bandwidth is considered. The reduction is equivalent to the negative of the bandwidth ratio (BWR) as given by:

$$BWR = 10 \log \frac{\text{Occupied baseband bandwidth}}{\text{Speech channel bandwidth (3.1 kHz)}} \text{ dBm0} \qquad (7.13)$$

Hence the noise power in a single speech channel can be given as

$$P_{m(total)}/\text{channel} = P_{m(total)} - BWR$$

As the receiver and multiplex are not ideal, and there is also intermodulation and crosstalk noise, the noise in the 3.1 kHz speech channel is also increased by the residual noise as measured when doing the NPR test.
Since

$$NPR = P_{m(total)}/\text{channel} - P_{residual} \text{ dB}$$

The noise in the channel is increased by $P_{residual}$ or $P_{m(total)}/\text{channel} - NPR$

Using the expression for $P_{m(total)}/\text{channel}$ above, the total noise in the channel becomes:

$$\text{Noise per channel} = P_{m(total)} - BWR - NPR$$

If the noise is referred to a zero test level point, the noise per channel becomes noise in dBm0 per channel, and as from equation 7.8, (*S/N*) ratio = $-$ noise in dBm0, to give:

$$(S/N) \underset{(unweighted)}{\text{ratio}} = NPR + 10 \log (\text{baseband frequency range})/3.1 \text{ kHz} - P_{m(total)} \quad (7.14)$$

and

$$(S/N) \underset{weighted}{\text{ratio}} = NPR + 10 \log (\text{baseband frequency range})/3.1 \text{ kHz} - P_{m(total)} + W \quad (7.15)$$

where W is the weighting factor as given in Chapter 1.

The following is a list of equations relating (*S/N*) ratio *NPR* for different system capacities.

Equations 7.14 and 7.15 can be approximated with an error of approximately 0.2 dB to:

For $N < 240$

$$(S/N) \underset{unweighted}{\text{ratio}} = NPR + 6 \log N + 2.1 \text{ dB} \quad (7.16)$$

$$(S/N) \underset{psoph.}{\text{ratio}} = NPR + 6 \log N + 4.6 \text{ dB} \quad (7.17)$$

$$(S/N) \underset{C\text{-}message}{\text{ratio}} = NPR + 6 \log N + 3.6 \text{ dB} \quad (7.18)$$

$$(S/N) \underset{FIAweight}{\text{ratio}} = NPR + 6 \log N + 5.1 \text{ dB} \tag{7.19}$$

For N $\geqslant$ 240

$$(S/N) \underset{unweighted}{\text{ratio}} = NPR + 16.3 \text{ dB} \tag{7.20}$$

$$(S/N) \underset{psophometric}{\text{ratio}} = NPR + 18.8 \text{ dB} \tag{7.21}$$

$$(S/N) \underset{C\text{-}message}{\text{ratio}} = NPR + 17.8 \text{ dB} \tag{7.22}$$

$$(S/N) \underset{FIAweight}{\text{ratio}} = NPR + 19.3 \text{ dB} \tag{7.23}$$

Example

To tie in the noise permissible from the CCITT/CCIR recommendations as briefly presented in section 5.10, we can say:

Total pWp = Thermal noise per hop + intrinsic noise per hop + intermodulation noise per hop + feeder noise per hop + coupler noise per hop.

From equation 7.11, (S/N) ratio is given as

$$(S/N) \text{ ratio} = 87.5 - 10 \log \text{pWp} \tag{7.11}$$

and from equation 7.17 for $N < 240$,

$$(NPR) \text{ ratio} = S/N - 6 \log N - 4.6 \text{ dB} \tag{7.17}$$

Hence

$$NPR = 82.9 - 6 \log N - 10 \log \text{pWp for } N < 240 \tag{7.24}$$

In a similar fashion, relationships between all of the other noise units may be obtained. Thus:

$$NPR = 87.9 - 6 \log N - 10 \log \text{pW for } N < 240 \tag{7.25}$$

$$NPR = 84.9 - 6 \log N - \text{dBrnC0 for } N < 240 \tag{7.26}$$

$$NPR = 76.9 - 6 \log N - \text{dBa0 for } N < 240 \tag{7.27}$$

For $N \geqslant 240$, the following expressions will have an error of no more than 0.2 dB:

$$NPR = 73.7 - 10 \log \text{pW}_{\text{(unweighted)}} \tag{7.28}$$

$$NPR = 71.2 - 10 \log \text{pWp}_{\text{(weighted)}} \tag{7.29}$$

$$NPR = 70.7 - \text{dBrnC0} \tag{7.30}$$

$$NPR = 62.7 - \text{dBa0} \tag{7.31}$$

Example

For an 11-channel system, the loading factor recommended by a particular manufacturer is:

$$P_{m(total)} = -10.5 + 7 \log N \text{ for } N < 24 \tag{7.32}$$

and the baseband level into the transmitter is -35 dBr, and out of the receiver is -5 dBr. What would be the noise power measured in pW in a 3.1 kHz slot at the receiver output if the NPR measured at this point was 47 dB?

We need to find the expression for the NPR in terms of the given parameters. As the unweighted noise is asked for (pW), equation 7.14 is applicable in this case. Thus

$$(S/N)\underset{unweighted}{\text{ratio}} = NPR + 10 \log (11 \text{ channels} \times 4/3.1) - (-10.5 + 7 \log N)$$
$$= NPR + 11.52 - (-3.21)$$
$$= NPR + 14.73 \text{ dB}$$

From equation 7.12

$$(S/N)\underset{unweighted}{\text{ratio}} = 90 - 10 \log \text{pW0}$$

Hence equating the above two equations;

$$NPR = 75.27 - 10 \log \text{pW0 for an 11-channel system} \tag{7.33}$$

From equation 7.33, the noise referred to 0 dBm0 can be determined, since *NPR* is given as 47 dB. Thus $10 \log \text{pW0} = 28.27$ dBm0.

The receiver signal, however, is -5 dBr; hence the absolute noise is 5 dB below the dBm0 value, so $10 \log \text{pW} = 10 \log \text{pW0} - 5 = 23.27$ dBm.

This means that the noise in pW in a 3.1 kHz slot in the baseband = 212.3 pW. (*Answer*)

Example

Assume that the multiplex equipment, in the above example which is in an air-conditioned room held at 17°C, adds 100 pW of noise, what would be (*S/N*) ratio out of the audio channel?

In a similar fashion to that given by equation 5.23, the total noise out of the multiplex equipment will be:

$$N_{T_0} = \text{Multiplex gain} \times \text{input noise} + \text{multiplex noise}$$

As the multiplex gain is the difference between the output dBr level of +7 dBr, and the receive baseband level of −5 dBr, the multiplex gain is 12 dB or 15.85. The input noise from the previous example was found to be 212.3 pW, and the multiplex noise is given as 100 pW. Hence

$$N_{T_0} = 15.85 \times 212.3 + 100$$
$$= 3465 \text{ pW} = -54.6 \text{ dBm}$$

The signal level out of the audio channel is +7 dBm, so that

$$(S/N) \text{ ratio}_{audio\ channel\ (flat)} = 61.6 \text{ dB } (Answer)$$

Out of interest if this is compared with the value of S/N before entering the multiplex equipment, we get 61.73 dB; thus the 100 pW noise introduced by the multiplex equipment has a negligible effect on the $(S/N)_{flat}$ ratio out of the channel modem (unless it is added before the multiplex amplifiers).

If the channel had been weighted with CCITT weighting, then there would be an additional 2.5 dB improvement over the 61.6 dB to give $(S/N) \text{ ratio}_{audio\ channel\ (weighted)} = 64.1 \text{ dB}$.

Table 7.3

Channel capacity N	High-pass kHz	Filters Low-pass kHz	Band-* stop kHz	Noise loading level, dBm0	Conversion factor (NPR to dBrnC0) NPR + dBrnC0
6	4	28	16 27	−5.0	74.1
11	4	56	16 27 56	−3.2	72.5
24	12	108	16 40 105	+4.5	77.6
60	60	300	70 185 270	+6.1	75.2
72	12	300	16 185 270	+6.4	74.7
120	60	552	70 270 534	+7.3	73.3
132	12	552	16 270 534	+7.5	73.1

* Center frequencies of available measuring channels
(Source: Reference 9)

Similar examples may be constructed using dBrnC instead of pW.

Table 7.3 gives some typical values of noise loading level used in low-capacity systems.

7.9 ECHO DISTORTION NOISE[3]

Microwave systems consisting of large numbers of system channels are vulnerable to a form of distortion which is caused by the addition to the modulated wave of an echo of it. This form of distortion may arise in a mismatched aerial feeder or through multipath transmission. The level of the resulting intermodulation distortion generated by this small echo depends on the echo amplitude and delay, and on the phase relation between the wanted carrier and the echo carrier, as well as on the modulation parameters, which are the maximum and minimum modulating frequencies, the RMS deviation and the baseband spectral distribution.

Figure 7.2 shows the distribution of echo distortion across the baseband when the echo is phased for maximum distortion for a 600-channel system. The three delay times chosen, 0.125, 0.25 and 0.5 μs corrrespond to air-spaced, non dispersive feeder lengths of 62.5, 125 and 250 ft. The distortion associated with long feeders tends to be rather uniform over the major portion of the baseband. The effect of pre-emphasis on long-feeder distortion is not important, and does not show any advantages or disadvantages on system performance.

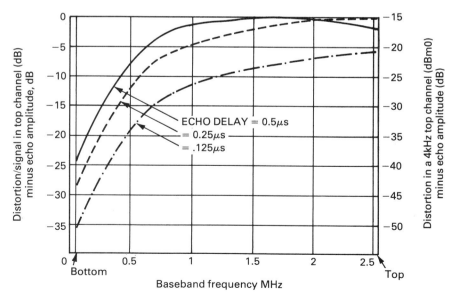

Figure 7.2   Varition of Echo Distortion over Baseband for a 600-Channel System. (Phased for max. distortion) (Courtesy of Electronic and Radio Engineer, Ref. 3)

Usually in working out the noise budget of a microwave system, echo distortion noise is allowed for. The signal-to-noise ratio of echo noise (S/N_e) in dBm0p can be expressed as:

$$-S/N_e = (\text{Distortion/signal ratio in the top channel} - \text{echo amplitude})$$
$$- (\text{antenna and waveguide return loss} + \text{equipment return loss})$$
$$- (2 \times \text{waveguide loss}) \text{ dBm0p} \tag{7.34}$$

Figure 7.3 provides the curves for different system channel capacities for equation 7.34. To obtain the noise due to echo distortion from Figure 7.3, for a particular channel capacity and echo delay time (or feeder run length), read off, for a particular feeder length or echo delay time, the value of distortion signal in the top channel minus the echo amplitude. This is otherwise known as the distortion, or the echo distortion noise minus the echo amplitude. The echo amplitude may be expressed as the negative value of the sum of the antenna and equipment return losses plus twice the feeder loss. Hence to obtain the distortion/signal ratio (echo distortion noise) these return and feeder losses must be subtracted from the reading taken from the curve. The return losses and feeder losses, which are required to complete the information in equation 7.34, must be determined from manufacturers' data, or actual tests on the equipment.

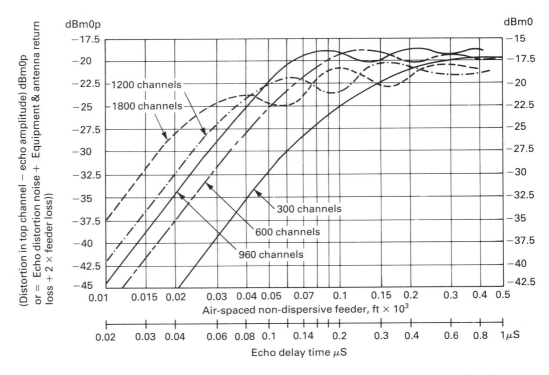

Figure 7.3   Echo Distortion Noise (Courtesy of Electronic and Radio Engineer, Ref. 3)

Therefore curve reading produces for a particular channel capacity, and feeder length (echo delay time):

(Distortion/signal in the top channel – echo amplitude)
= (Distortion in the top channel – echo amplitude)
= (Echo distortion noise (S/N_e) + equipment and antenna return loss
+ (2 × feeder loss)

Since Echo amplitude = −[equipment and antenna return loss + (2 × Feeder loss)].

7.10 RETURN LOSS (SEE SECTION 9.1.1)

The return loss can be defined as a measure of the match between the two impedances on either side of a junction point, and thus

$$\text{Return loss} = 20 \log \frac{Z_1 + Z_2}{Z_1 - Z_2} \text{ dB} \tag{7.35}$$

where Z_1 and Z_2 are the complex impedances of the two halves of the circuit.

An alternative definition of return loss is:

$$\text{Return loss} = 10 \log [(\text{Forward power})/(\text{reflected power})] \text{ dB} \tag{7.36}$$

A third definition is in terms of VSWR (voltage standing wave ratio), is:

$$\text{Return loss} = 20 \log \frac{\text{VSWR} + 1}{\text{VSWR} - 1} \text{ dB} \tag{7.37}$$

Example

Determine the echo distortion noise in pW for a 600-channel system, if the waveguide run is 200 ft, the waveguide loss is 2 dB, the equipment return loss is 20 dB and the antenna plus waveguide return loss is 24 dB.
From the right-hand side of Figure 7.3 (dBm0), for 600 channels and 200 ft of feeder, we get −17.6 dBm0. From equation 7.34 we get:

$$-S/N_e = -17.6 - (24 + 20) - 2 \times 4 = -69.6 \text{ dBm0}$$

Thus echo noise = 110 pW *Answer*

This value of 110 pW is the noise in the top channel without emphasis. If emphasis was used as is usually the case this figure would be reduced by a factor of 3 dB, bringing the noise down to 55 pW. Echo noise can be improved by increasing the equipment and antenna return losses.

7.11 THE WHITE NOISE GENERATOR[5,11]

7.11.1 Principles of Operation

As mentioned previously, if white noise occupying the traffic bandwidth is applied to a microwave radio system baseband at a suitable power level, a fully loaded telephone system will be very closely simulated. Furthermore, if a filter with a very narrow slot band, 3.1 kHz wide, is interposed between the white noise source and the system, the conditions then existing will be equivalent to a fully loaded system except for one quiet channel. A receiver tuned to the quiet channel may be used to indicate the noise level produced by intermodulation of the components of the white noise occupying the remainder of the frequency band. The method thus outlined is similar to a 960-channel system in which 959 channels are in use by subscribers, and the remaining channel is used to listen to the noise produced by the 959 telephone conversations.

Measurements are usually made at three or more widely separated channels (see Tables 7.1 and 7.3). The highest frequency is more sensitive to phase distortion and intermodulation in the RF and IF circuits under test. Undesired mixer frequency components (crossover distortion) are most noticeable at the middle frequency and non-linearity most noticeable at the lowest frequency. The effect becomes quite marked as the number of channels increase, because the channel frequencies being measured are further separated than for lower frequencies. To make a measurement, white noise of the appropriate baseband bandwidth and noise loading power level is applied to the transmitter baseband from the noise generator, with all slot filters 'out'. The far-end receiver baseband output is connected to the noise receiver, which is switch-tuned to the frequency of the band-stop filter that will eventually be used to produce the quiet channel. Usually the receiver samples all slots to ensure that the reading from each is the same overall. If it is not the same, equalizer potentiometers are adjusted for each slot to line them all up to the same reading. Figure 7.4 outlines the method used.

The receiver sensitivity should be adjusted to give a convenient meter reading when the noise generator is transmitting the simulated composite baseband noise, and has no band-stop filters inserted. Usually this convenient level on the meter is zero, with the receiver attenuators adjusted to produce the meter reading level. Once the noise generator switches in the narrow slot, the noise receiver input attenuator is again adjusted to restore the original meter deflection. The difference between the initial and final attenuator settings in decibels is the NPR. If the system is actually in use, and cannot be interrupted, CCIR Recommendation 398–3 applies. (See Section 7.13)

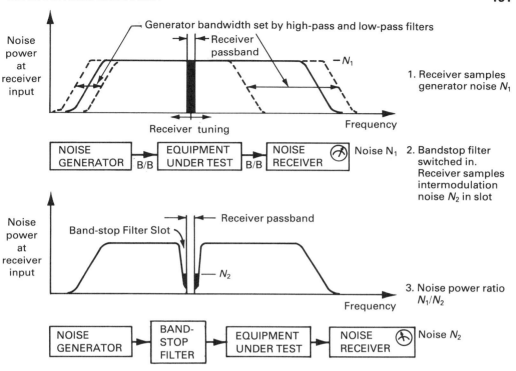

Figure 7.4 . NPR Measurement

7.11.2 Setting-up for Testing (Marconi TF 2091B)

The output of the noise generator must be terminated in 75 Ω when the instrument is being set up. Normally the equipment under test will provide this termination. Setting-up procedure:

1. Switch in a high-pass and low-pass filter to give the noise bandwidth required by the equipment under test as shown in Tables 7.1, 7.3 and 7.4*. Do not use more than one high-pass and one low-pass filter at a time. Switch out all the band-stop filters.
2. Set the required loading level as per equations 6.8, 6.9, etc., using the output attenuators in conjunction with the noise level control (potentiometer). Some typical loading levels at the baseband interconnection point are given in Table 7.2. Fractions of 1 dB can be set on the meter using the noise level control; maximum accuracy is achieved when the meter indicates within the 41 dB region. For levels

* See page 197

greater than 0 dBm, optimum accuracy is obtained with minimum attenuation of the fine attenuator.

Figure 7.5 is a pictorial diagram of the control panel of the white noise generator (Marconi TF 2091B). Some notes on the controls and other components on the panel are given below, the numbers of the items corresponding to the ringed numbers in Figure 7.5:

1. The meter indicates the available noise power in dB relative to 1 mW into a 75 Ω load at the output socket. The actual power output is given by the meter reading plus the algebraic sum of the attenuator dial settings.
2. Output attenuator controls adjust the output in 10 dB, 5 dB, and 1 dB steps. Positions marked in red indicate settings to be added to the meter reading, positions marked in black indicate settings to be subtracted.
3. The output socket is BNC, 75 Ω.
4. A noise level control regulates the noise power monitored on the meter by varying the gain of a high-level amplifier.
5. The noise on/off switch switches off the noise output. The lamp lights up when the noise is switched off.
6. A supply switch and pilot lamp are fitted.
7. High-pass and low-pass filters determine the noise bandwidth. The band-stop filters create slots in the noise band.

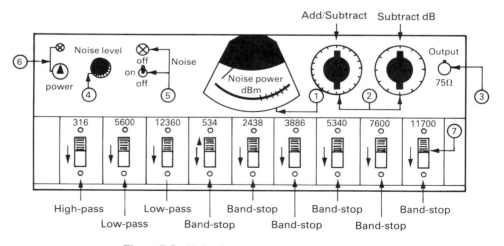

Figure 7.5   Noise Generator (Marconi TF 2091B)

Examples

1. To set −23.2 dBm noise level

Set the coarse attenuator to indicate 20 dB (shown in black on the knob skirting) and the fine attenuator to indicate 3 dB. Using the noise level control, set the meter to indicate −0.2 dBm.

2. To set +19.3 dBm (back-to-back) test

Set the coarse attenuator to indicate 20 dB (shown in red on the knob skirting) and the fine attenuator to indicate 0 dB. Using the noise level control, set the meter to indicate −0.7 dBm.

Note Equations 6.8 and 6.9 should be used to obtain the dBm0 power level setting. To obtain the actual power level to be set on the noise generator, the dBr level should be added to the dBm0 level, as dBm = dBr + dBm0. This means that the dBr level of the baseband must be known before the correct dBm noise level can be set.

7.12 THE WHITE NOISE RECEIVER (MARCONI TF 2092B)[5,11]

7.12.1 Setting-up for NPR Measurement

Figure 7.6 is a pictorial diagram of the control panel of the white noise receiver (Marconi TF 2092B). As for the generator (Section 7.11.2), the numbers in the following notes correspond with the ringed numbers in Figure 7.6:

1. The meter indicates the reference level for noise power ratio measurement. Graduations around the REF mark, which are in approximate steps of 0.2 dB, assist interpolation between the 1 dB steps of the fine attenuator. For the most accurate measurements, the attenuator must always be used to set the meter pointer as near to the REF mark as possible.
2. The set reference control adjusts the receiver gain to give a standard meter reading.
3. NPR/noise-input controls adjust the receiver input in 10 dB and 1 dB steps, and are calibrated in picowatts per channel and in noise power ratio. To facilitate direct reading, the coarse attenuator skirt is adjustable independently of the attenuator setting.
4. Record jack accepts two-pole telephone jack plug for connection to external indicator or recorder.
5. The input socket is BNC, 75 Ω.
6. A supply switch and power lamp are fitted.
7. Filters determine the receiver pass-band centre frequencies, which must coincide with the corresponding generator band-stop filter frequencies.
8. Frequency selector switch: positions 1–6 select the band-pass filter and the appropriate oscillator, providing the local oscillator (LO) frequency at the LO socket; Position 7 allows an external signal to be connected at the LO socket.
9. The operate/standardize switch selects input or standardizing noise source signal to feed into input attenuators. In the standardizing position, the input socket is terminated in 75 Ω.
10. Equalize range sensitivity controls allow each channel to be set to the same sensitivity.

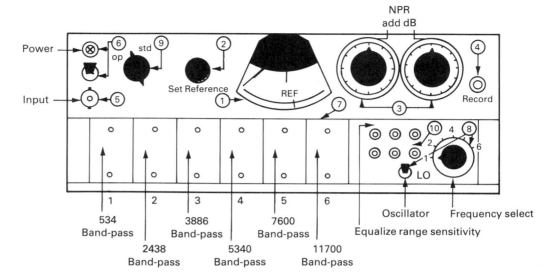

Figure 7.6   Noise Receiver (Marconi TF 2092B)

When the noise generator has been set up as in Section 7.11.2, and the noise receiver has been fitted with the appropriate band-pass filters and oscillators set out in Table 7.1, connect the equipment to the microwave receiver baseband output, as in Figure 7.4.

The noise receiver is set up to measure NPR as follows;

1. Turn all equalize range controls fully clockwise using a small screwdriver.
2. Turn the set reference control to mid-travel.
3. Choose one of the selected measurement frequencies by means of the frequency selector switch.
4. Switch the operate/standardize switch to operate.
5. Switch the fine NPR/noise-input control to indicate 0 dB, i.e. with the symbol 0 on the skirt vertical and coincident with the black arrow.
6. Adjust the coarse NPR/noise-input control until the meter reading is close to the REF mark.
7. Switch the frequency selector to the other selected frequencies in turn and note which gives the minimum meter reading. Leave the switch at the position of minimum reading.
8. Adjust the set reference control to give a meter reading on the reference mark.
9. Switch the frequency selector switch to the other frequencies in turn, and, without touching the set reference control, adjust the equalize range sensitivity controls for each frequency selector switch position, to give a meter reading at the same reference mark.
10. Leaving the control knob in the same switch position, adjust the calibrated skirt of the coarse NPR control so that the symbol 0 is vertical or coincident with the black indicating arrow. (The skirt is turned by pushing it towards the front panel

and rotating it. Clicks corresponding to each switch position as it is turned will be felt.)

With the coarse and fine controls set in this position, it will be noted that the indicated NPR is 0 dB. The receiver is now set up to measure NPR.

7.12.2 Measurement of NPR

The NPR of equipment or of a link, measured in the way detailed below, is the ratio of (1) the portion of the noise representing a multichannel signal that occurs in a narrow bandwidth to (2) the inherent and intermodulation noise in the same bandwidth when signals are not applied in that band, but are applied over the remainder of the multichannel frequency range. Part (1) of the ratio also includes inherent and intermodulation noise, but this is negligible in comparison with the noise representing the signal. Having set up the noise generator and receiver as in Sections 7.11.2 and 7.12.1, the NPR can be measured at any of the band-stop filter frequencies by the following steps:

1. Turn the frequency selector switch to the frequency at which the measurement is to be made. The switch positions are numbered and the first of six of these correspond to the band-pass filters fitted.
2. Switch in the generator band-stop filter of the same frequency as that selected in step 1 (see Figures 7.5 and 7.6). The generator meter reading should remain the same since automatic level control is applied. Only if second-order non-linear distortion dominates, should a correction be applied.
3. Adjust the receiver coarse and fine NPR/noise-input controls (set to 0 dB), so that the receiver meter indication is restored to the reference mark. The NPR is obtained by adding the dB readings, i.e. the sum of the figures indicated by the black arrows against the two attenuator skirts.
4. To make a measurement at other frequencies, follow steps 1 to 3, selecting another frequency and band-stop filter. If the receiver sensitivity was made equal on all chosen ranges, it will not be necessary to check the receiver level when changing frequencies.

It is not advisable to make an NPR measurement such that the final position of the NPR/noise-input coarse control switch is fully clockwise. In this position the signal is connected via the fine attenuator only to the input of the band-pass filter, and this is not such a good match for the system under test outside the pass-band frequencies. Under certain circumstances, e.g. low noise and high NPR, a reference mark low on the receiver meter should be used, say 30 per cent full scale deflection. However, the noise receiver should not be operated with the set reference control in the extreme counter-clockwise position, i.e. minimum gain. This would result in the need for a higher noise loading of the band-pass filters and would provide a non-optimum condition for inherent intermodulation distortion.

7.12.3 Measuring Residual or Thermal System Noise

The residual system noise [or baseband intrinsic noise ratio (BINR)] is a measure of the idle noise due to thermal noise, carrier leaks, spurious FM in oscillators, etc. It is defined as the ratio of the noise in a test channel with all channels loaded to the noise in the test channel with all loading removed from the baseband. The procedure is the same as that given in Section 7.12.1, followed by the switching off of the noise generator by switching the noise on/off switch to the off position and resetting the receiver meter to the reference mark using the receiver input attenuator. This is done for each slot. The difference between the two readings, that is with the system fully loaded and with the system unloaded, is a measure of the residual noise of the system. The difference between the measured NPR of Section 7.12.2, and the NPR of the residual system noise is a measure of the amount of spurious noise in the system caused by intermodulation and crosstalk. In addition to the measurement of residual noise, a useful check is to examine the baseband signal using a spectrum analyzer, or a selective level meter by which the baseband is slowly scanned. This is done to ensure that the residual noise measurement is not in error due to spurious signals arising from excessive carrier leaks, etc.

7.13 WHITE NOISE TESTING OF ACTIVE SYSTEMS [10]

The CCIR Recommendation 398–3 is specifically concerned with the white noise testing of a system which is carrying live traffic, and which cannot be withdrawn from service. At times noise measurements are required to be made on a live system in order that the quality of the system can be determined. The channels used for this kind of measurement are found outside the normal baseband signal, but close enough to the limits to effect meaningful results, without being so close as to result in expensive filters. Measurements made in channels above the baseband are generally more sensitive to changes of thermal and intermodulation noise in the radio-frequency and intermediate-frequency circuits of the equipment, whereas measurements in channels below this band are generally more sensitive to changes in the modulators and demodulators. The CCIR recommendation states that the noise should be measured both above and below the multiplex baseband, and that the centre frequencies of the noise measuring bands should be those given in Table 7.4.

7.14 RECEIVER NOISE QUIETING [6,12]

Receiver noise quieting is a test of the noise in a speech channel, for varying receiver input carrier levels. Refer to Figure 7.7, which is a typical noise quieting curve. When the receiver carrier level is below the noise threshold, no signal can be received, as

Table 7.4

System capacity N	Limits of band occupied by telephone channel, kHz	Frequency limits of baseband, kHz	Center frequencies of noise-measuring channels, kHz		
			Below	Above	
				(a)	(b)
24	12 – 108	12 – 108	10	116 or 119	?
60	12 – 252	12 – 252	10	304	?
	60 – 300	60 – 300	50	331	?
120	12 – 552	12 – 552	10	607	600
	60 – 552	60 – 552	50	607	600
300	60 – 1300	60 – 1364	50	1499	1549
	64 – 1296				
600	60 – 2540	60 – 2792	50	3200	3250
	64 – 2660				
960	60 – 4028	60 – 4287	50	4715	4765
900	316 – 4188	60 – 4287	270	4715	4765
1260	60 – 5564	60 – 5680	50	6199	6300
	60 – 5636				
1200	316 – 5564	60 – 5680	270	6199	6300
1800	312 – 8204	300 – 8248	270	9023	9073
	316 – 8204				
2700	312 – 12388	300 – 12435	270	13627	13677
	316 – 12388				

it is completely masked by the noise. Section 6.1 describes this threshold, and the equation defining it is again given here:

$$\left.\begin{array}{l} \text{Noise threshold} = -174 \text{ dBm} + 10 \log B + F \\ = 0 \text{ dB carrier/noise} \end{array}\right\} \qquad (6.1)$$

where B = the receiver IF bandwidth in Hz.
F = the receiver noise figure in dB.

As the signal strength is increased in power, it reaches the noise threshold to give a 0 dB carrier-to-noise ratio. As the signal strength is increased further, there is a decibel for decibel improvement in the reduction of the noise. This is indicated on the quieting curve by the upper dashed portion. The signal strength is increased to 10 dB or in some cases 12 dB above the noise threshold, and it triggers off a substantial improvement in the noise reduction. The improvement can be up to 20 dB, and is indicated by the lower dashed portion on the quieting curve. This improvement is the FM improvement threshold (Section 6.1), and is given by equation 6.2. Again for convenience, this equation is repeated here:

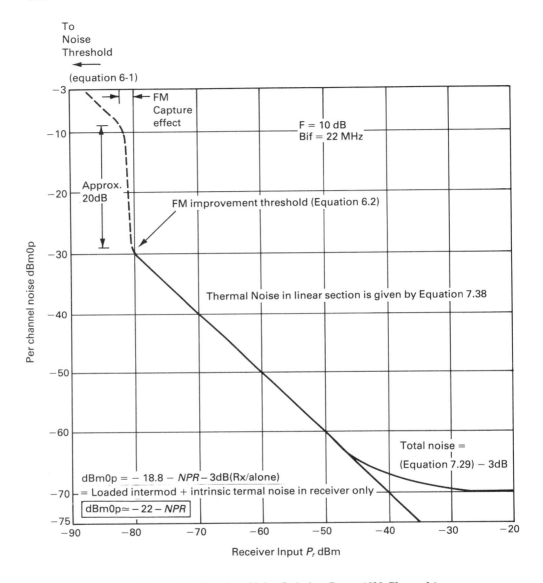

Figure 7.7 *Receiver Noise Quieting Curve (600 Channels)*

$$\left. \begin{array}{l} \text{FM improvement threshold} = -174 \text{ dBm} + 10 \log B + F + 10 \\ \text{Carrier-to-noise ratio} = 10 \text{ dB} \end{array} \right\} \quad (6.2)$$

A received carrier power stronger still produces, above the 20 dB improvement, again a decibel for decibel reduction in the noise. This is shown by the solid line in Figure 7.7. As the signal level increases, the noise improvement would also continue to increase if it were not for the receiver front-end amplifiers beginning to saturate. The saturation means that a further increase in signal into the receiver does not produce an output which is linearly related to that input. The result is intermodulation products. This is indicated in Figure 7.7 by the levelling off of the curve, which is asymptotic to the horizontal line representing the sum of the intermodulation and intrinsic (equipment) thermal noise. The reason why the receiver noise quieting curve is important in any in-station microwave radio tests is that it permits the isolation of equipment which may be causing a high noise reading in the system tests.

The system or white noise end-to-end baseband tests produce results on how the system is performing in respect of noise. The system comprises the transmitter, the receiver, and the radio path. If the results of these tests are unduly high, some action must be taken to isolate the offending equipment. The receiver noise quieting test assists in doing just that. It provides a means to determine whether or not the receiving equipment is operating correctly, or that the level of noise received is within specifications. If this test indicates that the receiving equipment is working to specifications, and there exists a high system noise, then it must be the transmitter or the radio path, and not the receiver that is causing the problem. In addition the quieting curve can indicate, if a fault exists, the item of equipment in the receiver which is causing the problem. i.e.:

1. If the top of the line curves too fast against the theoretical curve, then:
   (a) The receiver gain is lower than normal; or
   (b) The IF bandwidth is too large.
2. If the straight portion of the line departs from that calculated, then:
   (a) The receiver front end is noisy; or
   (b) The IF amplifier is noisy.
3. If the bottom portion of the line does not meet the specification, then the intermodulation of the system is excessive and may be caused by:
   (a) Excessive non-linearity; or
   (b) Excessive group delay; or
   (c) Interference.

To calculate the straight line portion of Figure 7.7, or to derive the noise in a speech channel resulting from the receiver equivalent input noise, for RF or carrier inputs above the FM threshold, equation 5.49 is used. That is at 17°C:

$$(S/N) \underset{unweighted}{\text{ratio}} = P_r + 20 \log (\Delta f/f_m) - F + D + 136.06 \, \text{dB} \qquad (5.49)$$

where P_r = the received carrier in dBm.

Δf = the RMS frequency deviation of the telephone channel due to a test tone.

f_m = the translated channel centre frequency in the baseband spectrum.

D = the de-emphasis factor in dB.

F = the receiver noise figure in dB.

To derive the other noise units from S/N, equations 7.9, 7.10 and 7.11 may be used (note that equation 1.22 gives the rounding-off of dBm0p $=$ dBrnC0 $-$ 90). Hence:

$$S/N_{(unweighted)} = -2.5 - N \text{ dBm0p} \tag{7.11}$$

Thus, noise N in dBm0p in a derived speech channel for receiver input level is given by

$$N = -P_r - 20 \log (\Delta f/f_m) + F - D - 138.56 \text{ dBm0p} \tag{7.38}$$

Similarly,

$$S/N = 88.5 - N \text{ dBrnC0} \tag{7.10}$$

Thus, noise N in dBrnC0 in a derived speech channel for receiver input level is given by

$$N = -P_r - 20 \log (\Delta f/f_m) + F - D - 47.56 \text{ dBrnC0} \tag{7.39}$$

Example

Determine the minimum acceptable receiver carrier level for a 600-channel system, if the receiver noise figure is 10 dB and the emphasis 4 dB, with a 24 MHz IF bandwidth. The maximum allowable limit on the noise out of the telephone channel is -35 dBm0p.

From equation 7.38 the value of P_r is

$$P_r = +35 + 22.1 + 10 - 4 - 138.56$$
$$= -75.46 \text{ dBm}$$

The FM improvement threshold is given by equation 6.2, and with the above values is:

$$\text{FM improvement threshold} = -174 + 10 \log 24 + 10 + 10$$
$$= -80.20 \text{ dBm}$$

As can be seen from the above results, if the noise out of the derived speech channel is the decisive factor, the minimum carrier level received is 4.74 dB higher than the FM improvement threshold. Table 7.5 gives the 20 log $(\Delta f/f_m)$ factors for the top baseband channel centre frequency.

As demonstrated in the example above, the FM improvement threshold is not necessarily the minimum acceptable received carrier. This is important to note, because it is the minimum received level which is the determining factor when considering the fade margin of the system. The *fade margin* is the amount the normal signal can fade before the radio system becomes inoperative.

Table 7.5[13]

Number of system channels	Δf	f_m (upper channel)	$20 \log (\Delta f / f_m)$
12 (type A)	35	58	−4.4
24 (type A)	35	106	−9.6
60	200	298	−3.5
120	200	550	−8.8
300	200	1298	−16.2
600	200	2538	−22.1
900/960	200	4026	−26.1
1260	200	5634	−29.0
1800	140	8202	−35.4
2700	140	12386	−38.9

Usually S/N out of a flat channel is specified as being 30 dB or better. If 30 dB is taken as the lower limit, then using the method outlined in the example, together with equation 7.38 or 7.39, the minimum acceptable received carrier can be determined. If the fade margin is specified as being not less than 35 dB, the transmitter power can be calculated using the information presented in Chapter 8. Four commonly used IF bandwidths are presented in Table 7.6.

Table 7.6

System channel number N	IF bandwidth, MHz
300	22
600	22
* 600 + TV	32
960	40
1260	40
1800	40

* The case where a TV channel is transmitted together with voice channels.

EXERCISES

1. It is intended that a noise generator be set up to perform noise tests across a link, which is designed to carry 1800 telephone channels. Assuming a baseband dBr level of −33 dBr, what should be the output power from the noise generator to simulate the live traffic composite baseband signal? (*Answer* −15.45 dBm)

2. What would be the cut-off frequencies of the high and low-pass filters used in the noise generator of Question 1? Assume testing is done on a non-active system. (*Answer* Refer to Table 7.1; H.P. = 316 ± 5 kHz, L.P. = 8160 ± 75 kHz)

3. If the NPR measured for an 1800-telephone-channel system was 45 dB in the 2438 kHz slot, what would the unweighted S/N be? (*Answer* 61.3 dB)

4. Using the above questions and answers; answer the following question: If a selective level meter was placed on the receiver baseband, and tuned to 2438 kHz, in the 3.1 kHz

* The case where a TV channel is transmitted together with voice channels.

bandwidth measuring mode, what would be the level of the noise read in dBm? (*Answer* −94.3 dBm)

5. Determine the echo distortion noise in pW for an 1800-channel system, if the waveguide loss is 5 dB, and the equipment return loss is 20 dB. The antenna plus waveguide return loss is 26 dB. (*Answer* 22.9 pW)

6. Assuming a de-emphasis factor of 4 dB, and a receiver noise figure of 10 dB, what would the noise in dBrnC0 be, in a derived speech channel, for a received carrier level of −60 dBm? The system channel capacity is 1800 channels. (*Answer* 53.84 dBrnC0)

REFERENCES

1. CCITT Recommendation G233, Orange Book, Vol. III–1, *Line Transmission*. VI Plenary Assembly, 1976.
2. CCIR Recommendation 380–3.
3. Medhurst, R. G., 'Echo-distortion in Frequency Modulation', *Electronic and Radio Engineer*, July 1959, p. 253.
4. CCIR Recommendation 399–3 (1978).
5. Tant, M. J., *The White Noise Book* (Marconi Instruments, 1974).
6. Microwave System Design: Noise Performance, Part 1 – Origins of Noise, Motorola R38–5–5, 1974.
7. Ratcliffe, P. M., NPR–How Accurate is it?, *Marconi Instrumentation*, Vol. 13, No. 2, 1971.
8. Tant, M. J., Further Advances in N.P.R. Measurement, *Marconi Instrumentation*, Vol. 13, No. 5, 1971.
9. Farinon LD–G Multiplex procedures, Farinon Electric, 1691 Bayport Avenue, San Carlos, California, 94070.
10. CCIR Recommendation 398–3.
11. Marconi White Noise Test Set – OA 2090B, Equipment handbook.
12. White, R. F., *Engineering Considerations for Microwave Communications Systems* (GTE Lenkurt, San Carlos, California, 1975).
13. *Reference Data for Microwave System Engineers* (Telettra Telefonia Elettronica e Radio SPA, Milano, 1976).

8 LINE-OF-SIGHT (LOS) RADIO LINK ENGINEERING

8.1 INTRODUCTION

A radio link system may be defined as one whose signals follow a straight line or 'line-of-sight' (LOS) path. The signal's propagation is affected by the 'free space' attenuation, objects within the path and precipitation. In the transmission of speech channels, the baseband width has to increase in accordance with the number of channels, or the system channel capacity. This in turn means a higher RF carrier. In LOS microwave systems the RF carrier is greater than 150 MHz, and the transmission of information is accomplished using angle modulation (frequency or phase modulation) rather than amplitude modulation. In the engineering of a radio link system, there are usually four sets of calculations. These are:

1. Frequency study and interference calculations
2. Antenna height calculations
3. Path calculations
4. Performance calculations

These calculations are interrelated to some extent. In this chapter the last three sets of calculations will be considered.

8.2 ANTENNA HEIGHT CALCULATIONS

It is assumed that, before these calculations are started, the type of traffic, the amount of traffic and the reliability specification have been laid down, and that the finance available for the project has been specified. Some idea of the type of equipment to be used has probably been thought about, and the frequencies for the system have either been allocated, or are pending until the antenna height calculations have been completed.

8.2.1 Sites

From maps with contours at 50 or 100 ft intervals a fair idea can be gained about the routing of a radio link system. The repeater, or terminal, sites are made as high as practicable and without any unnecessary obstructions, so that line of sight to the next repeater station from any particular station is possible.

8.2.1.1 Constraints

One constraint imposed upon the selection of suitable sites is the interference which may arise from links within the same system operating on the same frequency. This has to be considered when the initial site planning is being done. Another constraint is reflection paths between the repeater or terminal sites. If the radio path has a stretch of water, such as a lake, or a flat marshy stretch midway between two sites, link fading may become more noticeable than with an alternative route having no water at the link mid-point. Reflections causing fading occur not only in the vertical plane, but also in the horizontal. Large buildings or obstructions either side of the link path mid-point within the specified Fresnel radius can also instigate reflections which produce link fading. The cost of building the tower at a particular site is another factor. As the cost of the tower increases in proportion to the square of the height, the choice of a particular site may not be warranted if other higher sites are available, unless there is a very good reason for it.

8.2.1.2 Field Survey

Once the sites have been tentatively agreed upon, as the places where the towers will be erected, or where existing towers will be used, a field survey is usually carried out. The field survey covers the following considerations:

1. A full description of each site by its latitude and longitude to the nearest second, as well as a physical description of the site.
2. The existence of access roads if any, and the degree of difficulty due to the type of ground (granite, sandstone, etc.) of making such access roads.
3. Weather conditions such as wind velocities (for wind loading calculations in the tower design), formation of ice, rain, temperature, etc.
4. Average height of the trees or foliage, and the height of any high crop of trees or obstruction in the proposed link path, together with their distance from that site.
5. The relationship of sites to aeroplane flight paths, and airports, since aeroplanes can cause anomalous fading due to reflections. Also tower heights may be restricted due to Government regulations affecting the sites.
6. The height of the site above mean sea level or some reference such as LLWS (low-level water spring), and the height of the site after possible levelling.
7. Location of the nearest commercial electric power distribution system, and the addresses of the appropriate authority.
8. The location of nearby operating transmitters and if possible information on their output power and frequencies.
9. The heights of any existing towers nearby, details of any antennas mounted on them, and what authority they belong to.

10. Local building regulations, and environmental regulations.
11. Any other information which may affect the decision whether to confirm the proposed site as one of the final terminal or repeater station sites.

8.3 PATH PROFILING

Path profiling is usually completed before a field survey is made, and adjusted accordingly after the field survey to take the additional factors into consideration which were unknown on the initial attempt. The purpose of profiling is to permit the determination of the antenna heights, and therefore the tower heights. It is advisable to consider the complete microwave system when determining the antenna heights, rather than a single link in isolation, because on one tower there may be more than one antenna, each one at a different height, and with the height of an antenna depending on the height of its complementary far-end antenna.

A path profile is effectively a cross-section of the terrain along the line-of-sight path. Along the vertical axis is measured height in steps of 50 or 100 ft, or 20 or 50 m, and along the horizontal axis is distance, in miles or kilometres. On topological contour maps a straight line is drawn between the two proposed sites, and from the contours the height read off against the distance from a particular site. This information is then plotted on the path profile graph. Due to the refractive index of the air, some bending of the radio beam occurs; hence path profile proforma charts do not have a straight-line horizontal axis, but it is usually curved according to the k-factor. This factor, the refractive index and other aspects of the microwave beam are discussed in Section 8.3.1.

8.3.1 The k-factor[1,2]

Changes in refractive index n of the air of only a few parts in a million can play an important part on the effect of radiowave propagation. As the values of n are so close to unity (typically 1.00035), it is usually more convenient to work in parts per million *above* unity, i.e. in terms of the *refractivity N*, where

$$N = (n-1) \times 10^6 \tag{8.1}$$

Equation 8.1 expresses the relation of the refractive index to the refractivity. The dependence of atmospheric refractivity on the pressure P in millibars, and the temperature T in kelvins together with the water vapour pressure e in millibars, is given by

$$N = 77.6 \, P/T + 3.73 \times 10^5 \, e/T^2 \tag{8.2}$$

This equation is correct to within 0.5 per cent for atmospheric pressures between 200 and 1100 mb, for air temperatures between 240 and 310 k, for water vapour pressure less than 30 mb, and for radio frequencies less than 30 GHz. The term $77.6 \, P/T$ = N_{dry}, and the term $3.73 \times 10^5 \, e/T^2 = N_{wet}$.

The expression for the *maximum* possible (saturated) vapour pressure e_s at an air temperature $t°C$ is:

$$e_s = 6.11 \exp(19.7 \, t/(t - 273)) \tag{8.3}$$

This equation may also be used to relate e and the dew point temperature t_D in place of e_s and t.

For any relative humidity H per cent

$$e = He_s \tag{8.4}$$

It is sometimes preferable to consider the mass of water per unit volume (m) in the air. This mass is referred to as the *water vapour concentration*, and sometimes as the *vapour density* or *absolute humidity*. It is related to water vapour pressure and temperature by:

$$m = 216.7e/T \tag{8.5}$$

Two other parameters are in common use:

1. The *humidity mixing ratio* r, which is the ratio of the mass m_v of water vapour in a volume to the mass m_a of dry air in the same volume. We have:

$$\begin{aligned} r = m_v/m_a &= 0.622e/(P - e) \text{ gram/gram} \\ &= 622e/(P - e) \text{ gram/kilogram} \end{aligned} \tag{8.6}$$

2. The *specific humidity, moisture content* or *mass concentration* q is the ratio of the mass of water vapour in a volume to the mass of moist air in the same volume:

$$q = m_v/(m_v + m_a) = r/(r+1) \tag{8.7}$$

One of the most significant factors in the influence of radio propagation is the large-scale variation of refractive index with height, and the extent to which this changes with time. In practice, the measured median of the mean refractivity gradient in the first kilometer above ground in most temperate regions is about -40 N-units/km. The mean refractivity gradient in the first kilometer height dN is given by:

$$dN = N_s \left[1 - \exp(-1/h_0)\right] \tag{8.8}$$

where N_s = is the surface value of refractivity
$\quad\quad h_0$ = is a 'scale' height.

In temperate climates the monthly median values of N_s and dN vary from about 300 to 350, and -35 to -55 respectively, according to location and season.

The CCIR defines an average atmosphere as one in which $N_s = 315$, and $dN = -40$, i.e.

$$N(h) = 315 \exp(-h/7.36) \tag{8.9}$$

The consequence of the atmospheric refractive index changing with height is that, in terms of geometric optics, radio waves do not propagate in straight lines. For a spherically stratified medium, Snell's law in polar co-ordinates becomes:

$$n(h)(h + a) \cos B(h) = K \tag{8.10}$$

where $n(h)$ = the refractive index at height h above the earth's surface.
 a = the radius of the earth = 6.37×10^6 m.
 $B(h)$ = the ray angle with respect to the horizontal.
 K = a constant along a ray.

For a vertical gradient of refractive index dn/dh, the rays are refracted towards the region of higher refractive index with a radius of curvature r, so that

$$1/r = -(1/n).dn/dh.\cos B \tag{8.11}$$

When the refractive index gradient dn/dh may be assumed constant over a considerable height interval, a geometric transformation is often used to produce models for which either straight rays propagate above a model earth of an 'effective earth radius a_e', or rays of effective ray radius r_e propagate above a flat earth. The former model is more often employed for profiles.

For the straight-ray model:

$$1/a_e = 1/a - 1/r = 1/a + (1/n).dn/dh \cos 0°$$
$$= 1/a + dn/dh \tag{8.12}$$

Hence, if the effective earth's radius $a_e = k\,a$, we have:

$$k = \frac{1}{1 + a.dn/dh} = a_e/a = \frac{1}{1 + a.dN/dh \times 10^{-6}} \tag{8.13}$$

where k = the k-factor, or effective earth-radius factor.
 $\dfrac{dn}{dh}$ = gradient of radio refractive index with respect to height
 $\dfrac{dN}{dh}$ = gradient of refractivity per kilometer and is expressed in N-units/km.

Now $dn/dh = dN/dh \times 10^{-6}$, since $N = (n-1) \times 10^6$

When $dN/dh > -39$ N-units/km, it is said to be 'subrefracted', or bent down *less than normal*, in other words it rises towards the sky.

When $dN/dh < -39$ N-units/km, it is said to be 'superrefracted', or bent down *more than normal*, in other words it bends towards the ground more than normal.

When $dN/dh = -39$ N-units/km, it is said to be a $k = 4/3$ condition. As a large number of experimental values have been obtained for the rate of change of refractivity with height, and the average value has been found to be -39 N-units/km, this value is used to determine the normal refractive path of the radio beam through the atmosphere within 1 km of the earth's surface.

8.3.1.1 Modified Refractive Index
For convenience, the term M, or the modified refractive index, is sometimes used. Its relation to the refractivity N is:

$$M = (n-1 + h/a) \times 10^6 = N + h/a \times 10^6 \tag{8.14}$$

Using 6.37×10^6 m as the radius of the earth a, we find that the rate of increase of M with height is:

$$dM/dh = (dn/dh + 1/a) \times 10^6 = (dN/dh + 157) \ M\text{-units/km} \tag{8.15}$$

dM/dh is negative for gradients of refractivity per kilometer less than -157 N-units/km. This condition exists if ducting is to occur, and is consequently a good indicator to this effect. Atmospheric conditions are classified according to the shape of the M-profile, as shown in Figure 8.1. Curve B is a linear M-profile with standard gradient of 118 M-units/km (that is $dN/dh = -39$ N-units/km, or $k = 4/3$). This is referred to

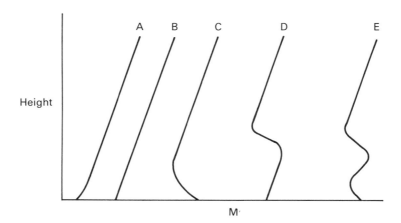

Figure 8.1   Standard Types of M-Profiles

as a *standard surface layer*. Layers with larger gradients are called *substandard surface layers*, and generally appear near the surface of the earth as in curve A. Curves C and D show radio ducts, with C being a surface duct, and D an elevated duct. Curve E shows a more extreme case of multiple ducts, which occur fairly often in practice.

8.3.1.2 Ducting

Negative values of dM/dh are called *M-inversions*. In such cases, the curvature of the rays is concave downwards on a plane-earth (flat-earth) diagram, although their true curvature is greater than that of the earth. Hence rays entering the duct at small angles are bent downwards, and will be trapped in the duct. Figure 8.2 shows how this happens for a surface duct. In the case where the transmitter is located in the duct, the angle of radiation β becomes critical. For any angle less than that shown by ray 1 in Figure 8.2, the signal is trapped within the duct, and may be propagated far beyond the horizon. In extreme conditions, it may be bent downwards so much that it is reflected, and propagated down the duct in a series of hops; see ray 2, Figure 8.2.

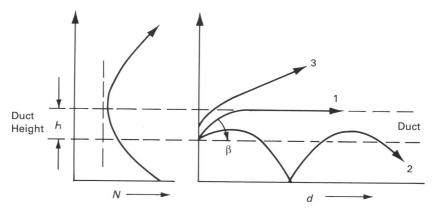

Figure 8.2 *Propagation in a Surface Duct (Courtesy of CCITT, Reference 2)*

Ray 3 illustrates a leakage signal escaping from the duct. A receiver located outside the duct would experience severe attenuation-type fading. It may also happen that several rays, leaving the transmitter at different angles, remain within the duct to cause interference-type fading at a receiver located in the duct. The conditions for ducting to occur are:

1. The refractivity gradient is equal to or more negative than −157 N-units/km – that is a super-refracting condition.
2. The refractivity gradient is maintained over a height of *many wavelengths*. Natural ducts do not have the sharp metallic cut-off of man-made waveguides, but they do have a wavelength cut-off above which waves will not propagate. Since the duct thickness (*t* meters) does not have a sharp limit, the cut-off wavelength (λ_{max} meters) is not sharp, but an indication between them is given by:

$$\lambda_{max} = 2.5 \times 10^{-3} \left(\delta N/t - 0.157\right)^{1/2} t^{3/2} \tag{8.16}$$

where δN is the refractivity change across the duct.

8.3.1.3 Ray Bending and Effective Radius of the Earth

Figure 8.3 shows pictorially the effects of bending on a microwave beam. In (*a*) the line marked $k = 1$ represents the radio path when the dielectric constant does not change with height, i.e. no bending occurs. When the dielectric constant increases with height, the ray bends upward, as shown by the line marked $k = 2/3$. Conversely, when the dielectric constant decreases with height, the ray bends downwards. It may even travel parallel to the surface of the earth as shown by the line marked $k = $ infinity. The line marked $k = 4/3$ represents an intermediate value, the so-called 'standard atmosphere', because it is the value to be expected most often during daylight hours.

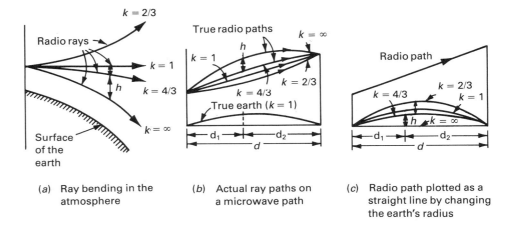

| (a) | Ray bending in the atmosphere | (b) | Actual ray paths on a microwave path | (c) | Radio path plotted as a straight line by changing the earth's radius |

Figure 8.3   Bending of Radio Waves in the Atmosphere (Courtesy of CCITT, Reference 2)

Figure 8.3(*b*) shows what actually happens as the rays bend on a microwave path. When $k = 1$ the ray reaching the receiver travels in a straight line, but for other conditions the true radio path is curved. When the rays curve up, the one reaching the receiver follows a path similar to the one marked $k = 2/3$; the result of this is to reduce the path clearance. Conversely, when the rays curve downward, the one reaching the receiver follows a path similar to the one marked $k = $ infinity; this increases the path clearance. The path for $k = 4/3$ falls half-way between these two extremes.

Figure 8.3(*c*) shows the same radio path, but with the ray drawn as a straight line and the curvature of the earth changed to give the same path clearance as in Figure 8.3(*b*). This method of drawing profiles gives rise to the concept of 'earth bulging', corresponding to the values of $k < 1$. As shown in (*c*), for $k = 2/3$ the earth appears to 'bulge' up into the path more than for the case of $k = 1$ (true earth). Conversely, when $k > 1$, the effective earth's curvature is less; this is called 'earth flattening'. The design range of k normally allowed for is from 2/3 to infinity; values of k outside this range should not occur often enough to be of serious concern. The 'earth bulge' at any point in a radio path is given by:

$$h = 4/51.d_1.d_2/k = 0.0785\ d_1.d_2/k\ \text{m} \qquad (8.17)$$

where d_1 and d_2 are the distances in kilometers to the near and far ends of the path respectively.

Unless special path profiling paper is used with the earth bulge already plotted for a specific k-factor, the actual 'earth bulge' must first be plotted from the horizontal axis, so that the microwave beam can be drawn as a straight line.

8.3.2 Effect of Path Clearance on Radio Transmission

8.3.2.1 Propagation Equation[4,10]

Propagation through the atmosphere is affected by the presence of the earth, causing the received signal to be weaker than expected in free space. The ground acts as a partial reflector and as a partial absorber, and both these properties affect the distribution of energy in the region above the earth. The ratio of received signal strength over plane earth (E) to that which would be received in free space (E_0) is given by:

$$E/E_0 = 1 \qquad\qquad +Re^{jD} \qquad + \ (1-R)Ae^{jD} + ... \qquad (8.18)$$
$$\text{direct wave} + \ \text{reflected wave} + \ \text{ground wave} + \text{secondary effects.}$$

where E = received signal strength over plane earth.
E_0 = received signal strength in free space.
R = coefficient of ground and is approximately equal to -1, when the angle θ between the reflected ray and the ground is small. The magnitude and the phase can be computed from the following equation:

$$R = \frac{\sin\theta - z}{\sin\theta + z} \qquad\qquad (8.18\text{A})$$

where z = $(e_0 - \cos^2\theta)^{1/2}/e_0$ for vertical polarization

$\qquad\quad = (e_0 - \cos^2\theta)^{1/2}$ for horizontal polarization
$\quad e_0 = e - j60\,\phi\lambda$
ϕ is the conductivity of the ground in mhos/meter, and λ is the wavelength in meters, and $j = \sqrt{-1}$.

A = attenuation factor of a surface wave, which depends upon the frequency, ground constants, and type of polarization. It is never greater than unity, and decreases with increasing distance and frequency.

D = the phase difference in radians resulting from the difference in the length of the direct and reflected rays. It is given by the expression below when the distance d between the two antennas is greater than about five times the sum of the two antenna heights.

$$D = \frac{4\pi h'_1 h'_2}{\lambda d} = \frac{f\pi h'_1 h'_2}{75000d}$$

where h'_1, h'_2 = heights of the antennas above the reflecting surface or tangent plane in meters.
d = the length of the path in km.
λ = the wavelength in meters.
f = the frequency in MHz.

The effect of the ground shown in equation 8.18 indicates that ground wave propagation may be considered to be the sum of three principal terms: namely, the direct wave, the reflected wave and the surface wave. The first two types are similar to the propagation of visible light, but the surface wave is pertinent only to microwave propagation. Since the earth is not a perfect reflector, some energy is transmitted into the ground and is absorbed. As this energy enters the ground, it sets up ground currents. The surface wave is defined as the vertical electric field for vertical polarization, and the horizontal electric field for horizontal polarization, that is associated with these ground currents. The practical importance of the surface wave is limited to a region above the ground of about one wavelength over land, and about 5 to 10 wavelengths over sea water, since for greater heights the sum of the direct and reflected wave is larger in magnitude. The surface wave is the principal component of the total ground wave at frequencies less than a few megahertz, but is of secondary importance in the VHF range (30 – 300 MHz); and it can usually be neglected at frequencies above 300 MHz.[11]

When D is greater than about 0.5 radian, and $R = -1$, equation 8.18 reduces to:

$$|E/E_0| = 2 \sin (n\pi)/2 = 2 \sin 2\pi h_1 h_2/d\lambda \qquad (8.19)$$

where n = the Fresnel zone number associated with a given path clearance.
h_1, h_2 = heights of the antennas above ground in meters.

Equation (8.19) is the sum of the direct and reflected waves in Figure 8.4(b) and can be interpreted as ground-produced interference fringes or lobes which cause the field intensity, at a given distance and for a given frequency, to oscillate around the free-space field as either of the two antenna heights is increased. For both the direct and reflected waves, the energy is travelling in the troposphere, where most of the effects of weather take place, so that propagation is affected greatly by meteorological conditions.

Using high-powered transmitters and sensitive receivers, at wavelengths less

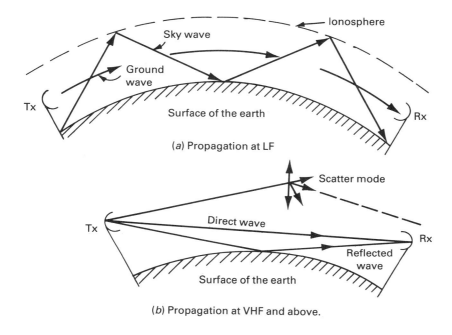

(a) Propagation at LF

(b) Propagation at VHF and above.

Figure 8.4   Effect of Path Clearance on Radio Transmission (Courtesy of CCITT, Reference 2)

than about 1 meter, we may detect usable signals at distances well beyond the optical horizon. This is due to the 'scatter' effect, where radio energy is reflected by changes in the refractive index of the air. Scatter links can operate over distances up to about 600 km.

Further enhancement of signals beyond the horizon can also be caused by ducting, where signals are trapped between layers in the atmosphere and conducted for long distances. This mode of propagation is not to be depended upon for reliable reception, for it is not a normal condition.

8.3.2.2 Huygens' Principle and Fresnel Zones

In path profiling, after the earth bulge has been attended to, another most important aspect is to ensure that the radio path is sufficiently high above any earth obstructions so that the radius, or part of it, clears these obstructions. This is done to prevent any reflected ray from the transmitter reaching the receiver and destructively interfering with the direct-path received signal.

Figure 8.5 shows a radio transmitter Tx, transmitting energy which travels outwards from the source in an expanding wavefront. Huygens' principle states that each element of this primary wavefront acts as a new source of radiation sending out a secondary wavefront, as at P′, P″, etc. The secondary radiation from all elements of the original wave add up to form a new wavefront, each element of which re-radiates in turn. This pattern is repeated indefinitely, so that the field strength at the

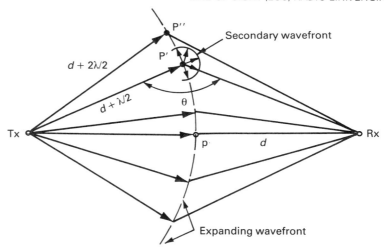

Figure 8.5   Illustration of Huygens' Principle (Courtesy of CCITT, Reference 2)

receiver Rx is the vector sum of the infinite number of tiny wavefronts set up by the transmitting antenna.

At P', only part of the new wavefront will reach the receiver, due to the dispersion of the energy of the new wavefront. The amount of energy received by the receiver depends on the distance from the transmitting antenna, and on the angle θ. For the shortest path, or the direct path d, between Tx and Rx, the angle θ will be 180°, but for any other path will be between 0 and 180 degrees. For any point, such as P', the cosine of the angle θ is a measure of the 'obliquity factor', which defines the magnitude of the component reaching the receiver. The smaller the angle θ, the less the energy received at Rx.

In Figure 8.5 energy arriving via P' will arrive a little later than energy arriving by the direct path d at the receiver Rx, due to the further distance it has to travel. If the path via P' is longer than the direct path d by one half-wavelength $\lambda/2$, the phase difference between the two signals will be 180°, and the two signals will subtract. If the length of the indirect path is increased by another half-wavelength, the signal arriving over this path (via P'') will add in phase with the direct signal. The length of the indirect path can be increased indefinitely, to define paths over which the signal will alternately subtract and add, or reinforce, the direct wave.

Figure 8.6 is another view of the same radio path. As before, the dashed line represents the wavefront expanding from Tx. The point P' again defines the indirect path for which the extra path length is one half-wavelength ($\lambda/2$). If P' is imagined to move around the circumference of the inner circle in Figure 8.6, this will define all possible paths with a length of $d + \lambda/2$. This inner circle is called the *first Fresnel zone*, with a radius F_1 given by

$$F_1 = 31.6 \sqrt{\left(\frac{\lambda d_1 d_2}{d}\right)} = 548 \sqrt{\left(\frac{d_1 d_2}{fd}\right)} \qquad (8.20)$$

where d_1 = the distance to the near end of the path, km.
$\quad\;\; d_2$ = the distance to the far end of the path, km.
$\quad\;\; d$ = $d_1 + d_2$ = total path length, km.
$\quad\;\; \lambda$ = wavelength, m.
$\quad\;\; f$ = frequency, MHz.

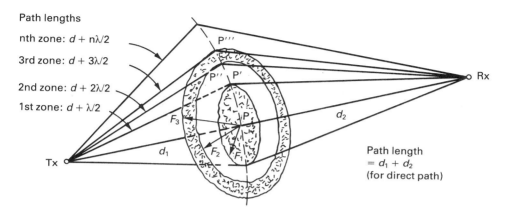

**Figure 8.6   Cross-section of Radio Path showing Fresnel Zones
(Courtesy of CCITT, Reference 2)**

The boundary of the second Fresnel zone is defined by the circle for which the point P″ traces all possible indirect paths with a length of $d + 2\lambda/2$. The radius of this second circle is F_2, where

$$F_2 = (2)^{1/2} . F_1 \text{ m} \qquad\qquad (8.21)$$

Similarly the radius F_n of the nth zone is given by

$$F_n = (n)^{1/2} . F_1 \text{ m} \qquad\qquad (8.22)$$

If we completely define the boundaries of the Fresnel zones, the result is the familiar cigar shaped surface, or ellipsoid, as shown in Figure 8.7. A cross-section, taken at right-angles to the path, shows the concentric circles of Figure 8.6. The corresponding view taken along the path shows a series of ellipses. The inner ellipse has its boundary defined by the radius of the first Fresnel zone. The next ellipse corresponds to the second Fresnel zone, etc. The area of each of the annular rings bounding the different zones, of which one is shown in Figure 8.7 is approximately equal to the area of the adjacent ring (not shown), so that the energy transmitted through each is roughly equal. The contribution to the field strength at Rx, from each zone, is proportional to the area of the zone and its obliquity factor, and is inversely proportional to the path distance.

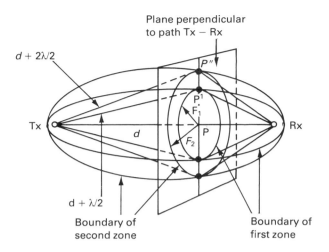

Figure 8.7   Fresnel Zones in Three Dimension (Courtesy of CCITT, Reference 2)

Since the area is nearly the same for each zone, the contributions at Rx from any two adjacent zones will tend to cancel, because of the phase delays. However, because of the obliquity factor, the contributions from the higher-order zones become progressively smaller. The net result is that the total field at Rx, from all zones, is approximately half of that from the first zone alone.

8.3.2.3 Effect of Path Obstructions [3]

The diffraction at obstacles as well as reflections by plane surfaces in the transmission path can be estimated with the aid of the calculated Fresnel zones drawn on the path profile. In practice, obstructions are mainly caused by rocky, or mountainous terrain, and by tall city buildings. In all cases, diffraction attenuation only occurs if more than $0.5 - 0.7$ of the first Fresnel zone is obstructed. Reflections from plane surfaces in the terrain, such as lakes, marshes, buildings, etc., or from the troposphere have an effect only if they reach the receiving antenna. The received signal at a particular frequency is cancelled if the amplitudes of the reflected and direct path signals are of equal magnitude and have a phase difference between them of 180°, or an odd multiple of 180°. Reflection from a surface however, does produce a phase discontinuity of about 180°; hence reflected signals reaching the receiving antenna are to be avoided if possible.

Figure 8.8 shows the variation in signal level as a function of clearance for four theoretical types of radio path. The Fresnel zones are plotted in terms of the ratio of the actual path clearance F to the first Fresnel zone radius F_1. The first two terrain types are: plane, or flat, earth and a smooth spherical surface, for both of which $R = -1$, that is the negative value for the reflection coefficient R. This indicates that a 180° phase shift has taken place for reflected energy at low angles. The third type is shown by the curve marked $R = -0.3$; this represents a more practical situation which may be experienced on many paths. In this, there is, on the average, a reflected signal with a 180° phase shift, but with an amplitude only 0.3 times that of

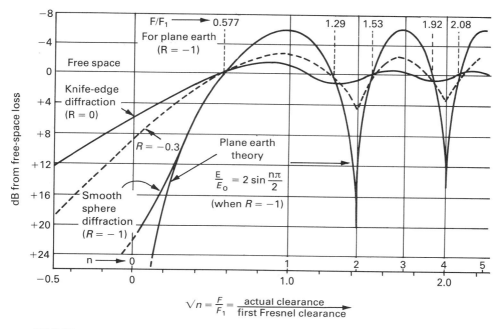

DO NOT USE THESE CURVES IN SYSTEM DESIGN AS THEY ARE REPRESENTATIVE ONLY.

Figure 8.8   Effect of Path Clearance on Radio Transmission (Courtesy of CCITT, Reference 2)

the incident signal, alternatively, there is a reflected signal whose amplitude is nearly equal to the incident signal, but whose phase is not shifted by 180°. The criterion for smoothness, and thus reflection, depends upon the Rayleigh criterion[2], which defines a surface to be smooth when;

$$H = 7.2 \, \lambda/a \tag{8.23}$$

where H = the height between the top and bottom of the surface irregularities, m.

λ = the wavelength of the radio carrier signal, m.

a = the angle at which the beam is incident from the horizontal, degrees.

This criterion can only be a general guide, since true specular reflection applies only when H and a tend towards zero. Equation 8.23 does however indicate that a surface is considered smooth to a radio beam, but rough to the eye, when the wavelength of the carrier is large, or the carrier frequency is low.

Finally, the last curve in Figure 8.8 when $R = 0$, represents knife-edge diffraction. Note that a free-space signal is reached at a path clearance of approximately 0.6 times the first Fresnel zone radius, for all four types of terrain. All the existing rules for determining necessary path clearance for good radio transmission are based upon this fact.

PLANE EARTH

For the case discussed in Section 8.3.2.1, the attenuation relative to free space can be given by:

$$\text{Attenuation} = -10 \log (E/E_0)^2 \tag{8.24}$$

Hence from equation 8.18 for $R = -1$, and for small grazing angles θ,

$$\text{Attenuation} = -6 - 10 \log \sin^2 (2h_1'h_2'\pi/\lambda d) \tag{8.25}$$

Alternatively, the attenuation may be expressed in terms of the difference between the sum of the reflected path distances and the direct path (dr), where

$$dr = \lambda D/2\pi = r_1 + r_2 - r_0 \tag{8.26}$$

Thus

$$\text{Attenuation} = -6 - 10 \log \sin^2 (\pi dr/\lambda) \tag{8.27}$$

where λ = wavelength of the radio signal carrier.

However, this diffraction loss, which is the attenuation with respect to free space, can also be expressed in terms of the clearance of the smooth earth, or any other obstacle, from the first Fresnel radius. From equation 8.22, the clearance F from the centre of the beam to the obstruction (see Figure 8.9) can be considered to be within the radius of the nth Fresnel zone; hence

$$\sqrt{n} = F/F_1 \tag{8.28}$$

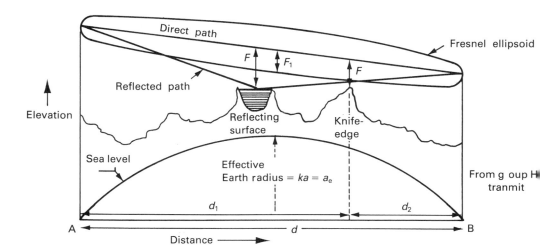

Figure 8.9   Path Profile

This provides the alternative expression

$$E/E_0 = 2 \sin\left[\tfrac{\pi}{2}(F/F_1)^2\right] = 2 \sin(n\pi/2) \tag{8.29}$$

resulting in the attenuation with respect to free space of:

$$\text{Attenuation} = -6 - 10 \log \sin^2(n\pi/2)$$
for the ideally reflecting flat, or plane, earth case. (8.30)

Equation 8.30 is shown in Figure 8.8, where the signal rises from zero (attenuation is infinity) at grazing incidence ($n = 0$), to its free-space level at $(n)^{1/2} = 0.577$, where the path length difference dr, as calculated from equation 8.26, is $\lambda/6$. When the reflection coefficient is not -1, the first free-space crossover point is not reached at $F/F_1 = 0.577$. When $R = 0$, the first free-space crossover point is reached at $F/F_1 = 0.707$. This is the case when the signal strikes well-wooded obstructions, such as forests, which do not act as knife edges. The initial analysis of path-test results always assumes a clearance of either $0.55F_1$ or $0.71F_1$ at the point where free space is first reached. Reference 6 gives details of VHF propagation in tropical forests.

SMOOTH SPHERE FOR WITHIN-THE-HORIZON RADIO PATHS
This case (Figure 8.10) is a modification of the plane earth condition; it involves both reflection and diffraction over a spherical surface. The reflection is assumed to occur on a plane tangent to the surface of the earth at the point of reflection. For this case, the path length difference dr is measured with reference to the tangent plane, and the modified antenna heights h' and h' are used, so that, from CCIR Report 338–3, for small grazing angles θ, and dr taken as approximately equal to $2h_1'h_2'/d$, together with equation 8.27, we have:

$$dr \simeq 2h_1 \sin(\theta) = h_1[(\phi_h^2 + 4h_1/(3a_e))^{1/2} + \phi_h] \tag{8.31}$$

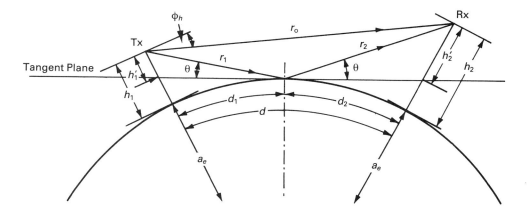

Figure 8.10   Geometrical Relationships for Radio Path – within the Horizon

where ϕ_h = the angle of elevation of the direct ray r_0 relative to the horizontal at the lower antenna, h_1, and may be positive or negative. If $\phi_h = 0$, $d_1 = 2h_1/(3(\theta))$, for the following conditions;

$h_1 \leqslant h_2$, $h_1 \leqslant 9a_e\,\theta^2/2$, and θ is small.

For equal antenna heights over a spherical earth of effective radius a_e;

$$dr = d(\sec\theta - 1) \tag{8.32A}$$

The greatest distance $d = d_0$, for which the attenuation = 0, may be obtained graphically from

$$2h_1^2/d_0 - h_1 d_0/(2a_e) + d_0^3/(32a_e^2) = \lambda/6 \tag{8.32B}$$

SPHERICAL EARTH FOR BEYOND-THE-HORIZON RADIO PATHS
For determining the diffraction attenuation A, relative to free space beyond the horizon, an approximate formula may be used over a smooth earth (CCIR Report 715, 1978):

$$A = G(X_0) - F(X_1) - F(X_2) - 20.5\ \text{dB} \tag{8.33}$$

$$\begin{aligned}
\text{where } X_0 &= d_0.B_0 & (8.34)\\
X_1 &= d_1.B_0 & (8.35)\\
X_2 &= d_2.B_0 & (8.36)\\
B_0 &= 670\,(f/a_e^2)^{1/3} & (8.37)
\end{aligned}$$

d_0 = total distance between the two antennas, km.
d_1 = distance from A terminal to its radio horizon = $(2a_e\,h_1)^{1/2}$ (8.38)
d_2 = distance from B terminal to its radio horizon = $(2a_e\,h_2)^{1/2}$ (8.39)
h_1, and h_2 are the antenna heights in km, a_e is the effective earth's radius in km, and f is the frequency in MHz.

The functions $G(X_0)$, $F(X_1)$, and $F(X_2)$ are plotted in Fig 8.11. The factor K in this figure depends on the frequency and the electrical characteristics of the earth. The curve labelled $K = 0$ is appropriate for horizontal polarization over water, or good and average grounds, at frequencies of 100 MHz or above. For horizontal polarization over poor ground, this same curve is applicable for frequencies of 600 MHz or above, and approximately so (within 2 dB) for frequencies down to 100 MHz. For vertical polarization over land, the curve labelled $K = 0.01$ is applicable for frequencies of 600 MHz or above. For vertical polarization over sea water, the curves $K = 0.01$, 0.1, 0.3 and 1.0 are applicable for frequencies of 3300 MHz, 120 MHz, 30 MHz and 7.5 MHz respectively, or less.

The error in the attenuation A will be less than 1 dB if

$$X_0 - X_1.d(X_1) - X_2.d(X_2) > 320 \tag{8.40}$$

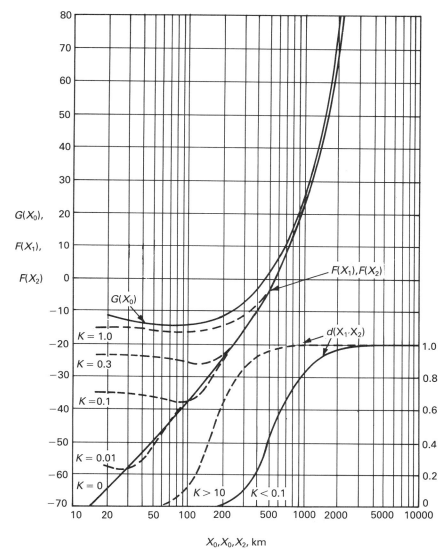

Note: For large values of X, $F(X) \approx G(X) - 1.356$
where $G(X) \approx 0.0575104\ X - 10 \log X + 2.066$

Figure 8.11 *The Functions of G and F in Equation 8.33 (Courtesy of CCIR, Reference 12)*

for horizontal polarization, or for vertical polarization on overland paths. For the error in A to be 1.5 dB or less for vertical polarization on oversea paths, the limit indicated as the right-hand side of equation 8.40 must be 320 for frequencies of 600 MHz or more ($K \leqslant 0.03$), 335 for 120 MHz ($K = 0.1$), and 115 for 7.5 MHz or less ($K \geqslant 1.0$).

CCIR Report 715, 1978 provides a series of nomographs to permit readily the diffraction attenuation for a spherical earth beyond line of sight for both land and sea links, with horizontal and vertical polarizations.

DIFFRACTION ATTENUATION DUE TO OBSTACLES AND IRREGULAR TERRAIN – KNIFE EDGE[12]

Usually in a propagation path where there are mountains, rock outcrops, buildings, etc., there can be at least one obstacle encountered. This obstacle may not warrant the increase in antenna heights which would be necessary to clear it, if the obstruction loss can be determined. To make such a calculation of obstruction loss it is necessary to idealize the form of the obstacle, or obstacles, either by assuming a knife edge of negligible thickness, or a thick smooth obstacle with a well-defined radius of curvature R at the top. As real obstacles are seldom encountered in the same forms as the idealizations, the calculations provided in this and the next section should be regarded as only approximate. These calculations are also valid if the wavelength is fairly small in relation to the size of the obstacle as given by equation 8.23, or if the carrier frequency is greater than 30 MHz.

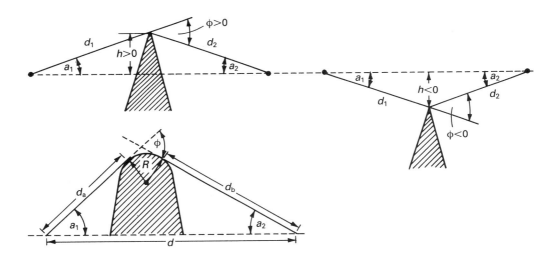

Figure 8.12 Diffraction over Obstacles (Courtesy of CCIR, Reference 12)

In the extremely idealized case shown in Figure 8.12, all the geometrical parameters are lumped together in a single dimensionless diffraction parameter denoted by v. This diffraction parameter assumes a variety of forms according to the geometrical parameters selected:

$$v = <n^{1/2} \tag{8.41}$$

$$v = h[2(d_1 + d_2)/\lambda d_1 d_2]^{1/2} \tag{8.42}$$

$$v = \phi \left[2d_1 d_2/\lambda(d_1 + d_2)\right]^{1/2} \tag{8.43}$$

$$v = [2h\phi/\lambda]^{1/2} \quad (v \text{ has the sign of } h \text{ and } \phi) \tag{8.44}$$

$$v = [2d \, a_1 a_2/\lambda]^{1/2} \quad (v \text{ has the sign of } a_1 \text{ and } a_2) \tag{8.45}$$

where h = the height of the top of the obstacle above the straight line joining the two ends of the path. If the height is below this line, h is negative.

d_1 and d_2 = the distances of the two ends of the path from the obstacle.

d = the length of the path: $d = d_1 + d_2$.

ϕ = the angle of diffraction in radians; its sign is the same as that of h. The angle ϕ is assumed to be less than about 0.2 radian or 12 degrees.

a_1 and a_2 are the angles between the top of the obstacle and one end as seen from the other end. Both are of the sign of h.

h, d, d_1, d_2 and λ (the wavelength) are all expressed by the same unit.

Figure 8.13 gives, as a function of v, the loss in decibels caused by the presence of the obstacle. For v more positive than -1, an approximate value can be determined from the expression:

$$A(v) = 6.4 + 20 \log \left[v + (v^2 + 1)^{1/2}\right] \text{ dB} \tag{8.46}$$

The error is within 0.5 dB.

DIFFRACTION OVER OBSTACLES AND IRREGULAR TERRAIN – ONE ROUNDED OBSTACLE

A propagation path with a single isolated rounded-top obstacle which provides the horizon for both terminals may be considered as a single diffracting rounded knife-edge between the terminals. For $\phi \geqslant 0$, the diffracting attenuation A with reference to free-space path loss may be evaluated from:

$$A = F(v) + G(\rho) + E(X) \tag{8.47}$$

(a) The Fresnel-Kirchhoff loss $F(v)$ is given in Figure 8.13 as a function of the diffraction parameter v, where

$$v = 2 \sin(\phi/2).[2(d_a + R.\phi/2)(d_b + R.\phi/2)/\lambda d]^{1/2} \tag{8.48}$$

In equation 8.48 λ is the radio carrier wavelength, d_a and d_b are the distances from each terminal to their horizons on the terrain feature, R is the effective radius of curvature for the terrain feature between horizons as determined by the product of the geometrical radius and the earth-radius factor k. All distances and lengths are in the same units. For $R = 0$, equation 8.48 reduces to equation 8.43.

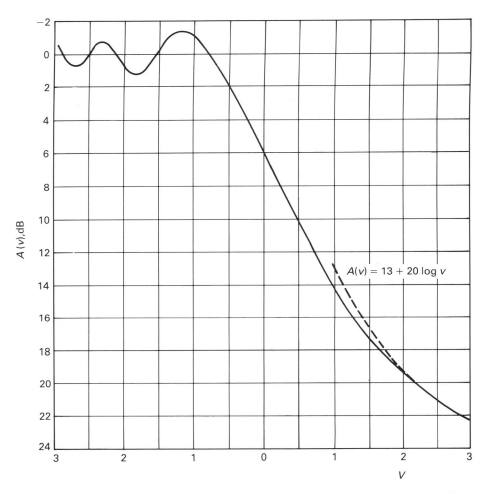

Figure 8.13   Knife-Edge Diffraction Loss (Courtesy of CCIR, Reference 12)

(*b*) $G(\rho)$, the loss in decibels for incidence upon the curved surface may be determined from:

$$G(\rho) = 7.192\rho - 2.018\rho^2 + 3.63\rho^3 - 0.754\rho^4 \tag{8.49}$$

where ρ is determined from:

$$\rho^2 = \frac{(d_a + d_b)}{d_a d_b} \cdot \frac{1}{[(\pi R/\lambda)^{1/3} \cdot 1/R]} \tag{8.50}$$

(*c*) $E(X)$, the dB loss for propagation along the surface between the horizons, is given by:

$$E(X) = \begin{cases} 12X \text{ for } X < 4 \\ \\ 17.1X - 6.2 - 20 \log X \text{ for } X \geqslant 4 \end{cases} \tag{8.51}$$

where $X = [\pi R/\lambda]^{1/3} \phi$ (8.52)

For $R = 0$, ρ and X go to zero and equation 8.47 reduces to the first term. For $\phi = 0$, $X = 0$ and equation 8.47 gives the loss for grazing incidence upon the rounded knife-edge or obstacle.

The solution of equation 8.47 results from the first term of an asymtotic formulation of the field in inverse powers of s. This assumes that the radio transmitting and receiver antennas are both remote from their horizon on the diffraction terrain feature. Thus:

$$s = (2\pi/\lambda).[d_a d_b]^{1/2} \geqslant 10 \tag{8.53}$$

and

$$[(\pi R/\lambda)^{1/3} /R]^2.d_a d_b \geqslant 0.1 \tag{8.54}$$

The above conditions are normally met for frequencies > 30 MHz and with any polarization, and for the values of ground constants normally encountered for irregular terrain. The solution is also applicable to a spherical earth, provided that the equations 8.53 and 8.54 are satisfied. An exception is the special case of vertical polarization over sea water, unless the frequency is more than 1 GHz.

8.3.3 Effect of Path Reflections

The point of reflection on a smooth path moves as k changes. Figure 8.14(a) shows the actual path of the microwave beam for three values of k, namely 2/3, 4/3 and infinity. The point of reflection for these three values is plotted, giving points A, B and C. These points are shown on a corrected earth basis in Figure 8.14(b). The dashed line through A, B and C shows how the reflection moves along the path as k varies. By drawing a line parallel to the direct ray, tangent to the true earth, the point G is determined at grazing incidence. By projecting this point on to the distance axis, d_1 can be read directly. If these four points are plotted on a path profile, and joined using a french curve, the value of d_1 can be read off for any other k. This provides a quick method of finding the reflection point, once the four points A, B, C and G are known. To determine initially these four reflection points (or more if necessary), two different methods may be used. These methods are to use nomographs[2], and to solve a cubic equation. Presented below is the method of calculating the reflection point distance d_1 from a given terminal, for any value of k.

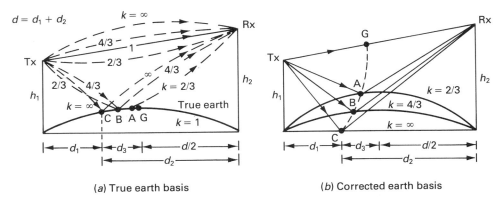

(a) True earth basis   (b) Corrected earth basis

Figure 8.14   Change in d_1 as a Function of k (Courtesy of CCITT, Reference 2)

8.3.3.1 Solution of a Cubic Equation to Determine the Reflection Point

METHOD 1

Refer to Figure 8.10. Given h_1 and h_2 the heights of the antennas above mean sea level, together with the distance between the two terminals d, and the k-factor, the reflection point distance d_1, from terminal Tx can be determined by solving directly the cubic equation:

$$d_1^3 - 3/2dd_1^2 + (1/2d^2 - a_e(h_2 + h_1))d_1 + a_eh_1d = 0 \tag{8.55}$$

where $a_e = k \times$ earth's radius $= k \times (6370 \text{ km})$.

METHOD 2

Three parameters are first evaluated. These are:

$$r = 6.37 \ (k/4)d(h_2 - h_1) \tag{8.56}$$

$$t = d^2/12 + 8.5(k/4)(h_1 + h_2) \tag{8.57}$$

$$\alpha = \cos^{-1}[r/t^{\frac{3}{2}}] \text{ degrees}$$

$$d_3 = 2(t)^{1/2} \cos(\alpha/3 + 240) \tag{8.58}$$

As $$d_1 = d/2 + d_3 \tag{8.59}$$

we have $d_1 = d/2 + [2(t)^{1/2}\cos(\alpha/3 + 240)]$ $\tag{8.60}$

where h_1 is the height of the antenna above sea level. If the reflection point is not at sea level but at some height h_r, the Tx terminal is on land, which is at a height H_a above sea level, and the antenna is at a distance h_a up the tower, then:

$$h_1 = h_a + H_a - h_r \tag{8.61}$$

Similarly,

$$h_2 = h_b + H_b - h_r \tag{8.62}$$

h_1, and h_2 are in meters, whilst d, d_1, d_2 and d_3 are in kilometers.

8.3.4 Radio Path Profiling and Approximations

Each radio path requires a profile, showing the elevation of all points. All obstructions and possible reflecting surfaces must be plotted on the profile as accurately as possible. These elevations are plotted up from the line representing the curvature of the earth, as in Figure 8.15. Trees are usually represented by arrowheads or vertical

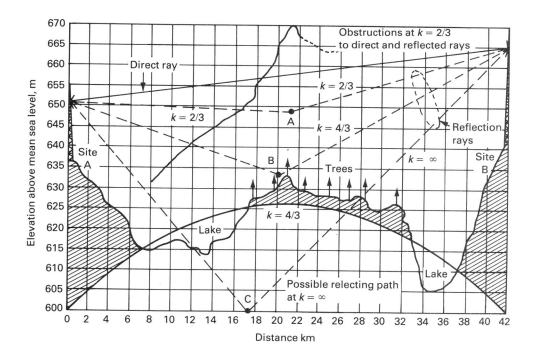

Figure 8.15 Plotting Profiles on Linear Graph Paper

arrows scaled accordingly and placed in the positions at which they occur on the profile. Figure 8.15 illustrates the most convenient way of plotting profiles, using rectangular graph paper. The line representing the curvature of the earth is drawn using a template, or if no template is available, drawn using the earth bulge equation 8.17. Figure 8.15 is plotted at $k = 4/3$ with the main obstructions also shown at $k = 2/3$, using dashed lines. A straight line between the antennas represents the path of the radio beam, and clearance over obstructions or reflecting surfaces can be scaled directly from the profile. Profiles are sometimes drawn on a 'flat earth' basis ($k =$ infinity) but standard practice is to draw profiles for $k = 4/3$, especially when analyzing propagation tests.

Figure 8.16 shows a common method of plotting profiles, using printed forms

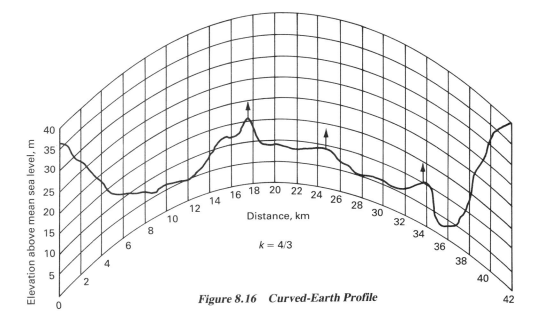

Figure 8.16   Curved-Earth Profile

with a curved-earth grid already printed on them. At higher frequencies than VHF, and for better accuracy, the method of Figure 8.15 is normally used. Figure 8.17 illustrates some of the approximations involved in plotting the profiles on rectangular graph paper. These errors are:

1. The surface of the earth is plotted as a parabolic arc, rather than as a circular arc, because of equation 8.17, which is rearranged to give:

$$(h - d^2/(51k)) = -4(d_1 - d/2)^2/(51k) \tag{8.17}$$

As can be seen, this is the equation of a parabola, with focus $1/(51k)$ and translated along the x-axis by $d/2$, and the y-axis by $d^2/(51k)$. The error is less than 0.3 per cent.

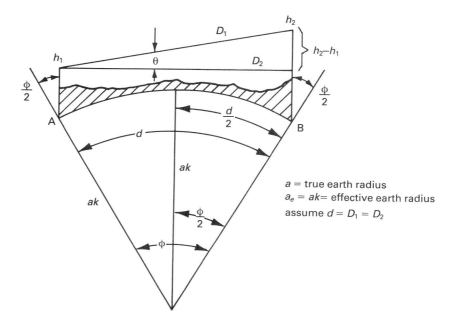

Figure 8.17 *Errors in Plotting Profiles on Linear Graph Paper (Courtesy of CCITT, Reference 2)*

2. The elevations are plotted along vertical lines, and not along the radial line from the center of the earth. This error is small because the angle $\phi/2$, even on a 50 km path, is:

$$\phi/2 = (d/2)1/a_e = 0.5397(d/2)(1/k) \text{ minutes} \qquad (8.63)$$

So for 50 km the angle $\phi/2 = 13.5/k$ minutes.

3. The true length of path d is not the distance measured along the surface of the earth. The length of path used in calculations is usually measured from a map, or calculated from map co-ordinates, and not corrected for the difference in elevation $(h_2 - h_1)$. The error is small, because the angle θ is seldom larger than 2 degrees. As a result it is assumed that $d = D_1 = D_2$.

4. Figure 8.18 shows another common error which also arises from the distortion in scales. The radii of the Fresnel zones are technically at right-angles to the line of sight, as shown by the dashed arrow. In practice, they are plotted along vertical lines, as shown by the solid arrow. From Figure 8.18

$$\theta = (h_2 - h_1)(\pi/180) \, 60/(1000 \, d)$$
$$= 3.4377 \, (h_2 - h_1)/d \qquad (8.64)$$

The error is usually insignificant, since θ seldom exceeds 2 degrees, but would, on the exaggerated scale of the profile, seem to be much larger. On a 50 km path

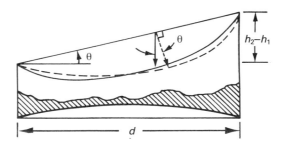

Figure 8.18   Error caused by Difference in Antenna Heights (Courtesy of CCITT, Reference 2)

with a height difference between the two antennas $(h_2 - h_1)$ of 1000 m, $\theta = 1.15$ degrees. With this profile plotted on scales of one cm equal to 2 km, and 1 cm equal to 50 m, θ would appear to be almost 38.7 degrees. It is obvious that plotting the radii of the Fresnel zones at right angles to the line of sight would cause large errors. Only when h_2 equals h_1 would there be no error.

Other errors which may arise when determining the elevations of points along a profile are;

5. A straight line drawn on a map does not represent the true radio path, which follows a great circle.
6. It is very difficult to locate a spot on a map as being the exact location of a radio site, or as being on the radio path.
7. Conditions change rapidly; existing maps are usually obsolete; new buildings are constructed; trees are cut down, etc.
8. Conditions change with the season; snow, flooding, tides, tree growth, etc.
9. Paths crossing over the slope of a hill make it extremely difficult to estimate the heights of construction. It is helpful if such points as shown on the profile are emphasized in some manner.
10. The value of k is assumed to be the same along the entire path. This is not usually true. Variations occur, especially if the elevations of the sites differ greatly, or the terrain changes rapidly. But in practice the assumption of constant k does work well.

8.3.5   Rules for Determining Antenna Heights

The rules are all based on the curves given in Figure 8.8, and on the need to limit obstruction fading at some assumed extreme low value of k, usually 2/3. The designer must use his own judgment for the rules take no account of the degradation due to reflections in a path; blind application of the rules may result in a path clearance equal to an even-zone radius over a reflecting surface. There are four rules in common use, as follows:

Rule 1. Allow a clearance of $0.577 F_1$ over obstructions at $k = 0.7$ (8.65)

Rule 2. Allow a clearance of $1.00F_1$ over obstructions at $k = 4/3$ (8.66)

Rule 3. Allow a clearance of $0.30F_1$ over obstructions at $k = 2/3$ (8.67)

Rule 4. Allow grazing clearance (zero F_1) over obstruction at $k = 2/3$ (8.68)

where F_1 is the first Fresnel zone radius.

Figure 8.19 illustrates how these rules provide varying degrees of clearance over four obstructions on a 60 km path, at 4 GHz. The profile is drawn for $k = 2/3$,

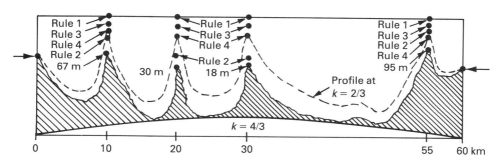

Figure 8.19   *Application of Clearance rules to a Radio Path 60 km long*
(Courtesy of CCITT, Reference 2)

to show what happens at such a low value of k. Note that the path clearance becomes negative for the obstructions at 10, 20 and 30 km, for $k = 2/3$; only at d_1 values in the valleys does it remain positive. Table 8.1 shows the estimated obstruction fading to be expected for each obstruction on this path at $k = 2/3$, for frequencies of 1, 2, 4 and 6 GHz. The losses were estimated from the profile, using the knife-edge diffraction curve of Figure 8.13. Inspection of Table 8.1 indicates that rule 1 is conservative in all cases, while rule 3 consistently limits the loss to only 3 dB or more than free space. Rule 2 would result in excessive losses at 2, 4, and 6 GHz, but would be acceptable at 1 GHz or lower. Rule 4 indicates a loss of 6 dB in all cases, but actual measurements indicate that the true loss at grazing ranges from 9 to 15 dB, except on true knife-edge paths. In fact, all the above estimates of loss are optimistic, as they are taken from the knife-edge curve. Hence, it is recommended to use Rule 4 on knife-edge diffraction paths only; Rule 3 on any path at any frequency up to and including 6 GHz; Rule 2 on any path at 1 GHz or less; and Rule 1 on any path at any frequency (noting that it is more conservative than Rule 3). Use only Rule 1 at frequencies of 7 GHz and upwards.

At frequencies in the 100–500 MHz range, it is usually necessary to operate with negative path clearance, and to overcome the extra path loss by using higher power or larger antennas. Increasing antenna heights usually means higher feeder loss, so that very high towers are not practical.

Table 8.1

Rule	6 GHz				4 GHz				2 GHz				1 GHz			
	d_1 km				d_1 km				d_1 km				d_1 km			
	5	10	20	30	5	10	20	30	5	10	20	30	5	10	20	30
1	+1	−1	−2	−2	+1	−1	−1	−2	+1	−1	−1	−1	+1	−1	0	−1
2	−6	−12	−15	−16	−4	−8	−12	−13	−1	−4	−7	−7	+1	−1	−3	−3
3	−3															
		for an obstruction at any distance, on any path, at any frequency														
4	−6															

* Note that loss is represented by a negative value in this table.

8.4 PATH CALCULATIONS

The purpose of this section is to determine the fade margin for a microwave link or hop. This is done by summing the total gains and losses between the transmitter and receiver to obtain the unfaded received signal level. The difference between the required signal level for a 30 dB signal-to-noise ratio, and the unfaded received signal level then provides the value of the fade margin. To determine if the resultant fade margin is sufficient, a subsection dealing with fading is included. This will enable the depth of fade expected on a particular path to be determined.

8.4.1 Free-space Path Loss

The loss in free space over a radio path is the reference to which all variations in path loss (fading) are compared. It is the transmission loss that would occur between the two antennas in free space, remote from the earth and its effects. In the ideal case, the ratio of the transmitted power to the power arriving at the receiver is given by:

$$\frac{P_t}{P_r} = \frac{(4\pi d)^2}{\lambda^2} \cdot \frac{1}{G_t G_r} \tag{8.69}$$

The path loss in free space (FSL) between these two antennas is simply 10 times the logarithm of this ratio:

$$FSL = 10 \log \frac{(4\pi d)^2}{\lambda^2} \cdot \frac{1}{G_t G_r} \tag{8.70}$$

where P_t = transmitted power in watts, or mW.
   P_r = received power in watts, or mW.
   G_t = power gain of the transmitter antenna.

G_r = power gain of the receiver antenna.
d = distance between the antennas in meters.
λ = wavelength of the carrier signal in meters.

By definition, the power gain of an isotropic antenna, i.e. an antenna which radiates energy uniformly in all directions, is unity, and the gains of other antennas are usually given relative to the isotropic.

Equation 8.70 for the path loss between any two antennas in free space simplifies to:

$$FSL = 32.4 + 20 \log d + 20 \log f - G_t - G_r \qquad (8.71)$$

where G_t and G_r are the gains of the transmitting and receiving antennas in dB.
d is in kilometers.
f is in MHz.

For isotropic antennas in free space, equation 8.71 reduces to:

$$A_0 = 32.4 + 20 \log d + 20 \log f \, \text{dB} \qquad (8.72)$$

where A_0 is the path loss in free space given in dB.

8.4.2 System Gains

8.4.2.1 Antennas

Antennas will be dealt with in depth in Chapter 9. It will suffice here to discuss briefly the dish antennas to obtain an idea of what gain to expect from such equipment.

A parabolic-reflector antenna, with a primary radiator arranged symmetrically about the vertex line, is an antenna with symmetrically fed dipoles, open waveguide, or horn radiators located at the focal point of the parabolic reflector. The field strength distribution over the aperture plane depends on the directivity of the primary radiator. The gain G, relative to an isotropic radiator, of a parabolic antenna with an aperture S_1 and an aperture efficiency η (ratio of effective area S_e to the geometric aperture S_1), which is determined by the illumination function, is given by the equation:

$$G = \frac{4\pi S_1}{\lambda^2} \cdot \eta = \frac{4\pi S_e}{\lambda^2} \qquad (8.73)$$

where λ = the wavelength in meters.
S_1 = geometric aperture area in meters.
S_e = effective area in meters.
η = the aperture efficiency.

The aperture efficiency usually ranges from 0.40 to 0.55 for 2–4 antennas in the 2 GHz range, giving a range of gains from 30 to 35 dB for each antenna. As there are two antennas in a system, i.e. one antenna at each end of the hop, from the system point of view, the gain of each must be added. Hence for antennas with 35 dB of gain each, the total gain from the antennas is:

35 dB + 35 dB = 70 dB

From equation 8.73, it can be seen that by increasing the surface area, or aperture size, the gain increases for a given fixed frequency. Similary, if the aperture size is kept constant but the frequency is raised, the gain also increases. This is similar to saying that the aperture size has increased relative to the radio signal's wavelength.

8.4.2.2 Transmitter Power
Usually expressed in dBm, this can be considered as a system gain relative to the free-space loss. Note that as it is an absolute unit (in this case dBm, but equally valid if in dBW), it permits an absolute level in the same units to be calculated at the receive input.

8.4.2.3 Total System Gain
This is the sum of the antenna gains and the transmitter power in dBm (or dBW).

8.4.3 System Losses

8.4.3.1 Diffraction Loss or Obstruction Loss
This has been discussed in Section 8.3.2.3, and the results of the losses calculated must be added here.

8.4.3.2 Transmission Line Loss
This loss is found from the manufacturer's data, and is expressed as a loss L per 100 meters. or something similar. The actual feeder loss is sometimes computed as 1.5 times the antenna height, in order to proceed with the system design. On the average this rule-of-thumb figure is fairly close, after considering the running-in of the feeder to the equipment from the tower, and the sometimes tenuous path traversed. Thus

$$\text{Feeder loss} = 1.5Lh \qquad (8.74)$$

where L = the loss in db/unit-length.
h = the height of the antenna from the base of the tower.

8.4.3.3 Connector Loss
This loss is the small loss associated with pressure windows, bends and flanges found in waveguides. The value of 0.5 dB per hop end is usually taken as a conservative estimate. Thus:

$$\text{Connector loss} = 1 \text{ dB per link} \tag{8.75}$$

8.4.3.4 Branching Loss

This loss is the circulator, isolator and duplexer, or if so equipped the quadruplexer, losses associated with the transmitter and receiver separately. Typical figures are 6.5 dB for the transmitter end, and 2.5 dB for the receiver end.

8.4.3.5 Radome Loss

This loss can vary according to the type of radome used. A typical figure for a 6 GHz link, unheated radome is 0.5 dB.

8.4.3.6 Safety-factor Loss

In some cases it may be prudent to add 0.5 dB at either end of the hop to cover additional losses not covered by those given above.

$$\text{Safety-factor loss} = 1 \text{ dB/link} \tag{8.76}$$

8.4.3.7 Free-Space Path Loss A_o

This loss is given by equation 8.72 and is again repeated here.

$$\text{Free-space loss } A_o = 32.4 + 20 \log d + 20 \log f \text{ dB} \tag{8.72}$$

where f = the centre frequency of the radio channel frequency band in MHz.
   d = the path length in km.

8.4.3.8 Total System Loss

The sum of all losses in sections 8.4.3.1 to 8.4.3.7 inclusive provides the total system loss. In some systems not all of these items will apply, and hence should be omitted when that is so.

8.4.4 Unfaded Received Signal Level

The subtraction of the total system gain from the total system loss provides the unfaded received signal level in dBm (or dBW). Due to the absolute level given by the transmitter power, the received level is also an absolute level in the same units.

$$\text{Unfaded received signal level} = (|\text{Total system loss}| - |\text{Total system gain}|) \tag{8.77}$$

If the transmitter power is omitted from equation 8.77, the resulting loss is called the *net loss*, i.e.

$$\text{net loss} = \text{Total system loss} - \text{antenna gains} \tag{8.78}$$

8.4.5 Practical Threshold

From equation 7.38 the practical received carrier level in dBm (P_r) can be determined as shown in the example on the use of this equation. This practical receiving level is the level at which the carrier must be received in order to satisfy the specified noise requirements in a derived speech channel.

8.4.6 Fade Margin

The difference between the unfaded received signal level and the practical received carrier level is the level which the carrier signal can fade from its normal level to that level at which the system becomes inoperative due to noise. This difference is known as the *fade margin*, and so

$$\text{Fade margin in dB} = \text{unfaded received carrier level in dBm} \\ - \text{practical threshold in dBm} \tag{8.79}$$

8.4.7   Figure of Merit

Equation 5.49 with a weighting factor W, gives a weighted signal-to-noise ratio in dB above the FM threshold out of a derived speech channel, i.e.

$$(S/N)_{(weighted)} = P_r + 20 \log (\Delta f/f_m) - F + D + W + 136.06 \text{ dB} \tag{8.80}$$

The *figure of merit* is the idealized value of the signal-to-loss dependent noise ratio which would be obtained with a lossless connection between the transmitter and receiver reference points, assuming the transmitter to be noiseless. The reference points are those used for the calculation of the path loss. Thus in equation 8.80, the P_r is replaced by P_t, and the weighted signal-to-noise ratio is replaced by the figure of merit M:

$$\text{System figure of merit} = M = P_t + 20 \log (\Delta f/f_m) - F + D + W + 136.06 \text{ dB} \tag{8.81}$$

where P_t = the output power from the transmitter in dBm (before filters and circulators).
$\Delta f/f_m$ = the RMS modulation index of the top FDM channel (see Table 7.5).
F = the receiver noise figure, dB.
D = the emphasis factor as per CCIR Recommendation 275–2, dB.
W = the weighting factor (2.5 dB for CCITT).
136.06 is $KT.2B$ in dBm, at 17° with $B = 3.1$ kHz.

It is apparent that if the transmitter power P_t has the net loss (equation 8.78) subtracted from it, the result is the unfaded received signal level P_r, i.e.

$$P_t - \text{Net loss} = P_r \tag{8.82}$$

Thus the ratio of the signal to the loss-dependent thermal noise as given by equations 7.38 and 7.39 is obtained. Note that, if a baseband combiner is used in a diversity system, there will be an improvement of 2 dB or more in the loss-dependent noise value, i.e.

$$(S/N)_{loss\ dependent} = P_t - \text{net loss} + 20 \log (\Delta f/f_m) - F + D + W + C + 136.06$$
$$(8.83)$$

where C is the baseband combiner improvement factor in dB.

Equation 7.8 represents a general equation relating signal-to-noise ratio to dBm0, i.e.

$$(S/N) = -(\text{noise in dBm0}) = 0 \text{ dBm0} - (\text{noise in dBm0})$$
$$(7.8)$$

For the specific case where psophometric weighting occurs, this expression becomes:

$$(S/N)_{weighted} = 0 \text{ dBm0p} - (\text{noise in dBm0p})$$

or

$$(S/N)_{weighted} = 10 \log(10^9 \text{pW0p}) - (\text{noise in dBm0p})$$

Hence:

$$(S/N)_{weighted} = 90 - 10 \log (\text{noise in pW0p})$$

or

$$\text{noise in pW0p} = 10^{0.1(90 - S/N_{weighted})}$$

relating this to equation 8.83 gives:

$$\text{Loss-dependent noise in pW0p} = 10^{0.1(90 - S/N_{weighted})}$$
$$(8.84)$$

8.4.8   Distance and Azimuth from Latitude and Longitude[13]

Often a repeater or terminal station is located on a map by having knowledge only of the station's co-ordinates, i.e. only the latitude and longitude may be given for each site. To determine accurately the distance between the two stations, and the direction in which the rigger should align each antenna with the aid of a compass, calculations involving the two stations' co-ordinates are required.

8.4.8.1 Calculated Distance between Two Station Sites
Referring to Figure 8.20, A and B are two places on the surface of the earth with given latitudes and longitudes, with B as the site of greater latitude (nearer a pole). The angles X and Y at site A and site B of the great circle passing through the two places and the distance Z between A and B along the great circle is calculated as follows:

$$L_A = \text{latitude of A}.$$

L_B = latitude of B.
C = difference in the longitudes of A and B.

Then

$$\tan(Y - X)/2 = (\cot C/2)\,\frac{\sin[(L_B - L_A)/2]}{\cos[(L_B + L_A)/2]} \tag{8.85}$$

and

$$\tan(Y + X)/2 = (\cot C/2)\,\frac{\cos[(L_B - L_A)/2]}{\sin[(L_B + L_A)/2]} \tag{8.86}$$

Equations 8.85 and 8.86 give the values of $(Y - X)/2$ and $(Y + X)/2$, from which:

$$(Y + X)/2 + (Y - X)/2 = Y \tag{8.87}$$
$$(Y + X)/2 - (Y - X)/2 = X \tag{8.88}$$

In equations 8.85 and 8.86, north latitudes are taken as positive and south latitudes as negative. If both places are in the southern hemisphere and $L_B + L_A$ is negative, it is simpler to call the place of greater south latitude B, and to use the above method for calculating bearings from true south and to convert the results afterward to bearings east of north.

The distance Z (in degrees) along the great circle between A and B is given by:

$$\tan Z/2 = [\tan(L_B - L_A)/2].[\sin(Y + X)/2]/[\sin(Y - X)/2] \tag{8.89}$$

The angular distance Z (in degrees) between A and B may be converted to linear distance as follows:

Z (in degrees) × 111.12 = kilometers
Z (in degrees) × 69.05 = statute miles
Z (in degrees) × 60.00 = nautical miles

In multiplying, the minutes and seconds of arc must be expressed in decimals of a degree.

8.4.8.2 Azimuth

The azimuth is the angle in the horizontal plane with respect to a fixed reference, usually true north, measured clockwise from the reference (magnetic north). Figure 8.20 shows simply the azimuth of an antenna at A, so that it may be aligned to point directly to site B. The angle X and $(360 - Y)$, as determined from equations 8.85 and 8.86, provide the azimuths of the A and B sites connecting line directly when $L_B > L_A$, B is in the northern hemisphere, and B is east of A. For other combinations differences occur. To construct the table below the following also applies: any latitude in the southern hemisphere or any longitude in the eastern hemisphere (East

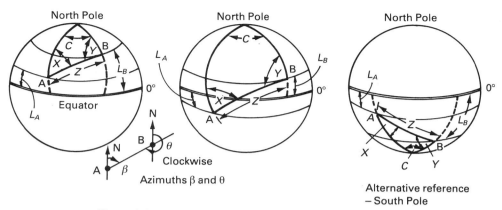

Figure 8.20   Great-Circle Distance Z and Azinmuths β, θ
(Courtesy of Howard W Sams, Reference 13)

of Greenwich), will make that latitude or longitude a negative quantity. This also applies for any A which is east of B.

L_A	L_B	*$l_A - l_B$	Azimuth at A	Azimuth at B
+	+	+	X	$360 - Y$
−	+	+	X	$360 - Y$
+	+	−	$360 - X$	Y
−	+	−	$360 - X$	Y
+	−	+	X	$360 - Y$
−	−	+	X	$360 - Y$
+	−	−	$360 - X$	Y
−	−	−	$360 - X$	Y

* l_A and l_B are the longitudes of sites A and B respectively.

Only the $l_A - l_B$ column correlates with the azimuth at A or the azimuth at B columns, by relating a positive sign against X, and a negative sign against $360 - X$, thus only if the longitude of the A terminal is east of the B terminal is there a change in the azimuth angles. Thus if all longitudes which are east are considered negative, and then $L_A - L_B$ is negative, the azimuth at A is $360 - X$, and at B is Y; otherwise if the difference in longitudes is positive, the azimuth at A is X, and at B is $360 - Y$. We may summarize the position as follows:

Rule 1. All northern latitudes are positive.
Rule 2. All southern latitudes are negative.
Rule 3. All western longitudes are positive.
Rule 4. All eastern longitudes are negative.
Rule 5. If longitude of site A − longitude of site B is positive, then azimuth at A
    = X and azimuth at B = $360 - Y$.
Rule 6. If longitude of site A − longitude of site B is negative, then azimuth at A
    = $360 - X$ and azimuth at B = Y.

8.4.9 Fading

Fading of radio signals may be divided into six categories. These are given in Sections 8.4.9.1–8.4.9.6.

8.4.9.1 Obstruction Fading

This type of fading occurs due to variations in the k-factor, which causes earth bulge effects. Thus obstructions in the radio path are made to vary in height. This variation in k-factor causes a varying obstruction loss to occur, which in turn causes the received signal to vary in level, and so obstruction fading may arise. Fading of this type seldom occurs on paths designed with the recommended clearance given in Section 8.3.5. When this type of fading does occur though, it affects all radio channels simultaneously. The method of calculating the obstruction loss for differing values of k-factor is given in Section 8.3.2.3.

8.4.9.2 Multipath Fading

The fading due to multipath effects is due to the refraction associated with the time-varying vertical gradient of the refractive index, and the formation of phase-interference patterns due to diffraction and reflection by the Earth's surface and to atmospheric refractive index discontinuities. The total signal is the sum of the direct ray and one or more indirect rays. The indirect rays arrive via paths above or below the direct signal; these paths change in length because of random changes in the air layers through which they travel. The rays adding to the direct signal and those sub-tracting from it will change constantly, and if the effective components are of the same order of magnitude, the most severe fading occurs. The net signal will fade relatively slowly with occasional fast deep fades in a stable atmosphere, but in a turbulent atmosphere will produce fast, but low-magnitude, fades which would cause fewer outages of the radio system. Rays of two different frequencies traversing the same path will differ in phase at the receiver, so that one channel of a radio system can be in a null whilst the other is at a maximum. Both types of atmospheric multi-path fading have a time-versus-depth fade distribution, where, for example, in poor propagation areas, a 40 dB fade margin could approach an outage time of one hour per year. Increasing the fade margin will reduce the outage time due to this type of fading, and as shown in Figure 8.21, for each 10 dB increase in fade margin, or decrease in fading depth, the outage period decreases by a factor of 10. Multipath fading is frequency selective, and all types of diversity will reduce the baseband out-age time. This type of fading is also sensitive to antenna orientation and size. The larger the antenna dish diameter, the better becomes the selectivity against off-path secondary rays. Multipath fading varies also with the length of the path and the type of terrain, but in the frequency range 2–6 GHz is not very dependent, unless the path length is over 100 km long. CCIR Report 338–3, 1978, states that the provisional curves as given in Figure 8.22 have been derived from UK and French test results. These curves give the distribution of fading depth during the worst month of a year, relative to free space, for various path lengths, for average terrain, and for the climate in north-western Europe. The frequency is 4 GHz. The curves fit the original data with an RMS error of about 4 dB, but great care should be exercised in applying these provisional curves to any other conditions of climate, terrain or path clearance.

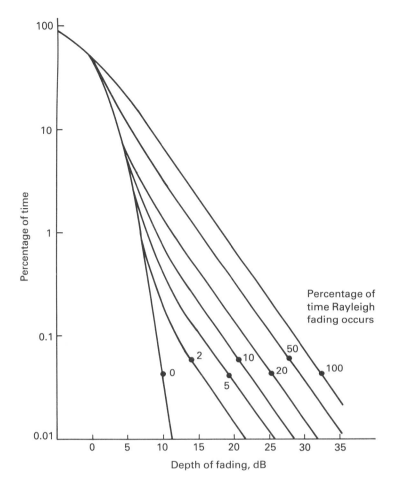

Figure 8.21   Composite Fading Distributions (Courtesy of IEE, Reference 6)

Reference 6 states that with adequate path clearance and in the absence of a singular specular reflection on a path, the very deep fading is due to multipath propagation through the atmosphere, which, over a single section, gives rise to a Rayleigh distribution of the received signal amplitude against time. Such a distribution is characterized by a slope of 10 dB per decade for fading depths greater than 10 dB. Reference 6 then goes on to say that if it is postulated that multipath fading (or Rayleigh fading, as it is usually termed) occurs for a proportion of the month, and, during the rest of the time, the fading depth has a much less severe distribution, the overall distribution for the month will display the characteristic Rayleigh slope for the deeper fades. Figure 8.21 shows combinations of Rayleigh fading for a given percentage of the time, with fading following a log-normal distribution, with a standard deviation of 3 dB, for the remainder of the time.

For designing a radio system conforming to CCIR recommendations, it is necessary to predict the probability of deep fades for very small percentages of the time (e.g. about 0.0002 per cent for an average hop of about 50 km). To overcome the difficulty which may be encountered in predicting fading depth for such a small percentage of the time, a method of utilizing the occurrence probability of Rayleigh fading depth has been developed[5,14]. The method developed relates to clear line-of-sight paths with negligible earth reflection. The fading probability $P_r(W)$ is given by the equation:

$$P_r(W) = K.Q.(W/W_0).f^B \, d^c \tag{8.90}$$

where d = path length, km.
      f = frequency, GHz.
      K = factor for climatic conditions.
      Q = factor for terrain conditions.
      W_0 = received power in non-fading conditions.
      $P_r(W)$ = probability that the received power is less than or equal to W.

Equation 8.90 applies only for fading exceeding about 15 dB.
    The values of the constants for differing parts of the world are as follows:

Japan, for the worst season:
$B = 1.2, C = 3.5, K = 0.97 \times 10^{-9}, Q = 0.4$ (over mountain) $= 1.0$ (over plain)

with h_1 and h_2 in meters; $Q = 101.82 \, (h_1 + h_2)^{-\frac{1}{2}}$ (over sea or coast)

N.W. Europe, for the worst month:
$B = 1,\quad C = 3.5, K = 1.4 \times 10^{-8}, Q = 1$

United States, for the worst month:
$B = 1,\quad C = 3,\quad K = 1.2 \times 10^{-6}$ for equatorial, maritime temperate regions.
                        $K = 9.0 \times 10^{-7}$ for maritime subtropical regions.
                        $K = 6.0 \times 10^{-7}$ for continental temperate climates.
                        $K = 3.0 \times 10^{-7}$ for polar climates or high dry mountainous regions.
$Q = (15.2/S)^{1.3}$ Where S is the terrain roughness measured in meters by the standard deviation of terrain elevations at 1 km intervals.
$Q = 3.35$ For smooth terrain (S is equal or less than 6 meters).
$Q = 1.0$ For average terrain ($S = 15.2$ meters).
$Q = 0.27$ For rough terrain (lower limit, where S is equal or greater than 42 meters).

From Figure 8.22, the attenuation value which is not exceeded for 80 per cent of the worst month is represented for N.W. Europe by the following CCIR 1974–78b formula as:

$$A = 10 \log [1 + d^2.f^{0.8}/8500] \tag{8.91}$$

where A = the attenuation in dB.
      d = the path length in km.
      f = the frequency in GHz.

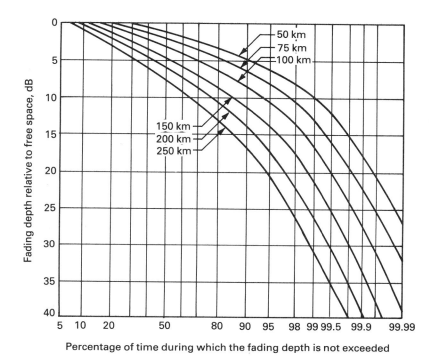

Figure 8.22 *Provisional Curves of Fading Depth not exceeded for Percentage of Worst Month (Courtesy of CCIRR, Report 338–3)*

For frequencies greater than 12 GHz, it is necessary to include an allowance for the attenuation by water vapour and oxygen.

DURATION OF FADES

The distribution of fade durations influences the type and design of diversity protection systems of the timing circuits of digital systems. Measurements made in the USSR show that the statistical distribution of fade durations approximates a log-normal law. The median value of fade duration decreases when fading depth increases, decreases with increasing path length and increases with relative path clearance. Measurements in the United States also show that the median fade duration distribution is log-normal characterized by a standard deviation of 5.6 dB for non-diversity paths. Long-term measurements on line-of-sight 40–70 km paths in the United States have shown that multipath median fade duration times $<t>$ can be expressed for a non-diversity signal as follows:

$$<t> = 56.6 \times 10^{-F/20} \times (d/F)^{1/2} \tag{8.92}$$

where d = path length, km, f = frequency, GHz, F = fade depth in dB for deep
        fades exceeding 20 dB.
    $<t>$ = median fade duration in seconds at fade depth F.

8.4.9.3 Duct Fading

Prolonged fading, or signal enhancement, can result from propagation through
ducts, especially when either the transmitter or the receiver is located within the
duct. Signals may be trapped within the duct and propagated far beyond the horizon.
Ducts may also cause multipath fading. Paths trapped in elevated ducts may exhibit
wide, slow, level changes, up to 30 dB above median in some cases, and thus may be
regarded as a waveguide coupling between the transmit and receive antennas. As
ducts are very narrow, space diversity can be a remedy. Frequency diversity is of little
value except as protection against the accompanying multipath fading.

Blackout fading may occur when low-clearance paths traverse areas supporting
super-refractive ground-based layers. In this case there may be a complete loss of
signal for periods of up to 24 hours. These areas are usually where there are river
deltas, marshes, paddy fields or hot wet climates. A rising atmospheric layer which
is usually not visible, but is sometimes associated with a visible steam-type fog
formed over warm water or moist ground, may intercept and trap the path. The path
failure is occasionally preceded by reflection fades occurring from signals reflecting
off of the layer, and obstruction fades. It has been found that increasing the fade
margin, adding frequency diversity links, and frequency band selection are of little
or no avail in avoiding this problem. To overcome this blackout type of fading,
avoidance of the blackout area, reduction in the spacing of repeater sites, or to
traverse the path above, or around the layer is suggested. These layers seldom
exceed 50 metres in elevation, so antennas may be chosen of sufficient height to clear
the obstruction. Widely spaced space-diversity antennas often provide a solution.

8.4.9.4 Rain Fading

Microwave radio propagation through the atmosphere shows attenuation due to
absorption and scattering by hydrometeors (rain, hail, snow and fog). This attenuation is usually negligible at frequencies below 5 GHz, noticeable at 6 GHz, and
serious above 10 GHz. Extremely heavy rains are required to cause complete failure
of a microwave link, but when fading does occur, it affects all channels in a link
equally. Therefore, in planning radio-relay systems to operate in the bands above
5 GHz, attenuation due to hydrometeors is a consideration whose importance
increases with rising frequency, since this may constitute the source of highest
attenuation. Fortunately, multipath fading is usually absent during heavy rain.
Fading due to cloud, fog, snow and hail is usually less important than fading due to
rain (CCIR Report 721). However, attenuation due to wet snow may be noticeable
especially if it builds up on the antenna. Figure 8.23 shows the theoretical curves of
attenuation as a function of rainfall rate.

Outages on a radio link are usually caused by rain cells (thunderstorms, etc.)
averaging 5 to 7 km in diameter and 5–15 min in duration. Automatic gain control
(AGC) recordings of receiver level on paths affected by rain show fairly slow erratic
level variations, with rapid path failure as the rain cells intercept the path. The fades

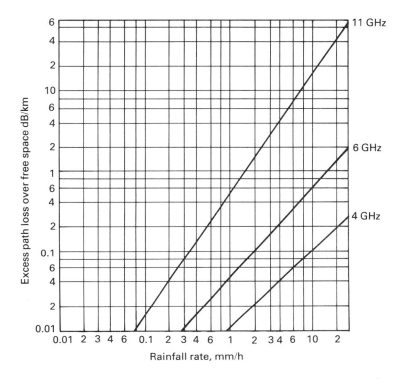

Figure 8.23   *Theoretical Microwave Attenuation as a Function of Rainfall Rate*
(Courtesy of CCITT, Reference 2)

are non-selective, and thus affect simultaneously all link radio channels in the band traversing the path. Whereas fading due to multipath propagation can be corrected by a vertical separation of several meters between the antennas, a choice of paths with a separation of several kilometers may reduce fading due to precipitation (raining). A path separation of 4 km has shown to give considerable improvement in a switched path system, but larger spacings did not provide a proportionately greater improvement. Increasing the fade margin to 50 to 60 dB may be some help, but if possible cross-band diversity would be more effective. The lower 6 GHz path is usually affected by atmospheric turbulence and may be stable when the upper 11 GHz is affected by rain cells.

8.4.9.5 Reflection Fading

Reflection fading or Fresnel-zone fading occurs when the reflected ray cancels the direct ray, i.e. whenever the path clearance equals, or is close to, an even-zone radius. The signal at the receiver may be reduced by many dB, depending on the reflection coefficient of the path. This is really another type of multipath fading, but often lasts much longer, especially on paths where a stable reflection occurs. The fade may be 10–20 dB and lasting for days or weeks. This is especially true when the path clearance is carelessly chosen to be close to an even-zone radius for normal atmospheric conditions.

ELEVATED LAYER REFLECTIONS

Reflections may also be caused by elevated layers. These elevated layers result from sharp variations in either temperature or humidity, and act as nearly perfect reflectors at microwave frequencies. These layers are usually quite stable, but can move slowly in a vertical plane with time. Fast, deep fades may occur during the night hours, and may be attributable to constantly changing undulating layers perhaps 3 m thick and 10 km long. Fading from layer reflections may be overcome by using large antennas with vertical polarization and 2 per cent frequency diversity or 100-wavelength-spacing space diversity.

GROUND OR WATER REFLECTIONS

On many paths it is impossible to avoid ground or water reflections. Some radio paths over treeless plains or ploughed fields have exhibited coefficients of reflection up to 0.90 corresponding to signal cancellations of over 20 dB.

Figure 8.24 shows a radio path, and the variation in signal strength vertically

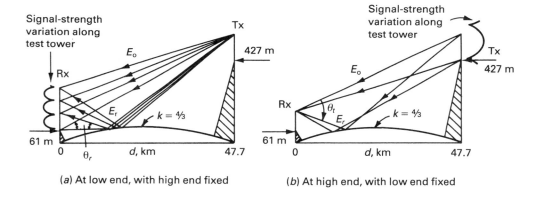

(a) At low end, with high end fixed   (b) At high end, with low end fixed

Figure 8.24   *Interference Patterns caused by Reflections (Courtesy of CCITT, Reference 2)*

along the transmit and receive towers. Figure 8.25(a) and (b) show the corresponding height/loss curves. These were obtained with one antenna at a fixed height on the test tower, and by taking measurements of signal strength at intervals up the tower at the other end of the path. The resulting signal strengths were then plotted relative to the free-space level for the path. Figure 8.25(c) shows the height/loss curve obtained when the signal strength is measured with both transmitting and receiving antennas moving in steps. This situation roughly corresponds to what happens when k changes; the reflection point moves and the path clearance changes simultaneously. On an overland path with a well-defined reflecting area, the reflection point would not ordinarily move. On paths with relatively smooth terrain, the reflections sometimes move along the path as the antennas are raised or as k changes. This makes it difficult to analyse the test results.

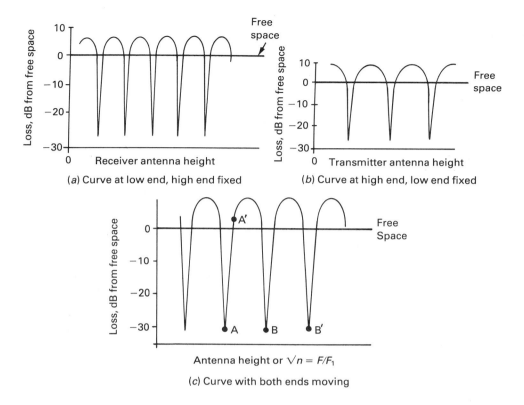

**Figure 8.25   Height/Loss Curves taken on the Path of Figure 8.24
(Courtesy of CCITT, Reference 2)**

From these height/loss curves it is evident that if the antenna heights are such that they correspond to one of the even-zone dips, the signal level will be reduced by many decibels. As k changes, the path clearance changes also and the vertical space pattern sweeps past the antenna, moving up as the earth bulges (k decreases) and down as flat earth is approached (k increases). This is accompanied by an accordion-like contraction and expansion because of the fact that the higher-order zones are spaced increasingly closer together. The result can be severe fading.

FREQUENCY SELECTIVITY OF REFLECTION FADING
Reflection-type fading is frequency-selective. The selectivity shows up only on paths with large clearance over the reflecting surface. Referring to Figure 8.25, let us take for example a chosen path clearance such that, at $k = 4/3$, channel 1 of a radio system operates at point A, whilst channel 6 operates at point A'. On most paths, this spread from A to A' is not significant, but when the path clearance is large enough, it is possible for two channels to operate in even-zone dips at the same time as shown by B and B'. The clearance F between the ground and the direct ray is the same for all

channels, but the radius of the first Fresnel zone F_1 changes with channel frequency. As a result, from equations 8.20 and 8.22, we have

$$n_1/n_2 = f_2/f_1 \qquad\qquad (8.93)$$

where n_2 and n_1 are the Fresnel-zone numbers for channels at frequencies f_2 and f_1.

For even zones relating n_1 and n_2, i.e. the zones at which these two channels can fade simultaneously, we get:

$$n_2 = n_1 + 2p \text{ for } p = 1,2,3, \qquad\qquad (8.94)$$

$$f_2/f_1 = n_1/n_2 = 1 - 2p/n_2$$

Hence $n_2 = 2p/(1 - f_2/f_1)$ and $n_1 = n_2 f_2/f_1 \qquad\qquad (8.95)$

If a particular microwave system is considered, there may be two adjacent channels having frequencies so related that both channels will be in even zones when the path clearance is such that n_2/n_1 equals 30/28, 60/56, 90/84, etc. Similar ratios may also be found for the other channels used in the band, but the smallest ratio of n_2/n_1 for any two channels means that the path clearance over possible reflecting surfaces on paths using this system should be restricted to less than the zone represented by the number in the denominator when k equals infinity.

To overcome this particular type of fading, an increase in the fade margin may help. In a non-condensing environment, the reflection point could be shielded by planting trees, erecting a screen, or by moving the antennas. Antennas are tilted slightly upwards to provide increased discrimination to the reflected ray. Large antennas, with smaller beam widths, especially at the low end of a high/low radio path could be used and space diversity, with optimum antenna spacing either computed or experimentally determined should also be considered. Vertical polarization is effective if the reflection angle is greater than 0.1°.

8.4.9.6 Inverse Bending Attenuation

In coastal areas, a substandard atmosphere generally associated with the formation of fog forming over cool water or land will cause the microwave beam to bend upwards with respect to the path under normal conditions. This can cause diffraction and the possibility of beam misalignment, resulting in a loss which is non-selective and affecting all radio channels in the same band. To overcome this problem, the fade margin may be increased, and the antenna heights raised. The lower bands are less susceptible to this form of fading than the higher bands, for a given amount of inverse bending. Where antenna heights cannot be raised, but frequencies can be changed, frequencies of 2 GHz or less are recommended.

Figure 8.26 shows typical AGC recordings for some of the different types of fading mentioned in Sections 8.4.9.1–8.4.9.6 inclusive.

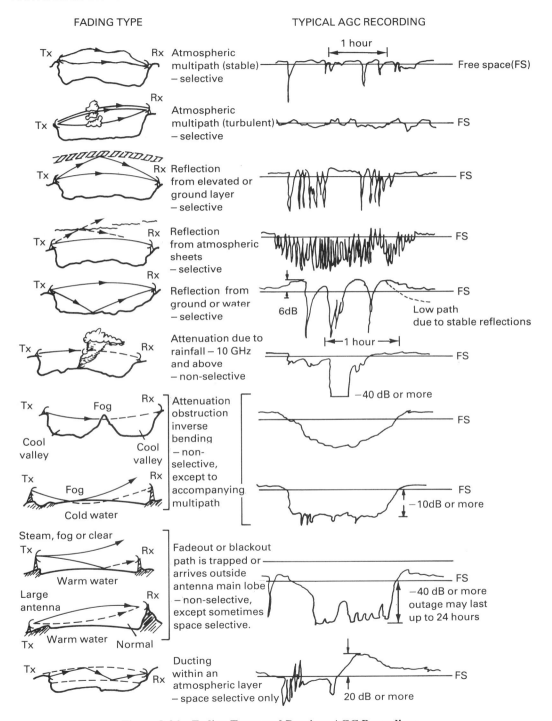

FADING TYPE   TYPICAL AGC RECORDING

Atmospheric multipath (stable) – selective

Atmospheric multipath (turbulent) – selective

Reflection from elevated or ground layer – selective

Reflection from atmospheric sheets – selective

Reflection from ground or water – selective

Attenuation due to rainfall – 10 GHz and above – non-selective

Attenuation obstruction inverse bending – non-selective, except to accompanying multipath

Fadeout or blackout path is trapped or arrives outside antenna main lobe – non-selective, except sometimes space selective.

Ducting within an atmospheric layer – space selective only

Figure 8.26   Fading Types and Receiver AGC Recordings

8.4.9.7 Optimum Antenna Size and Divergence[2]

DIVERGENCE

On a path over water or smooth earth, the curvature of the earth scatters the reflected ray, reducing the effective reflection coefficient to less than unity. This reduction can be expressed as the reflection coefficient multiplied by the *divergence factor D*, where

$$D = [1 + (5/16k).d_1^2.d_2/(d\,h_1')]^{-\frac{1}{2}}$$

$$= [1 + (5/16k).d_1 d_2^2/(dh_2')]^{-\frac{1}{2}} \tag{8.96}$$

where d, d_1 and d_2 are in kilometers, and h_1' and h_2' are in meters.

The value of the reflection coefficient R over water or smooth earth therefore becomes

$$R = R_0.D \tag{8.97}$$

where R_0 is the reflection coefficient over a plane surface. Note that D is a voltage ratio.

OPTIMUM ANTENNA SIZE

In practice, the fading caused by destructive wave interference and change in the k-factor is modified by the antenna itself. This is because the vertical dimension of the antenna aperture may be significantly large compared with the spacing between two successive even Fresnel zones. The reflected ray arriving at the bottom of the antenna will differ in phase from the ray arriving at the top. Figure 8.27 shows the simplified geometry. For a reflected ray, arriving off the center-line of an ideal rectangular antenna, the amplitude relative to the direct ray is given by:

$$\frac{E_r}{E_0} = \frac{\sin \pi a(2h_2'/\lambda d)}{\pi a(2h_2'/\lambda d)} = \frac{\sin \pi a(fh_2'/150d)}{\pi a(fh_2'/150d)} = \frac{\sin \pi a/a_0}{\pi a/a_0} \tag{8.98}$$

where a = the vertical dimension of the antenna aperture in meters.

      a_0 = the optimum dimension of the antenna aperture in meters.

          = $\lambda d/2h_2' = 150\,d/fh_2'$ at the lower end.

      a_0' = $\lambda d/2h_1' = 150\,d/fh_1'$ at the upper end.

      λ = the wavelength in meters.

      f = the frequency in MHz.

      d = the path length in meters.

      h_1', h_2' = the heights above the tangent plane in meters.

Then, if we designate by a' the antenna aperture at the upper end, and include the coefficient of reflection as given by equation 8.97, the relative magnitude of the reflected ray is given by

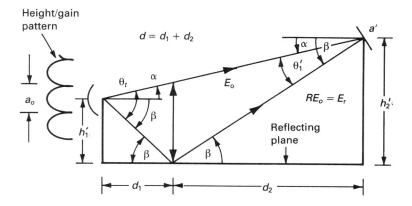

Figure 8.27   Simplified Path Geometry on a Reflecting Path (Courtesy of CCITT, Reference 2)

$$\frac{E_r}{E_0} = \frac{\sin \pi a f h_2'/150d}{\pi a f h_2'/150d} \cdot \frac{\sin \pi a' f h_1'/150d}{\pi a' f h_1'/150d} \cdot R$$

$$= \frac{\sin \pi a/a_0}{\pi a/a_0} \cdot \frac{\sin \pi a'/a_0'}{\pi a'/a_0'} \cdot R$$

$$= \frac{\sin \pi M}{\pi M} \cdot \frac{\sin \pi M'}{\pi M'} \cdot R \qquad (8.99)$$

The above equation indicates how the reflected ray may be suppressed, and the Fresnel-zone fading eliminated or reduced, if the vertical dimension a of the antenna is proportioned so that either $M = 1$, or $M' = 1$, i.e.:

$$\left.\begin{array}{l} a_0 = \lambda d/2h_2' = 150d/f h_2' \\[2mm] a_0' = \lambda d/2h_1' = 150d/f h_1' \end{array}\right\} \qquad (8.100)$$

where d, h_1', and h_2' are in the same units.

The received signal will then be independent of the reflected ray. These optimum antenna sizes, a_0 and a_0', occur when the vertical aperture is large enough to span a phase change of 360° from top to bottom. This corresponds to any two successive even, or odd, Fresnel zones. The optimum figure for antenna aperture holds only for one value of k; any change in k will alter the fading range accordingly. Usually k is chosen as 4/3, since the fading range from $k = 2/3$ to infinity will then be held to its least possible value. The extent of the fading may be calculated by taking 20 log $(1 - E_r/E_0)$ for even zone-fading and 20 log $(1 + E_r/E_0)$ for odd-zone fading.

The equations above apply only for an ideal rectangular antenna. In practice, the antenna will be a parabolic dish, or a horn-reflector type, and then equation 8.98 is modified to:

$$E_r/E_0 = (a_0/\pi a).2J_1(\pi a/a_0) = (2/\pi M).J_1(\pi M) \tag{8.101}$$

where J_1 is a Bessel function of the first order.

Figure 8.28 shows the curve of a/a_0 for both the rectangular and the circular aperture of the parabolic or horn-reflector antennas.

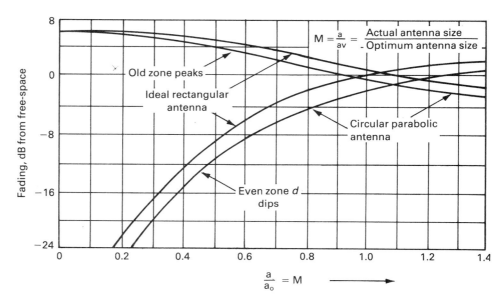

Figure 8.28   *Reflection Fading as a Function of M (Courtesy of CCITT, Reference 2)*

8.4.10 Optimum Length of Hops (Tandem Hops)*

A radio-relay link of a given overall length should generally be divided into a number of tandem hops. It is therefore important to select the optimum length for these hops. This optimum length may be defined, for instance, as the length which produces the shortest possible total outage time (as per CCIR Recommendation 557), or as the length which gives the highest possible mean value for the signal-to-noise ratio (as per CCIR Recommendation 395–2). It depends on the propagation and on the reliability of the equipment, including the power sources. It also depends on the structure of the radio-relay system (stand-by channels, diversity, etc.) and on the topography of the region.

* CCIR Report 784, 1978

With regard to the propagation, account should be taken of the correlation between fading on two successive hops. Rapid fading due to multipaths is generally uncorrelated. The same applies to attenuation due to rainfall when the length of the hop is clearly greater than the dimensions of the rain cells. On the other hand, if the length of the hop is about the same or less than the dimensions of the rain cells, there is a certain correlation, and hence the outage probability for the overall circuit is less than the sum of the outage probabilities for each hop.

Although there may be hops over 100 km long, the usual lengths are between 40 and 50 km in the frequency bands below 10 GHz. This length may still be used for the frequency band 10.7–11.7 GHz, at least with certain types of modulation, except in regions where the rainfall is very heavy. In the higher-frequency bands the length of the hops may have to be progressively reduced because of the rain, which at the same time reduces multipath fading. However, there is a limit to this trend, which is reached when the length of the hops becomes far less than the dimensions of heavy rain cells, i.e. a few kilometers.

8.4.11 Probability of Exceeding the Fade Margin during the Worst Month

The probability of Rayleigh fading such that the fading depth W_0 is reached during the worst month of the year is given by (CCIR Report 338–3):

$$P_r(W) = K.Q.(W/W_0).f^B.d^c \tag{8.90}$$

The fade margin F in dB is given by:

$$F = \text{Unfaded received carrier level in dBm} - \text{practical threshold level in dBm} \tag{8.79}$$

Thus

$$F = 10 \log W_0 - 10 \log W \text{ dB}$$

or

$$-F = 10 \log (W/W_0) \text{ dB}$$

and

$$(W/W_0) = 10^{-F/10} \tag{8.102}$$

Hence the probability of reaching the fade margin due to Rayleigh fading, or the fraction of the time during the worst month in which the hop is at the fade margin or is unavailable, is given by:

$$P(F) = K.Q.f^B.d^C.10^{-F/10} \tag{8.103}$$

where K, Q, f, d, B and C are as given in Section 8.4.9.2.

8.4.12 Percentage Availability of Hop

Equation 8.103 gives the 'outage time' or, if multiplied by 100, the *percentage unavailability*. Subtracting from unity the value found from equation 8.103, and multiplying by 100, will give the percentage availability during the worst month. Thus:

$$\text{Percentage availability } A \text{ during the worst month} = 100\,[1 - P(F)] \quad (8.104)$$

8.4.13 Diversity Improvement Factor[7]

Diversity, as described in Section 8.5, reduces the effect of noise on the quality of the transmitted signal due to fading and other causes, and so reduces the time which a specified operational noise level is exceeded. Since the fading on line-of-sight links is additive in time, rather than in the depth of fade, the diversity improvement factor is given as a time factor:

$$\text{Diversity improvement factor} = \frac{\text{Time with excessive fading without diversity}}{\text{Time with excessive fading with diversity}} \quad (8.105)$$

The diversity improvement factor on typical line-of-sight paths with multipath fading, and without substantial ground reflections, is given approximately as[15,16]:

$$I_S = 1.2 \times 10^{-3}.S^2.f/d.10^{(wf)/10} \qquad \text{space diversity} \qquad (8.106)$$

$$I_f = (1/12d).\Delta f\,/f^{\,2}.10^{(wf)/10} \qquad \text{frequency diversity} \qquad (8.107)$$

where I_S = the space diversity improvement factor.
       I_f = the frequency diversity improvement factor.
       S = the antenna separation for space diversity in meters.
       Δf = the frequency separation for frequency diversity in MHz.
       f = the radio frequency in GHz.
       W_f = the fading depth below the no-fade signal level in dB.
       d = path length in km.

Equations 8.106 and 8.107 are valid approximations under the following conditions:

$\Delta f \leqslant 500\text{MHz}$	$f = 2{-}11\text{GHz}$	$W_f = 30{-}50\text{dB}$
$S \leqslant 15\,\text{m}$	$d = 30{-}70\,\text{km}$	$I_S \text{ or } I_f \geqslant 10$

Hence with diversity, the probability of reaching the fade margin due to Rayleigh fading as given by equation 8.103 will be modified by the diversity improvement factor I, using I_S or I_f whichever is applicable, and where $W_f = F$, to give:

$$P(F) = (1/I).K.Q.f^{\,B}.d^C.10^{-F/10} \qquad (8.108)$$

If only one or two protection channels are available for several operating channels, reduced diversity effect will result, since several operating channels can request the protection channel simultaneously. Reference 7, p. 94 deals with this subject in more detail.

8.5 DIVERSITY

To combat reflection or multipath fading on a line-of-site radio link various diversity techniques have been developed. The concept is to select the highest received signal-to-noise ratio from two or more signal channels which carry the same information, but which are taken from separate receivers. As long as the signal channel received levels fade independently, or have a low cross-correlation, the effect of signal fading on the received signal can be reduced. In most diversity systems, either two receiving channels operate at different radio frequencies (frequency diversity), or they are fed from two antennas spaced apart (space diversity). The combining of both space and frequency diversity produces a system known as quadruple-channel diversity.

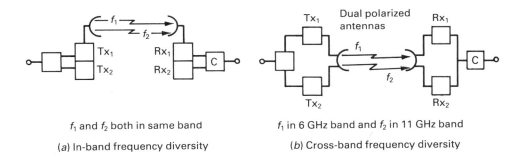

f_1 and f_2 both in same band

(a) In-band frequency diversity

f_1 in 6 GHz band and f_2 in 11 GHz band

(b) Cross-band frequency diversity

Figure 8.29   Frequency – Diversity Systems (Courtesy of CCITT, Reference 2)

8.5.1 Frequency Diversity[15]

Frequency diversity is used to take advantage of the fact that deep multipath fades in a given path tend to occur at different times on frequencies which are sufficiently separated. Figure 8.29 shows in-band and cross-band frequency diversity systems. In-band frequency diversity normally requires a spacing of 2 per cent in frequency shift from the adjacent radio channel to be reasonably effective. A spacing of 5 per cent would be needed to provide uncorrelated fading, but is not usually available because of assigned frequency bands and frequency allocations. A drawback to frequency diversity is that the frequency spectrum is overcrowded, and therefore to use one radio channel for protection may be considered extravagant. The cross-band diversity system utilizes 11 GHz channels for normal traffic, and 6 GHz channels for diversity. This protects against multipath and rain fading and at the same time conserves the frequency spectrum in the more crowded 6 GHz band.

The spacing between diversity frequencies using only one antenna at each end can be determined from:

$$\Delta f = 75.d/(h_1' h_2') \tag{8.109}$$

where h_1' and h_2' = the antenna heights above the reflection point tangent plane in meters.
d = the path distance between the two antennas in meters.
Δf = the diversity frequency spacing in MHz.

8.5.2 Space Diversity[16]

The space diversity arrangement shown in Figure 8.30 can provide full equipment redundancy when automatically switched hot stand-by transmitters are used, but does not provide a separate end-to-end operational path as does frequency diversity. This method uses two receiving antennas, mounted at different heights, to reduce the correlation fading of the two signals, which traverse different paths over the space between the transmitting and receiving antennas. Because of the requirement for additional antennas and waveguides, it is more expensive than the frequency diversity arrangement. However, it provides a more efficient use of the frequency spectrum, and extremely good diversity protection, which can often be better than that obtainable by frequency diversity, especially when the latter is limited to small frequency spacing intervals (less than 2 per cent).

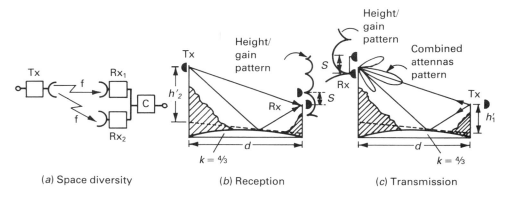

Figure 8.30 Space and Complementary Space Diversity (Courtesy of CCITT, Reference 2)

Complementary space diversity is intended to combat reflection fades. The correct spacing between the two antennas on a tower is different at each end of the path. The height/gain pattern at the receiving end shows that one antenna is placed at a signal maximum, whilst the other is at a minimum. As k changes, the height/gain pattern will move up and down, so that the optimum spacing will change. If the path clearance is very large the spacing may be such as to put both antennas in a minimum as k approaches infinity. This possibility should be considered when the calculations for the spacing are done. In order to have on the same tower one antenna at a maximum on the height/gain pattern, and the other at a minimum, the antenna spacing should be half that of the optimum antenna size as given by equation 8.100. Thus, for either end of the hop, the antenna spacings are given by:

$$\left.\begin{array}{l} S = \lambda d/4h_2' = 75\ d/(fh_2')\ \text{at the lower end} \\[2mm] S = \lambda d/4h_1' = 75\ d/(fh_1')\ \text{at the upper end} \end{array}\right\} \qquad (8.110)$$

For the transmitting antenna, if the antennas are connected in parallel, the resulting array will be such that the direct path will be on the main lobe, whilst the reflected ray is in a null in the antenna pattern, as shown in Figure 8.30(c).

Conventional space diversity is used to protect against multipath fading and it involves selecting the stronger of two signals from antennas spaced far enough apart to give zero or very little correlation between the fading on the two signals. Normal spacings are around 100–200 wavelengths. At 200 wavelengths extremely good protection against multipath fading results. In most cases, a spacing chosen for complementary diversity will also be large enough to reduce significantly the correlation of the multipath fading. Conversely, a spacing based on reducing the effects of multipath fading will normally be sufficient to protect against reflection fading.

8.5.3 Hybrid Diversity

This is a specialized form of diversity which consists of a standard frequency diversity

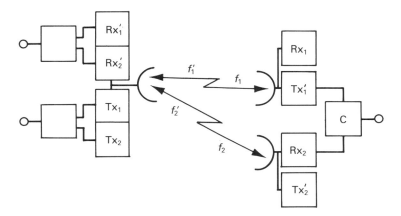

Figure 8.31   Hybrid Diversity

path, in which the two transmit and receive pairs at one end of the path are separated from each other, and connected to separate antennas which are vertically spaced as in the space diversity case. The arrangement provides a space diversity effect in both directions; in one direction because the receivers are vertically spaced, and in the other direction because the transmitters are vertically spaced. This configuration combines the operational advantages of frequency diversity with the improved diversity protection on reflective or difficult paths, of space diversity. Figure 8.31 shows the arrangement of the transmitter and receiver pairs. This arrangement has most of the advantages of both frequency and space diversity, and uses only half the number of antennas in a space diversity system. It is especially good in cross-band applications, since it uses two relatively inexpensive single-band antennas with better performance than the expensive dual-band antennas.

8.5.4 Diversity over Mountain Paths

Long mountain diffraction paths over the ocean, in the 800 MHz to 2 GHz range, encounter rapid multipath fading. At the receiver there exists a horizontal pattern, in a direction normal to the path. Two antennas are used, one at a maximum and the other at a minimum, in a manner similar to that for complementary diversity, but in a horizontal direction. The normal spacing is given by:

$$S = \lambda d_2/2L = 150d_2/f\, L \qquad (8.111)$$

where d_2 = distance from diffraction ridge to receiver in meters.
   λ = wavelength in meters.
   L = distance between the two diffracting points on the ridge in meters.
      for $0.5F_1 \leqslant L \leqslant F_1$
   f = the frequency in MHz.
   S = the spacing of the two antennas in meters.

In practice, if S is 150–200 wavelengths the correlation factor for fading on the two antennas will be approximately 0.6.

8.5.5 Space Diversity as Compared with Frequency Diversity

Either type of diversity provides good protection against multipath fading, but in areas where spectrum congestion is not a problem, frequency diversity has several advantages:

   Only one antenna and waveguide is required, instead of two.
   The towers are lower and more lightly loaded.
   It is easier to obtain redundancy of equipment, as no transmitter switching is required.
   A second complete electrical path is available to make maintenance easier.
   One stand-by channel can act as protection for two or more working channels.

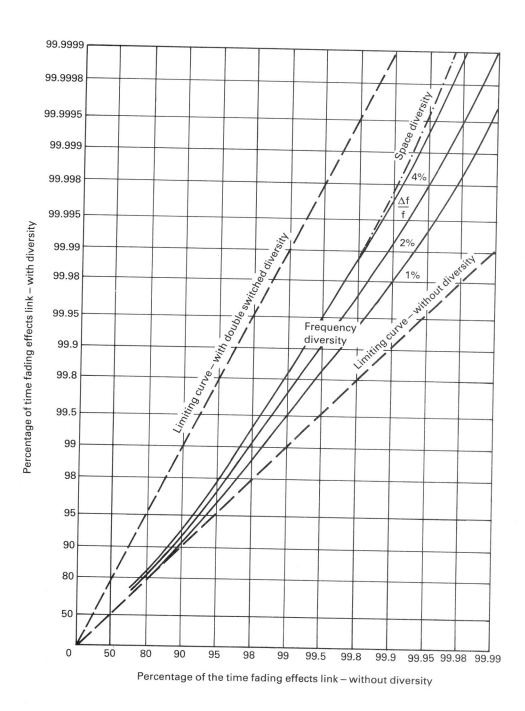

**Figure 8.32   Improvement Obtained by Space or Frequency Diversity
(Courtesy of CCIR Report 338–3)**

On the other hand, it is seldom possible with frequency diversity to obtain more than 2 per cent diversity, and frequency assignments may be scarce, while space diversity provides more protection against 'Earth bulging' or obstruction fading.

The amount of improvement against interference fading, multipath or reflection selective fading, which will be provided by diversity, depends on the distribution of the fading on each diversity half, and on the degree of correlation between the two distributions. If each diversity half has Rayleigh-distribution-type fading, the correlation coefficient between the two halves can be any value from 0 to 1, as shown in Figure 8.21. A correlation coefficient of 0 would mean completely independent fading on the two halves, and a correlation coefficient of 1 would mean that the two halves faded identically. The coefficient of correlation, when zero, means completely uncorrelated or independent fading, and this is often referred to as a 'Rayleigh-squared' distribution, since then, the probability that both halves will simultaneously fade below a given level is equal to the square of the probability that either half alone will fade below that level.

Figure 8.32[17] indicates the improvement obtained with space or frequency diversity. These curves are valid for the least favorable month on links seriously affected by fading, for frequencies between about 2 and 10 GHz. The improvement would probably be slightly better for lower frequencies. With space diversity, it is assumed that the vertical spacing between the antennas is greater than about 150 wavelengths.

8.6 PERFORMANCE CALCULATIONS

The performance calculations are completed so that the noise in a planned system can be compared with some objective before the system is implemented. Usually the calculated noise performance is equal to or better than that set down by the objective. If the planned noise performance of the hop or hops does not meet the objective, then serious consideration should be given to re-design of the system. One such objective is that given by CCIR Recommendation 395. This is reproduced in part in Section 8.6.1, and in Table 5.1.

8.6.1 CCIR Noise Performance Objectives

The CCIR recommends* that:

I. In circuits established over real links which do not differ appreciably from the hypothetical reference circuit, the psophometrically weighted noise power at a point of zero relative level in the telephone channels of a frequency-division multiplex

* In CCIR Recommendation 395–1

radio-relay system of length L, where L is between 280 and 2500 km, should not exceed:

1. $3L$ pW mean power in any hour (under review).
2. $3L$ pW one-minute mean power for more than 20 per cent of any month.
3. 47 500 pW one-minute mean power for more than $(L/2500) \times 0.1$ per cent of any month; it is recognized that the performance achieved for very short periods of time is very difficult to measure precisely and that, in a circuit carried over a real link it may, after installation, differ from the planning objective.

II. Circuits to be established over real links, the composition of which for planning reasons differs substantially from that of the hypothetical reference circuit, should be planned in such a way that the psophometrically weighted noise power at a point of zero relative level in a telephone channel of length L, where L is between 50 and 2500 km, carried in one or more baseband sections of frequency-division multiplex radio links, should not exceed:

A. *For 50 km $\leqslant L \leqslant$ 840 km*
1. $3L$ pW + 200 pW mean power in any hour (under review).
2. $3L$ pW + 200 pW one-minute mean power for more than 20 per cent of any month.
3. 47 500 pW one-minute mean power for more than $(280/2500) \times 0.1$ per cent of any month when L is less than 280 km, or more than $(L/2500) \times 0.1$ per cent of any month when L is greater than 280 km.
B. *For 840 km $< L \leqslant$ 1670 km*
1. $3L$ pW + 400 pW mean power in any hour (under review).
2. $3L$ pW + 400 pW one-minute mean power for more than 20 per cent of any month.
3. 47 500 pW one-minute mean power for more than $(L/2500) \times 0.1$ per cent of any month.
C. *For 1670 km $< L \leqslant$ 2500 km*
1. $3L$ pW + 600 pW mean power in any hour (under review).
2. $3L$ pW + 600 pW one-minute mean power for more than 20 per cent of any month.
3. 47 500 pW one-minute mean power for more than $(L/2500) \times 0.1$ per cent of any month.

III. The following notes should be regarded as part of the recommendation:

Note Noise in the frequency-division multiplex equipment is excluded. On a 2500 km hypothetical reference circuit the CCITT allows 2500 pW mean value for this noise in any hour.

Examples

1. What is the objective for a 156 km hop length, in terms of maximal fractional time, of not exceeding the 47 500 pW one-minute mean power recommendation?

   As the length of the hop is 156 km, from A.3 above, the fraction of the month which the noise should not exceed 47 500 pW is given by:

   $(280/2500) \times 0.1$ per cent $= 11\,200 \times 10^{-8}$, which is the objective.

2. What is the objective in terms of the maximal fractional time, of not exceeding the 47 500 pW on a system of length 420 km?

From A.3. above, the 47 500 pW one-minute mean noise power should not be exceeded for more than $(L/2500) \times 0.1$ per cent of the month where $L > 280$ km. Hence the objective is $420/2500 \times 0.1$ per cent $= 16\,800 \times 10^{-8}$

   CCIR Recommendation 393–1 allocates 10 000 pWp (psophometrically weighted) noise on a 2 500 km hypothetical reference circuit carrying frequency-division multiplex telephony circuits. Hence the total noise accumulation on a per-kilometer basis is given as:

   10 000 pWp/2500 km = 4 pWp/km

Of the 10 000 pWp, the amount allocated to terminal equipment such as multiplex equipment, etc., is 2500 pWp, and 7500 pWp to line equipment. Thus for radio-link systems, the total noise accumulation on a per-kilometer basis is given by:

   7500 pWp/2500 km = 3 pWp/km

Table 5.1 provides a summary of the CCIR general and special objectives.

3. Using the same hop length as that of Example 1 above, find the total noise objective in dBm0p.

   The length of the hop (or hops) is 156 km; from Table 5.1, the total line noise is given by

   $(3L + 200)$ pWp $= (3 \times 156 + 200)$ W $= 668$ pWp

   Hence the objective = 61.75 dBm0p.

4. Using a hop length of 420 km as in Example 2 above, find the total line noise objective in dBm0p.

   As the length of the hop is 420 km, from Table 5.1, the total line noise voltage is given by:

   $(3L + 400)$ pWp $= (3 \times 420 + 400)$ W $= 1660$ pWp

   Hence the objective = 57.8 dBm0p.

8.6.1.1 Listing of Performance Calculations
The performance calculations are based on equipment planning parameters which

are to be expected under normal operating conditions. Usually the calculations are given for the top slot. In order to verify that the specified limits can be met, a separate basic noise breakdown is given for the other two slots. Most of the listed parameters have already been given or derived elsewhere in previous chapters, and it is the purpose of this section to bring together all of these parameters in order to illustrate their use in an actual system design format. The following itemized list is the same as an actual listing of the parameters as they would appear in a typical design.

1. Path length. As calculated from equation 8.89
2. Number of hops in this section
3. Number of modem sections included
4. Number of through and drop/insert stations
5. Tx/Rx thermal and intermodulation noise in pW0p, as given by equations 7.25 or 7.29 for top slot
6. Interference noise. This is the allowance for unpredictable mutual influences between radio links of the planned system (pW0p)
7. Echo intermodulation noise. As given by Section 7.9 (pW0p)
8. Modem basic and intermodulation noise. As given for the top FDM channel for the equipment loaded with an equivalent white noise load as per CCIR Recommendation 399–3 (see Table 7.2). Modem basic noise is the residual equipment noise in the no-load condition, given for the top FDM channel (pW0p)
9. The insert contribution noise in the main baseband. This is due to the possible insertion at the repeater stations of in-band traffic (pW0p)
10. Noise contribution due to auxiliary traffic in the main baseband. This is due to possible insertion of wayside traffic and auxiliary traffic (pW0p)
11. Total fixed noise. This is the sum of the fixed noise contributions (items 5 to 10) (pW0p)
12. Loss-dependent thermal noise. As given by equation 8.84 (pW0p)
13. No-fade total noise. The sum of the total fixed and loss-dependent thermal noise (sum of items 11 and 12; pW0p)
14. The equivalent signal-to-noise ratio is given by:

$$\text{Equivalent } S/N = 90 - 10 \log (\text{no-fade total noise} - \text{item 13}) \tag{8.112}$$

15. Assumed simultaneous fade over all the hops of the same section. As given by equation 8.91 or additional information (dB)
16. Loss-dependent thermal noise under the conditions given in item 15. As given by equation 8.84 with the additional loss of item 15 included
17. The total faded noise is the sum of the fixed noise contributions and the total faded loss-dependent noise as given in item 16
18. The equivalent signal-to-noise ratio is given by equation 8.112 with the no-fade total noise replaced by the total faded noise as per item 17
19. Noise objectives in accordance with CCIR or other body. As given by Table 5.1 or other values adopted
20. Probability of exceeding 47 500 pW0p. This value is given by equation 8.103 with the fade margin (F) given by the unfaded received carrier level (equation 8.77) − 43.2 (As $S/N = 90 - 10 \log \text{pW0p} = 90 - 10 \log 47\,500 = 43.2 \text{ dBm0p}$)
21. Objective of reaching 47 500 pW0p as per CCIR Recommendation 395. This is found near the beginning of Section 8.6.1
22. Probability of exceeding outage. This is given by equation 8.103 with the fade margin determined from the practical threshold
23. Worst hop with 30 dB fade. This is the hop with the smallest fade margin to 47 500 pW0p
24. Total loss-dependent thermal noise with 30 dB fade on the worst hop. As per equations 8.79 and 8.84

25. Total noise with 30 dB fading is the sum of items 24 and the total fixed noise of item 11
26. The equivalent signal-to-noise ratio is given by:

Equivalent $S/N = 90 - 10 \log$ (total noise of item 25)

27. The noise objective is in accordance with the relevant specification

8.6.1.2 Basic Noise Breakdown

A breakdown of the psophometrically weighted basic noise power is usually given for three narrow measuring channels. The column 'fixed noise' gives the basic noise for the Tx/Rx and modem, with no noise loading applied to the system.

A separate breakdown is also made for the loss-dependent noise. The totals of the noise breakdown are guaranteed per modem section.

EXAMPLE OF PATH CALCULATIONS AS PER ABOVE LISTING

1.	Section length	154.5 km
2.	Number of hops	4
3.	Number of modem sections	1
4.	Number of through or drop/insert stations	5
5.	Tx/Rx thermal noise plus intermodulation	92 pW0p
6.	Interference noise	40 pW0p
7.	Feeder echo noise	20 pW0p
8.	Modem basic and intermodulation	35 pW0p
9.	Insert contribution	--
10.	Auxiliary traffic contribution	--
11.	Total fixed noise	187 pW0p
12.	Loss-dependent thermal noise	87.8 pW0p
13.	No-fade total noise	274.8 pW0p
14.	Equivalent S/N	65.6 dBm0p
15.	Assumed simultaneous fade	5 dB
16.	Loss-dependent thermal noise	277.6 pW0p
17.	Total noise	464.6 pW0p
18.	Equivalent S/N	63.3 dBm0p
19.	Objective	62.0 dBm0p
20.	Probability of exceeding 47 500 pW0p	597.1×10^{-8}
21.	Objective	6180.0×10^{-8}
22.	Probability of exceeding outage	55.4×10^{-8}
23.	Worst hop with 30 dB fades	2
24.	Loss-dependent thermal noise	28359.5 pW0p
25.	Total noise	28546.5 pW0p
26.	Equivalent S/N	45.4 dBm0p
27.	Objective	41.6 dBm0p

BASIC NOISE BREAKDOWN

	Measuring slot kHz			Measuring slot kHz		
	7600	3886	534	7600	3886	534
	(Fixed noise, pW0p)			(Loss-dependent noise, pW0p)		
Tx/Rx basic noise	52	52	20	–	–	–
Modem basic noise	15	20	30	–	–	–
Total basic noise	67	72	50	87.8	69.6	2.1

8.7 VARIATION OF K-FACTORS IN DIFFERENT PARTS OF THE WORLD

As seen from Section 8.3, the k-factor is one of the more important parameters in the antenna height calculations. Because the tower height is higher or equal to the antenna height, and its cost is approximately proportional to the square of the height, it is just as important not to over-design as it is not to under-design the tower height. As the maximum and minimum values of k-factor are different in different areas of the world, a knowledge of the range of k-factor values for a particular area becomes important for near optimum design. In the next few sections are provided some values which have been used in various parts of the world and which have been proved successful.

8.7.1 Continental Europe[7,17]

Figure 8.33 (CCIR Report 338–3, after References 7 and 17) shows the minimum effective value of k which is exceeded for approximately 99.9 per cent of the time, as a function of path length, and for only 1 per cent of the time. These curves then provide information on the maximum and minimum values of k to be expected in a continental temperate climate, such as that of Europe.

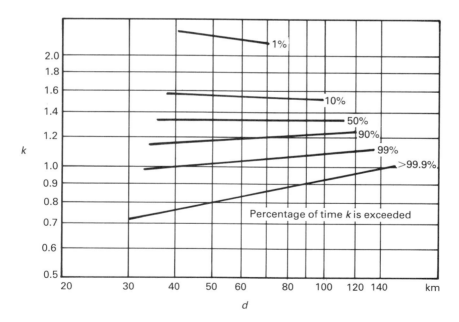

Figure 8.33   Minimum Effective Value of k with a Continental Temperate Climate (Courtesy of Heyden & Son, Reference 7)

8.7.2 Continental USA

Table 8.2 is taken from Reference 8, p. 188, and provides a guide to k-factors in the USA for varying propagation conditions.

Table 8.2 (for 99.9–99.99 per cent path reliability)

	Propagation conditions				
	Perfect	*Ideal*	*Average*	*Difficult*	*Bad*
Weather	Standard atmosphere	No surface layers or fog	Substandard light fog	Surface layers, ground fog	Fog moisture over water
Topology	Temperate zone, no fog no ducting, good atmospheric mix day and night	Dry, mountainous no fog	Flat, temperate, some fog	Coastal	Coastal, water, tropical
k-factor	1.33	1–1.33	0.66–1.0	0.66–0.5	0.5–0.4

8.7.3 Ivory Coast[18]

Figure 8.34 shows the cumulative distribution of the k-factor for Abidjan, and illustrates the variability of the effective earth's radius. For February (worst of the four months) the k-factor varies from approximately 0.5, exceeded for about 0.5 per cent of the observations, to infinity, not exceeded by the k-factor becoming negative for 95 per cent of all observations. The wide variation in k-factor suggests the increased probability of serious fading and even fade-out. In the Niger delta area, the January/February period is the period when the Harmattan winds bring the dust down from the Sahara desert, and is predictably the worst month for radio propagation. In this period there is a wide diurnal variation of temperature and humidity.

Details relating to Figure 8.34 are as follows:

Abidjan/Port Bouet, Ivory Coast (West Africa)
05–15N, 03–56W   15 meters MSL
Data: Radiosonde   0600Z (0600 LST)
                                                  8/58–8/62, 2/63–5/63
Temperature, °C   January 31/23; July 28/23
Mean dewpoint, °C   January 24; July 22
Precipitation, centimeters   Annual 207.0; June 55.2; January 3.1

Located on the north shore of Ebrie lagoon; separated from the Atlantic Ocean by a sand bar. A humid tropical maritime climate.

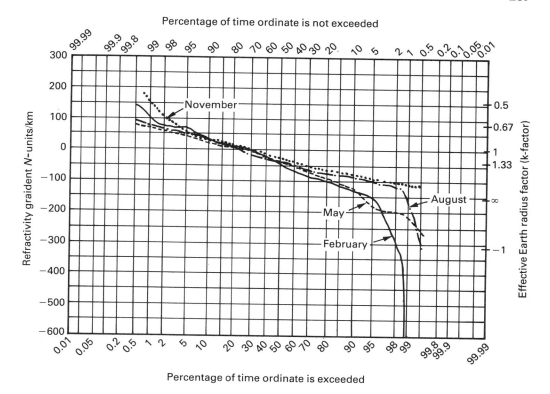

Figure 8.34 *Cumulative Distribution of the Observed Radio Refractivity Gradient in the Ground – based 100 m layer – Abidjan*

8.7.4 Nigeria

Radiosonde data showing the vertical structures of temperature, humidity and atmospheric pressure for the period from 1971 to 1975 were collected through the courtesy of the Research and Training Institute, Meteorological Department, Oshodi, Lagos, Nigeria and processed into the refractivity gradient dN/dh.

Figure 8.35 shows the cumulative probability distribution curve (solid line) for March, obtained from the data for 1971, 1973 and 1974, whereas Figure 8.36 shows the cumulative probability distribution for October 1971 to October 1975.

In these figures are also drawn the approximate normal exponential-combined-distribution curves (dotted lines). Both the estimated and actual curves agree to a fair degree. It is noted that the actual curve for October is much smoother than that of March, since the number of samples for October is greater (126) than those taken for March (58). The March curve is considered as the worst among the twelve curves for diffraction loss due to earth bulge. From the March curve, the percentage of time that the k-factor falls below a certain value is given as follows:

Percentage of time	dN/dh N-units/km	k-factor k
50	−48	1.44
10	−28.5	1.22
1	−13	1.09
0.1	−2	1.01
0.01	+8.0	0.95
0.001	+16.5	0.90

Note: $k = 157/(157 + dN/dh)$ (See Section 8.7.5) Seasonal variations of k-factor (mean value) are shown in Table 8.3.

Table 8.3 Seasonal k-factor (mean) Variations – Oshodi, Lagos, Nigeria

Month	Jan	Feb	Mar	Apr	May	June	July	Aug	Sept	Oct	Nov	Dec
−dN/dh	−50.8	−48.9	−48.1	−47.3	−51.0	−53.9	−51.5	−47.7	−49.7	−49.6	−49.5	−48.0
Standard deviation for dN/dh N-units/km	16.4	10.5	15.2	12.2	12.6	10.1	8.2	8.2	9.0	11.5	9.9	13.6
k-factor	1.48	1.45	1.44	1.43	1.48	1.52	1.49	1.44	1.46	1.46	1.46	1.44
No. of samples	50	53	58	57	56	86	112	119	109	126	87	49

8.7.4.1 Radio Path Clearance Criteria

The analysis of the data shows that the k-factor may take a value below one less than 0.1 per cent of the time for the worst month (March), however a design assumption would be that the k-factor would fall below 2/3 for a very small percentage of the time. For normal atmospheric conditions, the k-factor is distributed between 1.40 and 1.53 (one standard deviation). The radio path clearance criteria which may possibly be used, taking into account both the performance and economy of the system, are:

FOR 2 GHz AND 6 GHz LINKS
$0.577F_1$ for $k = 4/3$
$0.3F_1$ for $k = 2/3$

FOR A 450 MHz LINK OVER A SMOOTH EARTH
$0.3F_1$ for $k = 4/3$

FOR A 450 MHz LINK OVER A KNIFE-EDGE OBSTACLE
$0.0F_1$ for $k = 2/3$

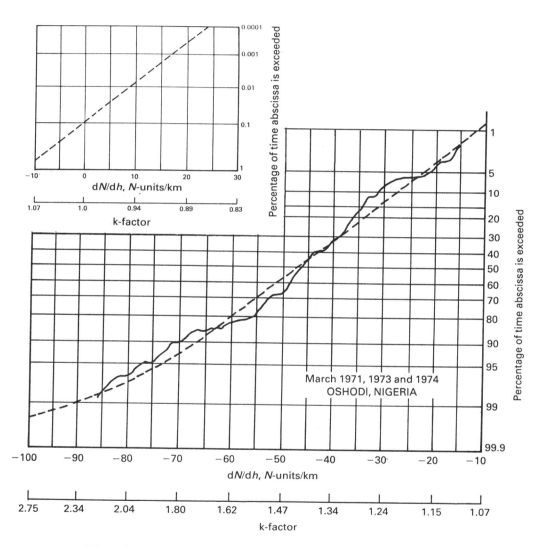

Figure 8.35   Cumulative Probability Distribution of Refractivity Gradient and k-Factor (March)

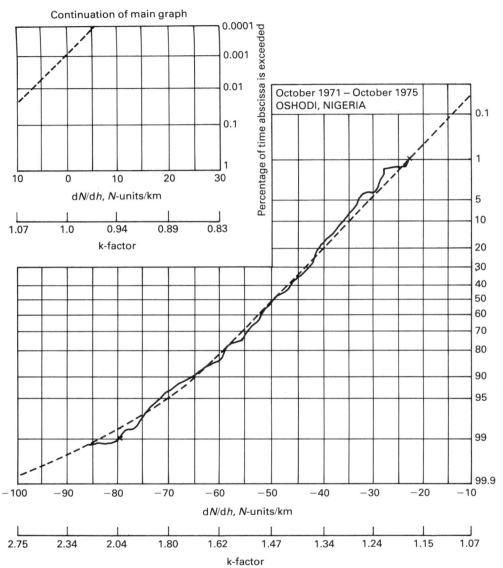

Figure 8.36 *Cumulative Probability Distribution of Refractivity Gradient and k-Factor (October)*

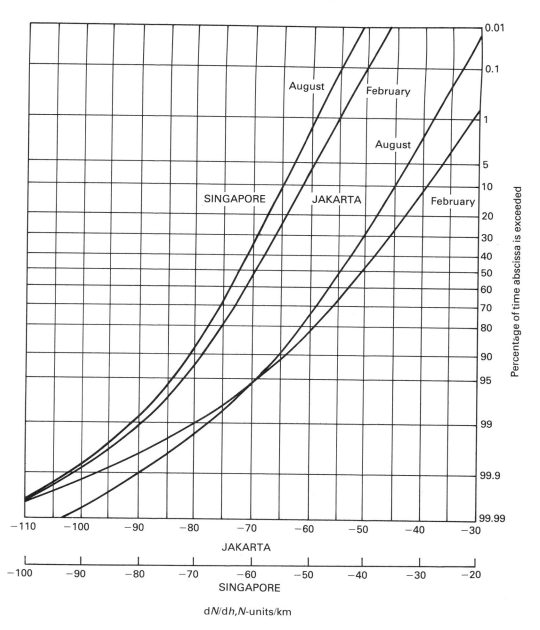

Figure 8.37 **Cumulative Distribution of dN/dh in South-East Asia**
(Courtesy of IECE of Japan, Reference 9)

8.7.5 South East Asia[9]

Atmospheric refractivity gradients in Indonesia and Singapore as well as in and around Japan have been analyzed in order to estimate the variation of refractivity gradient in South East Asia. Figure 8.37 provides the cumulative distribution of refractivity for both Singapore and Indonesia.

Using equation 8.13, where the Earth's radius a is 6.37×10^3 km, the conversion between dN/dh and k-factor becomes:

$$k = 157/(157 + dN/dh) \tag{8.13}$$

Table 8.4 Typical Values of k-factor for various Atmospheric Conditions

k-factor	dN/dh (N-units/km)	Atmospheric conditions	Microwave propagation
5/12	220	Humidity inversion (large positive gradient)	Extreme earth's bulge
1/2	157	Moderately sub-refractive	Moderate earth's bulge
2/3	80	Slightly sub-refractive	Slight earth's bulge
1	0	Uniform density	No refraction
1.25	−30	Dry atmosphere	Standard (mountainous)
4/3	−40	Standard atmosphere	Standard
1.6	−58	Humid atmosphere	Standard coastal
0	−157	Moderate negative gradient	Flat earth
−1	−314	Steep gradient	Possible fade-out
−0.5	−470	Extreme gradient	Fade-out

EXERCISES

1. What would be the earth bulge for a k-factor of 2, at a point 15 km from an antenna, if the path length is 40 km? (*Answer* 14.72 m)

2. What is the first Fresnel radius about the main beam, 15 km from an antenna on a 40 km path? The radio carrier centre frequency is 1 GHz. (*Answer* 53 m)

3. At a frequency of 1 GHz, is an irregularity in the terrain of height 1 meter, rough or smooth, when viewed 15° from the horizontal? (*Answer* Rough, since an irregularity would appear smooth only if it were less than 0.16 m high))

4. Determine the free-space path loss of the radio hop of Question 2. (*Answer* 124.44 dB)

5. What is the system figure of merit for an 1800-channel system with a de-emphasis factor of 4 dB, a receiver noise figure of 9 dB, a transmitter power of 30 W, and CCITT weighting? (*Answer* 142.93 dBm)

6. If the received signal level of Question 5 was calculated to be -60 dBm, what would the net loss be, and what would be the loss-dependent noise in pW0p? (*Answer* 104.77 dB, and 152 757 pW0p)

7. If the probability of reaching the fade margin due to Rayleigh fading is one part in one hundred million during the worst month, and the radio link is in North-West Europe, determine the value of the fade margin for the link of Question 2. (*Answer* 57.53 dB)

REFERENCES

1. Hall, M. P. M., *Effects of the Troposphere on Radio Communication* (Peter Peregrinus, 1979).
2. Economic and technical aspects of the choice of transmission systems, Appendix to Section B.IV.3 of the handbook *Propagation* (ITU, 1871), pp. 7–74.
3. Beckmann, P. and Spizzichino, A., *The Scattering of Electromagnetic Waves from Rough Surfaces* (Pergamon Press, 1963).
4. Bullington, K., 'Radio Propagation at Frequencies above 30 Megacycles', *Proc. IRE*, 1947, **35**, pp. 1122–1136.
5. Morita, K., 'Severe Temperature Inversion Layer Regions in the World', *Review of the Electrical Communications Laboratories*, 1980, **28**, No 11–12, pp. 1053–1056.
6. Pearson, K. W., 'Method for the Prediction of the Fading Performance of a Multisection Microwave Link', *Proceedings IEE*, 1965, **112**, No. 7, pp. 1291–1300.
7. Brodhage, H. and Hormuth, W., *Planning and Engineering of Radio Relay Links* (Siemens and Heyden, 1977).
8. Freeman, R. L., *Telecommunication Transmission Handbook* (Wiley, 1975).
9. Akiyama, T. and Hatano, K., 'Radio Refractivity Gradient in South-East Asia', *Trans. IECE of Japan*, 1977, **E60**, No. 1, pp. 21–22.
10. Chatterjee, B., *Propagation of Radio Waves* (Asia Publishing House, 1963).
11. Vincent, W. R., 'Comments on the Performance of VHF Vehicular Radio Sets in Tropical Forests', *IEEE Trans. Veh. Technol.*, 1969, **VT–18**, pp. 61–65.
12. CCIR Report 715, 1978, Propagation by Diffraction.
13. ITT, *Reference Data for Radio Engineers*, 6th Ed. (Howard W. Sams, 1981), pp. 28–9.
14. Morita, K., 'Prediction of Rayleigh Fading Occurrence Probability on Line–of–Sight Microwave Links', *Rev. Elec. Comm. Lab., NTT, Japan*, 1970, **18**, pp. 11–12.
15. Barnett, W. T., 'Microwave Line–of–Sight Propagation with and without Frequency Diversity', *BSTJ*, 1970, **49**, p. 8.
16. Vigants, A., 'Space–Diversity Engineering', *Bell System Technical Journal*, 1975, **54**, p. 1.
17. Boithias, L. and Battesti, J., 'Protection against Fading on Line–of–Sight Radio Relay Systems' (in French), *Ann. des telecomm.*, September-October, 1967.
18. Samson, C. A., US Department of Commerce, Office of Telecommunications, *OT–Report* 75–79, 1975.

9 ANTENNAS AND ANTENNA SYSTEMS

9.1 ANTENNAS

9.1.1 Voltage Standing Wave Ratio and Return Loss

An antenna is both a transmitter and receiver of carrier wave or signal power. Its impedance can be measured only when it is used as an instrument to transmit power from the transmitter to the intervening medium between it and the receiving antenna. To measure the antenna impedance a slotted line matched to the transmission line is connected between the transmission line and the antenna, and power from the transmitter applied. The slotted line then permits the measurement of the reflected power. If the reflected power is zero, the voltage standing wave ratio (VSWR) is unity, and the antenna is said to be perfectly matched to the transmission line. The relationship between VSWR and forward power P_f and reflected power P_r can be derived by the ensuing discussion:

$$P_f - P_r = \text{Power in watts delivered into the antenna} \qquad (9.1)$$

Section 7.10 provided alternative definitions of return loss. Of interest in this section is equation 7.36, which for convenience is repeated below:

$$\text{Return loss} = 10 \log (\text{Forward power})/(\text{reflected power}) \text{ dB} \qquad (7.36)$$

In terms of the forward signal voltage V_f, and the reflected signal voltage V_r, the negative return loss factor is given by:

$$P_r/P_f = \phi = (V_r^2/Z)/(V_f^2/Z) = (V_r/V_f)^2 \qquad (7.36)$$

where $10 \log \phi = $ The negative return loss in dB, or the return gain in dB.

When a standing wave is set up, the reflected voltage wave in its two extreme modes can be either opposing the forward wave, or in phase with it. The VSWR is thus the ratio of these two extremes, i.e.

$$\text{VSWR} = (V_f + V_r)/(V_f - V_f) = (1 + V_r/V_f)/(1 - V_r/V_f)$$

$$\text{VSWR} = (1 + \phi^{\frac{1}{2}})/(1 - \phi^{\frac{1}{2}}) = \rho \qquad (9.2)$$

$$\phi = (\rho - 1)^2/(\rho + 1)^2$$

$$\text{i.e. Return Loss RL} = 20 \log \frac{\rho + 1}{\rho - 1} \text{ dB} \qquad (7.37)$$

9.1.2 Effective Aperture and Gain of an Antenna[1,7]

The effective area or effective aperture of an antenna in a particular direction describes the property of an antenna when it is used as a receiver. With the antenna matched to its transmission line, and receiver, the antenna will deliver to the receiver an amount of power which it extracts from the incident plane wave coming from a particular direction, and having an energy density in watts/meter.[2] The amount of power which is delivered to the receiver, if divided into the energy density of the incident plane wave, produces an 'effective antenna receiving area' in the direction from which the plane wave is incident, and at the receiver frequency.

An isotropic antenna is one which radiates uniformly in all directions over the full 4π solid angle about it. Such an antenna is purely theoretical for it is not possible to construct one in practice, but it is used as the reference to which all antenna gains are usually referred. The *gain* of an antenna is a transmitting parameter, and is defined as the ratio of the power radiated per unit solid angle in a particular direction (usually the principal direction), to the power which would be radiated by a matched *isotropic* antenna with the same power applied to it. The gain is usually expressed in decibels.

$$\text{The effective area of an isotropic antenna} = \lambda^2/4\pi \qquad (9.3)$$

The equation relating the gain of any antenna in any particular direction to the effective area in the same particular direction is given by:

$$G(\theta,\phi) = 4\pi S_e(\theta,\phi)/\lambda^2 = S_e(\theta,\phi) \text{ referred to isotropic} \qquad (9.4)$$

where $G(\theta,\phi)$ = the gain in the direction defined by the angles θ, and ϕ.
$\qquad \lambda$ = the wavelength of the carrier.
$\qquad S_e(\theta,\phi)$ = the effective cross-sectional area of the antenna in the direction defined by the angles θ and ϕ.

The gain of a receiving antenna is the same as the gain of a transmitting antenna, and thus their polar patterns are the same. For any antenna, large or small, placed in an isotropic radiation field, will collect the same amount of energy. This means that a large antenna will collect more energy in its principal direction than in other directions, but will collect the same energy as a smaller antenna which collects less in the principal direction than the larger antenna, but more in the other directions. The larger the antenna, the greater will be the reception of power in a particular direction, and less in other directions; this is because the total energy collected by integrating over a complete sphere cannot exceed (but must equal) the total energy collected by integrating over a complete sphere and received by an isotropic antenna. More simply put, if an inflated spherical rubber balloon represents the total energy received (or transmitted) by an isotropic antenna, a larger high-gain, directional antenna can only change the shape of that balloon by stretching it out like a long pencil, without changing the amount of air (energy) contained by that balloon.

9.1.3 Aperture Efficiency[6]

The power transfer between two antennas in the absence of any absorption in the medium between them is given by:

$$P_r = P_t.G_t.S_e/(4\pi d^2) \tag{9.5}$$

where P_r = the received power.

P_t = the transmitted power.

G_t = the gain of the transmitting antenna in the direction of the receiver.

S_e = the effective area of the receiving antenna in the direction of the transmitter.

d = the distance between them.

Compare this equation with equation 8.69, in the determination of the path loss. To achieve good side-lobe suppression, the antenna is usually tapered at the edges, so that illumination is tapered. This in turn reduces the effective area of the antenna, and leads to the concept of aperture efficiency:

$$S_e = \eta_a.S \tag{9.6}$$

where η_a = the aperture efficiency.

S_1 = the physical area of the antenna.

With the tapering of the illumination, the edges of the antenna do not contribute as much to the gain of the antenna in its principal direction as they may otherwise contribute. With this reduction in the effective area comes the reduction in gain. Aperture efficiencies of large antennas vary between 50 and 90 per cent.

The effective area of small antennas (or antennas whose area is small compared to λ^2) does not bear any significant relationship to physical area, and so equation 9.6 does not hold.

9.1.4 Effective Temperature

The effective temperature of an antenna is the temperature of the radiation source which it sees. The side lobes of the antenna are an important consideration when the antenna is observing the sky at 4 K with its main beam. The side lobes, if observing the ground at 300 K, may easily make a disproportionately large contribution to its effective temperature, especially if the side lobes are not excessively far down on the main beam. To overcome this problem, it is common to cover the ground around certain antennas with metallic mesh, in order that it may reflect the temperature of the sky (See Section 5.6.1).

9.1.5 Approximations to Beam Width, Gain, and Angle between Nulls

The beam width θ_1 may be estimated to give an answer which is within a factor of 2 of the correct beam width from the following equation:

$$\theta_1^2 = 4/G_{max} \tag{9.7}$$

where G_{max} is the maximum gain of a circular dish antenna.

The gain G_{max} is given by:

$$G_{max} = 4\pi S_e/\lambda^2 = 4\pi\eta_a\pi a^2/\lambda^2 = (2\pi a/\lambda)^2.\eta_a$$

$$G_{max} = 20 \log d - 20 \log \lambda + 10 \log \eta_a + 9.943 \tag{9.8A}$$

where a = the radius of the circular dish
a, d and λ are in the same units.

Assume that a commercially available antenna has an aperture efficiency η_a of 55 per cent; then the antenna gain can be expressed in dB as:

$$\left.\begin{array}{l} G_{max} = 20 \log d + 20 \log f + 17.8 \text{ where } d \text{ is in meters} \\[2mm] {}_{max} = 20 \log d + 20 \text{ lof } f + 7.5, \text{ where } d \text{ is in feet} \\[2mm] G \end{array}\right\} \tag{9.8B}$$

In equations 9.8B f is in GHz.

Equations 9.8B show that, if the frequency of the radiated signal is increased, the gain increases, and similarly, as the diameter of the dish is increased, the gain increases. This means, of course, that if an antenna is to be used at a higher frequency with a constant gain, a smaller antenna diameter, and therefore a less costly antenna may be used.

The angle between the nulls on either side of the main beam is given as

$$2\theta_1 = 2\lambda/(\pi a(\eta_a)^{1/2}) = 4\lambda/(\pi d(\eta_a)^{1/2}) \text{ radians} \tag{9.9}$$

where a is the radius of the antenna, and d is its diameter

The precise answer depends upon the illumination pattern of the antenna. For a uniformly illuminated circular dish where the aperture efficiency is 55 per cent, the half-power width of the beam is given by:

$$
\left.
\begin{aligned}
2\theta_1 &= 1.7168\ \lambda/d \text{ radians} \\
&= 98.3674\ \lambda/d \text{ degrees} \\
&= 29.5/f.d \text{ where } d \text{ is in meters} \\
&= 96.6/f.d \text{ where } d \text{ is in feet}
\end{aligned}
\right\}
\tag{9.9A}
$$

9.1.6 Polarization

In line-of-sight radio links, the electric or E field is either horizontal, or parallel to the earth's surface or vertical. This gives rise to what is referred to as *horizontal* or *vertical polarization*. The feed to a dish antenna is in some cases a simple dipole, which is either orientated horizontally, to give horizontal polarization, or vertically. In some cases the feed may consist of both horizontally and vertically orientated dipoles, permitting the antenna to transmit and receive different frequencies on different polarizations at the same time. For interference studies, a knowledge of how much radiation a single polarized antenna radiates on the other polarization in accordance with its cross-polarization pattern is important. If a radiating antenna has a poor cross-polarization discrimination at the angle looking toward a receiving antenna on another system or on another part of the same system, no amount of discrimination of that receiving antenna will reduce the component it receives from the cross-radiation of the transmit antenna. Isolation of 20 dB or better may be expected between polarizations in non-precipitation conditions.

9.1.7 Side-lobe Attenuation and Back-to-Front Ratio

The side-lobe attenuation is the ratio in decibels of the power density transmitted or received in the principal direction to the power density at a given angle up to a maximum of 90° from that principal direction. The front/back ratio is the ratio in decibels of the power density transmitted or received in the principal direction to the power density at a given angle in the range 90–270° from the principal direction.

9.1.8 Antenna Bandwidth

The desired radiation characteristic of an antenna, as well as the impedance match to the feeder, can be achieved only over a limited range of frequencies. For this reason, the ratio of the upper to the lower limit of the operating frequency is referred to as the relative bandwidth of an antenna. In some antennas, the dipole feeding, say, a dish antenna is tuned to the centre frequency of the radio channel. Normally the bandwidth of the antenna is, however, much greater than the RF bandwidth of the transmitter, and may accommodate several radio channels in the same band.

9.2 ANTENNA TYPES

Considerations in selecting a microwave antenna are: frequency band, gain, radiation patterns, VSWR, polarization, mechanical and tower requirements, antenna costs, and costs of installation.

Standard antenna diameters are 2.0, 4.0, 6.0, 8.0, 10, 12, and 15 ft, or
                        0.6, 1.2, 1.8, 2.4, 3.0, 3.7, and 4.6 m.

Side and back angle radiation patterns for parabolic antennas may be improved by the use of a shield which takes the form of a cylindrical rim shrouding the front of the dish. Antennas used at frequencies below 3 GHz are usually fed by coaxial feeds which may be foam-filled to prevent the ingress of foreign particles, and to assist in maintaining the rigidity of the feed. Above 3 GHz, however, the feeds are terminated in a rectangular waveguide flange which is so designed that it shapes the beam so as to illuminate the reflector efficiently whilst maintaining a sharp cut-off at the reflector edge. The waveguide is usually pressurized with dry air to protect against condensation or the ingress of water.

Radomes are used to protect against the accumulation of ice, snow, dust, etc., and may also be used to reduce the wind loading of a solid (against a grid) antenna. The radome may reduce the antenna gain, and is usually included in the path calculations as a separate loss in decibels.

9.2.1 Parabolic Reflector Antennas

As mentioned above, these antennas are particularly useful for microwave bands where high directivity is required. A prerequisite is that the tolerance in the contour of the reflecting surface must not exceed roughly one-twentieth of the operating wavelength.

Figure 9.1 shows a cross-section of a parabola. The primary radiator, or a feed,

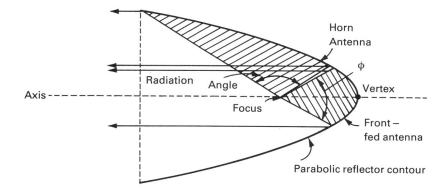

Figure 9.1   Illumination of a Parabolic Reflector

is located at the focal point, and is of such design that the radiation at the edges of the parabolic reflector is about 10 dB below the radiation which is directed towards the vertex, or centre of the dish. Unfortunately, in practice, the primary radiator is not of negligible area in relation to the reflector area, and so a shadow appears in the reflected radiation, which reduces the antenna gain and increases the side-lobe magnitude. The primary feed in addition to causing the shadow also acts as a receiving antenna to the reflected radiation, which reduces the matching capabilities of the antenna.

Two types of antenna in common use are constructed from the parabola. The first is the usual front-fed parabolic reflector antenna, and the second is the horn antenna.

9.2.1.1 Front-Fed Antenna

This antenna is fed from the focal point with its radiation directed towards the vertex of the parabolic reflector. The primary radiator may be a rectangular waveguide with its walls flared to form a horn or, for the lower frequencies, it may be a simple dipole arrangement if linear polarization is desired. Square or circular waveguides flared at the end to form a horn are used for circularly polarized waves or for simultaneous transmission of two orthogonally polarized waves. Horn feeds with either square or circular cross-section and which are excited in their fundamental mode exhibit different radiation patterns in the two orthogonal principal planes of their field of radiation, preventing the same illumination pattern from existing at the reflector edges, and hence producing a different gain for each of the two polarizations. The cross-polarization coupling is greater in this case than for symmetrical radiation patterns. To produce a better rotational symmetry of the radiation patterns produced by the square or circular horn, feeds fins, diagonal excitation, quarter-wave chokes and various other techniques may be employed.

9.2.1.2 Cassegrain Antenna

Figure 9.2 is a schematic of a Cassegrain antenna. It is based upon the multiple

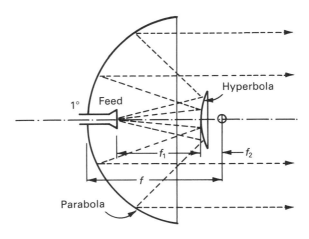

Figure 9.2   Cassegrain Antenna

reflector systems sometimes used in optics, where if the diameters of the reflectors are large with respect to the radiation wavelength a better overall illumination results. The primary feed emerges directly without any bends from the vertex of the main reflector (of focal length f) and is flared to form the primary radiator. A secondary- or sub-reflector which is a convex hyperboloid is mounted so that one of its foci coincides with the focus of the primary radiator (focal length f_1), and the other (focal length f_2) with the focus of the main reflector (paraboloid). The result is to produce an effective focal length f_w for the multiple reflector system given by:

$$f_w = (f_1/f_2).f \qquad (9.10)$$

Since $f_1 > f_2$, the result is an extension of the focal length and a reduction in the shadow loss of that of the direct front-fed parabolic reflector. A very high aperture efficiency can be achieved by suitable correction of the primary and secondary reflector contours.

9.2.2 Horn Reflector Antennas

The horn reflector (cornucopia) antenna has a section of a very large parabola, mounted at such an angle that the energy from the feed horn is simultaneously focused and reflected at right-angles as shown in Figure 9.1. As the gain reduction caused by the obstructing primary feed of a front-fed reflector is due to it being directly in the reflected radiation field, it follows that this gain reduction may be reduced if the primary radiator is directed to some part of the main reflector other than the vertex. The effect of this redirection is also to reduce the effects of antenna impedance mismatch caused by the obstructing primary radiator. As the feed horn is located outside the aperture so that blocking cannot occur, for reasons of practical construction, both radiator and reflector are formed as a single constructed unit. This enclosed design assists in achieving a high front/back ratio of the order of 70 dB or more, which permits operation from two directions or more from a station operating on the same frequencies.

The side lobes due to the construction are also at a very low level. Because of the reduction in the antenna mismatch features, there is a corresponding improvement in the VSWR of this antenna when compared with the front-fed type. With suitable coupling networks multiband operation may be used on both polarizations. The bandwidth is substantially larger than with Cassegrain or front-fed types, because of the horn forming the primary radiator. The radiation pattern and the match meet the requirements imposed on wide-band antennas used for microwave relay systems operating over several octaves. The limits of the frequency range are solely determined by the feeding waveguide, provided the radome is thin enough. The lower frequency limit is given by the waveguide cut-off frequency, and the upper limit by the possible occurrence of parasitic wave modes, which if uncorrected can cause severe distortion. The disadvantages of this type of antenna are that it is very big, heavy, and complex in its mounting. Also it is expensive, not only in the cost of the antenna itself, but in its installation and tower costs. There is only a limited choice of sizes.

Table 9.1 shows the availability of this type of antenna, with details from one manufacturer.*

Table 9.1

Frequency, GHz	Type number	Aperture, ft (m)	Gain, dBi ±0.5 dB			Beam width, degrees	F/B dB	VSWR max
			Bottom	Mid-band	Top			
3.7–4.2	SHX10A	10 (3.0)	39.3	39.8	40.3	2.1	80	1.02
5.925-6.425	SHX10A	10 (3.0)	43.1	43.5	43.8	1.38	90	1.02
7.725-8.275	SHX10A	10 (3.0)	45.2	45.5	45.8	1.10	90	1.02
10.7-11.7	SHX10A	10 (3.0)	47.3	47.7	48.1	0.80	90	1.02
12.7-13.25	SHX10A	10(3.0)	48.7	48.9	49.1	0.70	90	1.02

9.2.3 Cross-Band Parabolic Antennas

These are antennas similar to those described in Section 9.2.1, with feeds designed to permit operation in two widely separated frequency bands, such as 6 GHz and 11 GHz. Because of the complex and critical feed assemblies, these antennas have a slightly reduced gain and a poorer VSWR than single-band antennas.

9.3   DIRECTION-CHANGING ANTENNAS

As it is not always possible to select repeater sites to obtain a line-of-sight transmission, passive reflectors or repeaters may be used to change the direction of the radio beam in order to avoid obstructions which would otherwise cause excessive obstruction loss. As passive reflectors are expensive and difficult to adjust and construct, they should only be considered when no other viable alternative exists. Passive reflectors may consist of two parabolic-reflector antennas joined by waveguide or coaxial cable, or by one plane metal reflector. With paraboloidal reflectors of dimensions comparable with those used at the transmitting and receiving stations, the resulting path losses are often intolerable, so that preference is given to the more expensive plane reflector, provided that the deflection angle does not become too obtuse. If the deflection angle is too obtuse, then two plane reflectors may warrant consideration. Calculations depend on whether the beam direction is changed in the near field or in the far field. The boundary between these field zones is given approximately by:

$$\text{Near field} < 2L^2/\lambda \tag{9.11}$$

* Extracted from the 1981 catalogue of the Andrew Corporation.

where L = the maximum linear dimension of the reflector surface.
λ = the wavelength of the radiation in the same units as L.

For far-field direction changing it is necessary that reflectors are placed so that strong reflections off the surrounding terrain do not cause intolerable intermodulation noise. This intermodulation noise caused by reflections is akin to 'echo distortion' noise, and may be determined using the same theory (see Section 7.9).

The signal-to-noise ratio for intermodulation noise can be calculated if the difference in attenuation δA and the difference in delay δt between the wanted and unwanted signals can be estimated.

For delay differences $\geqslant 1\ \mu s\ S/N = \delta A + 22$ dB $\qquad\qquad$ (9.12)

For delay differences $< 1\ \mu s$, Figure 7.3 must be used.

9.3.1 Calculation of Path Loss using Passive Reflectors

9.3.1.1 Path Loss between Two Parabolic Antennas

The basic transmission loss A_0 can be expressed as:

$$A_0 = 20 \log 4\pi d/\lambda \qquad\qquad (9.13)$$

The power gain G of a parabolic antenna has been given in equation 9.4:

$$G = 4\pi S_e/\lambda^2 \qquad\qquad (9.4)$$

and if both transmitting and receiving antennas are considered, the path loss is given by:

$$A_0 - (G_t.G_r) = 20 \log 4\pi d/\lambda - 10 \log (4\pi/\lambda^2)^2.S_{e_t}.S_{e_r}$$

$$= 20 \log (4\pi d/\lambda)(\lambda^2/4\pi)/(S_{e_t}.S_{e_r})^{1/2}\quad \text{dB}$$

Thus

$$\text{Path loss} = 10 \log d^2\ \lambda^2/(S_{e_t}.S_{e_r})\ \text{dB} \qquad\qquad (9.14)$$

where λ = the wavelength of the radiated carrier in meters.
d = the distance from the antenna to the reflector in meters.
S_{e_t} = the effective area of the transmitting antenna in square meters.
S_{e_r} = the effective area of the receiving antenna or reflector in square meters.

This equation shows that to determine the total path loss between the transmitter, and the receiver, the path loss between the transmitter and the reflector is first calculated, and the decibel value added to the calculated path loss between the reflector and the receiver. This of course means that with back-to-back passive parabolic repeaters, the path loss between the transmitter and the receiver will be quite excessive.

9.3.1.2 Passive Repeater using one Plane Reflector
The gain of a flat (plane) reflector is given by:

$$G = 10 \log\left[\frac{4\pi}{\lambda^2} S \cos\frac{\phi}{2}\right] \tag{9.15}$$

where S' = the effective area of the flat reflector = $S \cos\frac{\phi}{2}$
  S = the true area of the flat reflector in the same units as λ
  ϕ = the angle between the incident and the reflected ray.

A flat reflector may be more effective than a parabolic antenna if the angle between the incident and reflected ray is not so obtuse as to reduce the effective area below that of the effective area of a parabolic antenna. The path loss of a plane reflector assuming free-space propagation conditions is given by:

$$\text{Path loss} = 20 \log d_1.d_2\lambda^2/(S_e.S') \tag{9.16}$$

where d_1 and d_2 = the distances in meters of the reflector from the transmitter and receiver respectively.
  S_e = the effective area of the parabolic transmitting or receiving antenna.
  S' = the effective area of the plane reflector.
        = actual area of the reflector × $\cos\frac{\phi}{2}$ $\qquad$ (9.17)
        = the projection of the geometric reflector surface onto a plane perpendicular to the beam direction.

If equation 9.16 is compared with equation 9.14, the ratio of the path losses is:

Path-loss (2 parabolic reflectors)/Path-loss (1 plane reflector) = $20 \log S'/S_{e_2}$ $\quad$ (9.18)

This equation shows that if the effective area of the plane reflector is larger than the effective area of the parabolic reflector, the path loss using the parabolic reflectors will be greater, by 20 log of the effective-area ratio, than the path loss using a plane reflector.

The beam width of the reflector in the vertical plane can become a limiting factor on how large the size of the reflector can become. This happens when the beam width in the vertical plane is too small, so that ordinary changes in the k-factor cause the main signal to change its angle of arrival by more than half the beam width of the reflector, thus producing a disalignment effect in the reflected beam towards the receiver. A further drawback to having a narrow beam width is the need to make the structure more rigid, in order to resist the deflection under wind and ice loads. Figure 9.3 provides the effective aperture of a plane reflector leading to a beam width of 1 degree, plotted as a function of frequency[1].

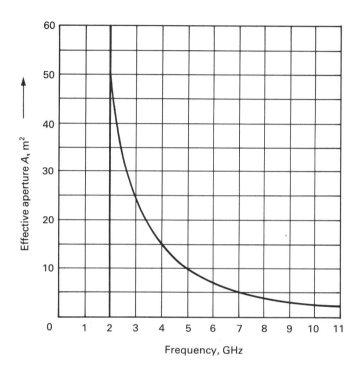

Figure 9.3 **Effective Aperture of Passive Reflectors for 3 dB Beam Width of 1°**
(Courtesy of Heyden & Son, Reference 1)

Figure 9.4 shows the path configurations using two parabolic passive repeaters, and one plane reflector. Figure 9.5 shows the path configurations using two plane reflectors in a path.

CONDITIONS OF PLANARITY FOR REFLECTORS
Irregularities in the reflector surface cause rays to reach the receiver via different paths. The attenuation caused by an irregularity of height H is given by:

$$\text{Attenuation} = 20 \log \cos [360° \, (H/\lambda)] \text{ dB} \qquad (9.19)$$

where H and λ are in the same units. Normal tolerance holds H to be less than $\lambda/8$.

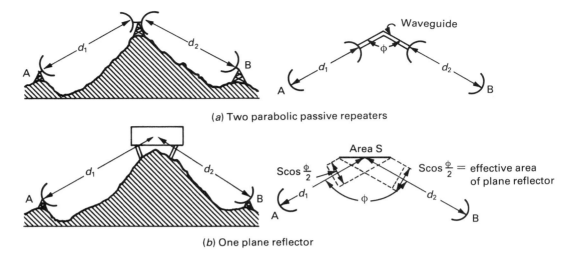

(a) Two parabolic passive repeaters

(b) One plane reflector

Figure 9.4   Single Passive Repeater Types

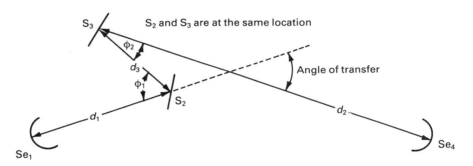

(a) Double plane reflectors at one location

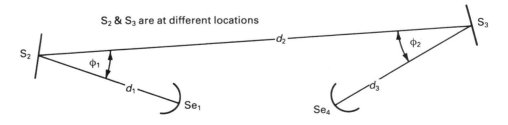

(b) Double plane reflectors at different locations

Figure 9.5   Double Plane Reflector Configurations

9.3.1.3 Two Plane Reflectors at one Location

In the arrangement of Figure 9.4 if the angle ϕ between the incident and reflected ray is greater than or equal to 140°, consideration should be given to the mounting of two plane reflectors of equal size, taking care that the two reflectors do not obstruct each other. Alternatively, instead of taking the angle ϕ as a criterion, the angle of transfer as shown in Figure 9.5(a) may be taken, and the requirement for the use of two plane reflectors is that this transfer angle should be less than 40°. In order to reduce the loss, the angles ϕ_1, and ϕ_2 should be kept as small as possible. The path loss may be calculated from equation 9.20. Note that the antenna gains have been taken into account, and should not be added again to the loss as determined from equation 9.20. This condition also applies to equations 9.14 and 9.16.

$$\text{Path loss} = 10 \log \{(\lambda^2 d_1 d_2)^2/(S_{e_1} S_2' S_3' S_{e_4})\} + \alpha_3 \text{ dB} \tag{9.20}$$

where S_{e_1} and S_{e_4} = the effective areas of the parabolic antennae in (meters)^2
$\quad$ S_2' and S_3' = the effective areas of the plane reflectors, i.e. the physical areas
$\quad\quad$ $\times$ cos (half the angle between incident and reflected ray).
$\quad$ S_2' $\quad$ = $S_2 \cos (\phi_1/2)$
$\quad$ S_3' $\quad$ = $S_3 \cos (\phi_2/2)$
$\quad$ λ $\quad$ = the wavelength in meters.
$\quad$ d_1 $\quad$ = the distance in meters between the first parabolic antenna and the first reflector.
$\quad$ d_2 $\quad$ = the distance in meters between the second reflector and the second parabolic antenna.
$\quad$ d_3 $\quad$ = the distance between the two reflectors in meters.
$\quad$ α_3 $\quad$ = the path loss between the two reflectors caused by d_3.
$\quad\quad$ = $20 \log f [2\lambda d_3/(S_2' S_3')^{1/2}]$, with f in GHz
$\quad\quad$ α_3 is usually no greater than 1.5 dB.

This system should be used only in extreme cases, for it is expensive, difficult to align and requires high transmitter power. To minimize the path loss, both the reflector areas and both the angles between incidence and reflection should be the same.

9.3.1.4 Double Plane Reflectors at Different Locations

The arrangement shown in Figure 9.5(b) permits two or more obstructions in the direct path between the transmitter and receiver to be bypassed. Again assuming that the reflectors are in the far field (see equation 9.11), the path loss (including the antenna gains) may be calculated from the following equation

$$\text{Path loss} = 20 \log (\lambda^3 . d_1 d_2 d_3)/(S_{e_1} S_2' S_3') \text{ dB} \tag{9.21}$$

Assuming $S_{e_1} = S_{e_4}$
The path loss in this case is increased by the full loss of the third-path section.

9.3.1.5 Periscope Antennas[2]

If a suitably designed reflecting antenna is placed in the near field, an additional gain over that provided by a parabolic antenna is possible. The geometric area of the plane reflector must be greater than twice the geometric area of the parabolic antenna. In addition to this condition for additional gain, the performance of the arrangement depends upon the height h of the reflector above the antenna as shown in Figure 9.6 and on the wavelength λ. Figure 9.7 shows the additional gain G_x of the antenna reflector as a function of the ratio $h\lambda/S'$, with the ratio of the geometric areas of the reflector (S'), and of the parabolic antenna (S_{geom}) as parameters.

Figure 9.6   Periscope Antennas with Plane Reflectors

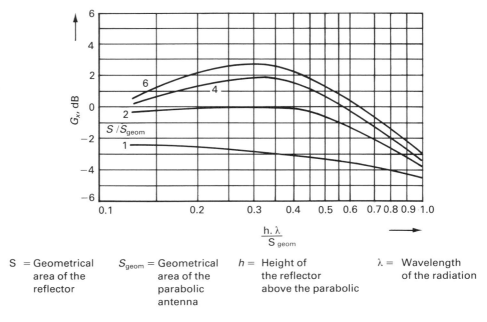

| S | = Geometrical area of the reflector | S_{geom} = Geometrical area of the parabolic antenna | h = Height of the reflector above the parabolic | λ = Wavelength of the radiation |

Figure 9.7   Additional Gain Due to Reflector (Courtesy of Heyden and Son, Reference 1)

Periscope antennas are used to reduce waveguide losses and thus reduce the effects of echo distortion, as well as to reduce the waveguide expense. Periscope antenna systems do not have particularly good off-beam discrimination characteristics, and in high-density systems or congested areas are not particularly suitable. In addition they require stiffer towers and suffer from poor discrimination against interference. The requirements for flatness or trueness of the reflector face, and for allowable deflections, become much more rigorous as the frequency is increased, and are quite severe in the 11 GHz and higher bands.

One potential problem which may require attention is that of the 'speak' echo path which may exist between the illuminating dish and the distant end. Unless this path is blocked by the intervening terrain, a signal may get through directly, particularly under super-refractive conditions, and if its magnitude is sufficient it can cause intermodulation problems. Since echo signals with long delays can cause significant intermodulation problems, even if they are 50–60 dB below the main signal, it is apparent that a good deal of blocking is required. In some cases the placing of metal shields on the path side of the illuminating dish may be required to cut down on the direct signal.

As can be seen from Figure 9.7, although the theoretical maximum gain is 6 dB, practical sizes of antenna and reflector limit this to 2–3 dB.

9.3.1.6 Diffractors as Passive Repeaters

A diffractor is employed to improve performance on paths which normally use mountain diffraction. It is especially useful at high frequencies, such as 11 GHz, where large, flat reflectors are expensive and difficult to construct. It is a microwave version of the optical Fresnel lens, and there are two types: the screen type and the dielectric type. The screen type acts by blocking off the wave components which would cancel the received field, while the dielectric type shifts their phase to add to the received field. They are placed on the ridge blocking the line of sight in a diffraction path, and give an effective gain over the knife edge alone.

Figure 9.8 shows the position of the diffractor on the ridge. The dielectric type

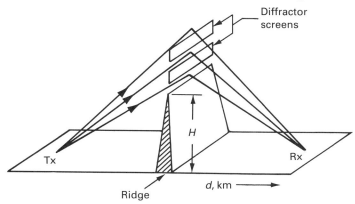

Figure 9.8   Microwave Diffractor (Courtesy of CCITT, Reference 7)

is much more efficient than the screen type, and so it can be made much smaller. The screen type has no requirement for surface flatness, and neither type needs to be adjusted after installation; both of these characteristics present distinct advantages over flat reflectors. A further advantage is that the supporting structure need not be rigid. The dielectric type has been successfully used to provide a gain of 10 dB on a knife-edge path at 11 GHz.

9.4   WIRE ENTRANCE LINKS

Sometimes it may be difficult to locate a station containing the multiplex or carrier equipment close to the antenna, as when the tower is sited on a high rocky outcrop. The building containing the carrier equipment may by necessity be located from 100 ft to 1 mile or so away. To run a feeder from the station to the top of the mast is obviously impractical. To overcome this problem, the radio equipment is located at the mast site, and the baseband signal which is of a lower frequency is run from the carrier station building to the antenna site. Due to the distance that the baseband cable has to run, a *wire entrance link* interfaces the baseband cable to the carrier equipment. Such links are operated on a four-wire basis, with separate cables for each direction of transmission. The wire entrance links usually contain cable equalizers of the fixed and variable type, line amplifiers, attenuator units, impedance matching transformers, lightning protectors and power supplies, each adjusted to the prevailing conditions. With heterodyne systems where a repeater does not demodulate down to baseband, but only down to IF it is also possible to operate wire entrance links at the IF, using coaxial cable and equalizers, etc. The heterodyne type of FM microwave equipment is preferred to equipment which demodulates down to baseband, when TV or long-haul traffic with no drop or insert facility is required, since it contributes the least amount of noise and distortion.

9.5   TRANSMISSION LINES AND RELATED EQUIPMENT[4,5]

Antenna transmission lines or feeders are employed to transmit the signal from the transmitter to the antenna with as little change to the original signal as practically possible. The longer the feeder, the more difficult this becomes due to the increased attenuation, and to the reflections produced by impedance mismatches at the feeder ends, as well as by discontinuities in the feeder itself. At certain frequencies, and with particular feeder constructions, spurious wave modes may occur which also adds to the degradation of the signal-to-noise ratio. Section 7.9 deals with the problem of 'echo distortion', which is a function of feeder length and channel capacity, etc.

Above frequencies of 2 GHz, waveguide may be used as the feeder, and above 4 GHz the use of waveguides is mandatory due to factors of availability, attenuation and VSWR. Below 3 GHz coaxial cable is usual. The factors affecting the choice of a particular feeder are mainly the frequency, the attenuation per unit length, the polarization (or polarizations) required, availability and cost.

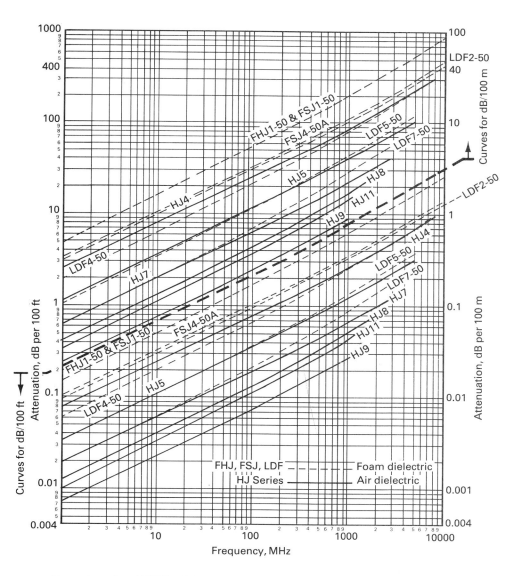

Curves based on; VSWR = 1.0; temperature = 24°C; atmospheric pressure; dry air

Figure 9.9   Cable Attenuation Versus Frequency
(Courtesy of Andrew Corporation, Reference 4)

9.5.1 Coaxial Cable

Coaxial cable that is commercially available has an impedance of 50 Ω. It is constructed either as an 'air dielectric', where the inner conductor is a hollow tube, and separated from the corrugated outer conductor by a spiral spacer made of dielectric material, or as a 'foam-dielectric', where a solid inner conductor is separated from the outer corrugated conductor by a solid dielectric filler. Both types have a plastic jacket to provide protection against corrosion and mechanical damage. Figure 9.9 provides information on air-dielectric and foam-dielectric coaxial cables.*

9.5.1.1 Attenuation

The manufacturer usually specifies the attenuation at a particular temperature (25°C) and applied power. If this temperature varies from that specified, a correction factor must be applied to obtain the correct attenuation factor. For example, if the temperature rises to 100°C, or if full average power rating is applied, the attenuation may rise to 13 per cent above that given in Figure 9.10.

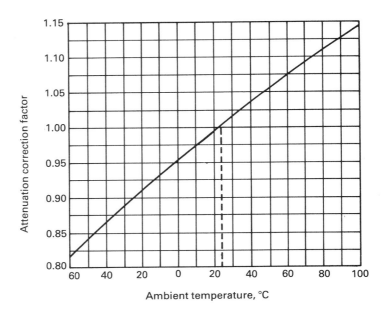

***Figure 9.10   Variation of Attenuation with Ambient Temperature
(Courtesy of Andrew Corporation, Reference 4)***

9.5.1.2 VSWR Effect on Transmission Loss

When the transmission line is attached to an antenna, the VSWR of the antenna increases the transmission loss of the line. Figure 9.11 shows this increase in loss plotted against VSWR.

* Extracted from the 1981 catalogue of the Andrew Corporation.

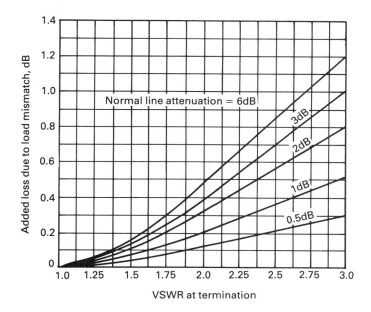

Figure 9.11   Effect of Load VSWR on Transmission Loss
(Courtesy of Andrew Corporation, Reference 4)

9.5.1.3 Power Rating Considerations

Peak and average power ratings are both required to describe fully the capabilities of a given transmission line. Typically, peak-power ratings limit the amplitude modulation or pulsed usage while average-power ratings limit the high-frequency usage.

PEAK-POWER RATING

The peak-power rating of a transmission line is limited by voltage breakdown between the inner and outer conductors. The voltage breakdown is independent of frequency, but varies with line pressure and the type of pressurizing gas. Peak-power ratings are therefore generally stated for the following standard conditions: VSWR = 1.0, zero modulation, and one atmosphere absolute dry air pressure (0 lb/in^2 or 0 kPa gauge) at sea level. The peak-power rating of the selected cable must be greater than the following expression in addition to satisfying the average-power handling criterion:

$$P_{pk} = P_t (1 + M)^2 \times \rho \qquad (9.22)$$

where P_{pk} = the cable peak-power rating in kilowatts.
$\quad\quad\ P_t$ = the transmitter power in kilowatts.
$\quad\quad\ M$ = the amplitude modulation percentage expressed as a decimal fraction.
$\quad\quad\ \rho$ = the voltage standing wave ratio.

From this relation it can be seen that 100 per cent amplitude modulation increases the peak power in the transmission line by a factor of 4. Also, the peak power in the transmission line increases directly with the VSWR. The transmission-line peak power rating can be significantly increased by pressurization. Since voltage breakdown in a transmission line carrying considerable power frequently results in permanent damage to the system, an adequate safety factor on peak power is necessary. This safety factor is usually taken as two on voltage, or four on peak power, and is intended as a provision for transmitter transients, and high-voltage excursions due to the unforeseen. Voltage breakdown will occur at approximately the same values for similarly sized air-dielectric cables; therefore, significantly higher peak-power ratings for similarly sized cables which may be given by manufacturers are possibly the result of operating with a lower safety factor, and should be carefully investigated.

The peak-power ratings of the Andrew Corporation HELIAX type of transmission line are determined according to:

$$P_{pk} = [(E_p \times 0.707 \times 0.7)/2]^2/Z_0 \tag{9.23}$$

where E_p = the DC production test voltage.
     0.707 = the RMS factor.
     0.7   = the DC to RF empirically derived factor.
     2   = the safety factor on voltage.
     Z_0 = the characteristic impedance of the cable.

Typical DC production test voltages for common sizes of air-dielectric cables are:

Nominal size, in	$\frac{7}{8}$	$1\frac{5}{8}$	3	$3\frac{1}{8}$	4	5
E_{pk}, kv	6	11	16	19	20	25

Foam-dielectric cables have a greater dielectric strength than air-dielectric cables of similar size and for this reason might be expected to have higher peak-power ratings than air cables. Higher peak-power ratings usually cannot be realized, however, because the commonly used connectors for foam cables have air spaces at the cable/connector interface which limit the allowable RF voltage to 'air cable' values.

AVERAGE-POWER RATING

Average-power ratings of transmission lines are governed by the safe long-term operating temperature of the inner conductor and dielectric. The maximum permissible inner conductor temperature varies with the type of dielectric and is based upon consideration of the softening temperature of the dielectric, oxidation of conductors, and the life of the dielectric limited by oxidation. All plastic dielectrics soften at elevated temperatures, and dielectric ageing is accelerated. The following are considered as safe, long-term temperatures:

Foam dielectric	100°C
Air dielectric, in $\frac{1}{2}, \frac{7}{8}, 1\frac{5}{8}, 5$	100°C
Air dielectric, in 3, 4	121°C

The average power ratings are based upon the inner conductor temperatures as above, with VSWR equal to 1.0, and an ambient temperature of 40°C. Figure 9.12 shows the variation of average power rating with ambient temperature.

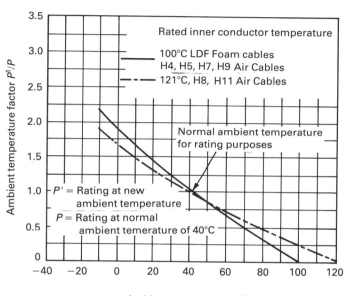

Figure 9.12 *Variation of Average Power Rating with Ambient Temperature (Courtesy of Andrew Corporation, Reference 4)*

The average power rating of the cable to be selected must be greater than P_{av} as given by the following expression, in addition to satisfying the peak-power handling criteria.

$$P_{av} = \frac{P_t \times [(1 + \rho^2) + F'(\rho^2 - 1)]}{ATF \times PF \times TCF \times 2 \times \rho} \tag{9.24}$$

where P_{av} = the cable average-power rating in kilowatts under standard conditions.

P_t = the transmitter power in kilowatts.

ρ = the voltage standing wave ratio.

F' = a derating factor for the VSWR that varies with frequency and line size, selected from Figure 9.13.

ATF = the ambient temperature factor, selected from Figure 9.12.

PF = the pressurization factor, selected from Figure 9.14.

TCF = the transmission characteristic factor: for AM or FM, $TCF = 1.0$; for Television use $TCF = 1.2$.

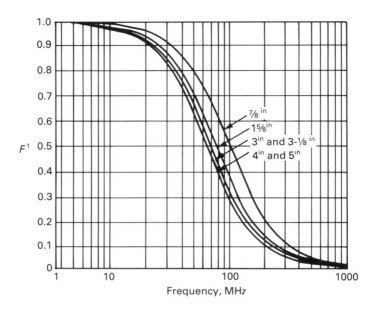

**Figure 9.13   Derating Factor for Average Power due to VSWR
(Courtesy of Andrew Corporation, Reference 4)**

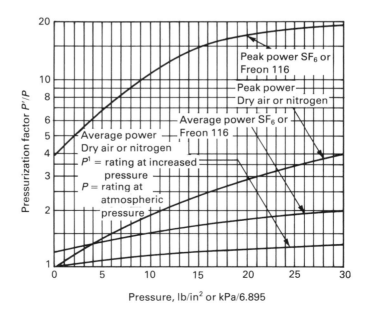

Figure 9.14   Pressurization Factor (Courtesy of Andrew Corporation, Reference 4)

9.5.1.4 Efficiency

The efficiency of a transmission line depends on its length and attenuation. The efficiency is defined as the percentage of transmitter power which reaches the antenna. It can be calculated as

$$\text{Efficiency} = 100 \times 10^{-A_l/10} \tag{9.25}$$

where A_l is the total attenuation of the transmission line at the frequency of interest in decibels. The remaining power is dissipated as heat in the transmission line.

9.5.2 Waveguides

Waveguide propagation in the fundamental propagation mode is possible only above the cut-off frequency of the waveguide, which is given by:

$$\text{Fundamental mode cut-off frequency} = c/2a \times 10^{-9} = 150/a \text{ GHz} \tag{9.26}$$
for a rectangular waveguide

where c = the speed of light (3×10^{11} mm/s).
   a = the widest inside dimensions of the waveguide in millimeters.

For a circular waveguide the fundamental mode cut-off frequency
$= c/1.706d \times 10^{-9} = 175.85/d$ GHz   (9.27)

where d is the inside diameter of the waveguide in millimeters.

These two equations are important because the fundamental (TE_{10}) mode is the propagation mode used for antenna feeders.

As can be seen from these two equations, at frequencies above 3 GHz the waveguide has dimensions which are not unmanageable, whereas coaxial cable has an unacceptably high attenuation. This is the main reason for the preference given to the use of waveguides above 3 GHz. Where straight runs and low VSWR are required, rigid waveguide is preferable to flexible waveguide. If the run, however, is not straight, but consists of many bends and turns, the additional flanges and elbows may cause sufficient reflections to raise the VSWR to the point where flexible waveguide with its ease in installation may be preferable, despite its higher VSWR.

9.5.2.1 Rectangular Waveguide

Rectangular waveguide consists not only of straight runs of waveguide, but the whole host of accessories which must come to connect the straight runs together. The straight sections are provided usually in standard lengths of 5 and 10 ft, and in non-standard lengths of up to 20 ft. As the waveguide is usually pressurized, there are pressure inlets and pressure windows to consider. For bending and twisting and rotating the polarization, there are elbows, flexible twist sections and rigid twist sec-

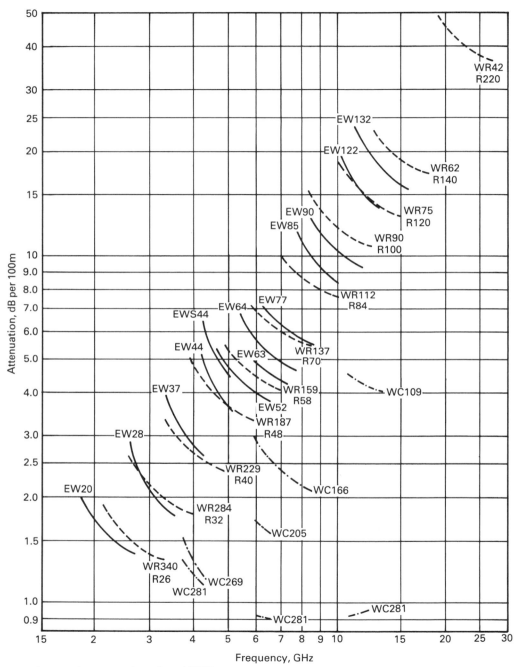

Figure 9.15   Microwave Waveguide Attenuation
(Courtesy of Andrew Corporation, Reference 4)

tions. For connecting the sections together whether of rigid or flexible waveguide, there are cover flanges, gaskets, etc. Table 9.2 and 9.3 list some rigid waveguides and various components used to make up the waveguides, with expected values of VSWR for them. Figure 9.15 provides information on the attenuation per 100 meters for different frequencies, and for different types of waveguides.

Table 9.2 Rigid Waveguide Frequency Bands

Waveguide size	Operating band, GHz	Low VSWR standard band, GHz
WR229	3.30 – 4.90	3.540 – 4.200
		3.700 – 4.200
WR187	3.95 – 5.85	4.400 – 5.000
WR159	4.90 – 7.05	5.925 – 6.425
WR137	5.85 – 8.20	5.925 – 6.425
		6.425 – 7.125
		7.125 – 7.750
WR112	7.05 – 10.00	7.125 – 7.750
		7.750 – 8.500
WR90	8.20 – 12.40	10.700 – 11.700
WR75	10.00 – 15.00	
WR62	12.40 – 18.00	
WR42	18.00 – 26.50	

The maximum VSWR ratings listed in Table 9.3 apply to the entire operating bands as given in Table 9.2.

Table 9.3 Rigid Waveguide Component VSWRs

Waveguide component	Standard type, maximum VSWR	Low-VSWR type, maximum VSWR
Straight section	1.05	1.02
Flex-twist (except WR42)	1.10	1.03
Flex-twist WR42	1.35	—
90° E or H plane elbow	1.05	1.02
90° twist	1.06	1.02
Premium 90° step twist	—	1.02
Premium flex (no twist)	—	1.03
Pressure inlet	1.01	1.01
Pressure window	1.04	1.01
Pressure-window/inlet	1.04	—

9.5.2.2 Circular waveguide

Circular waveguide is used for long vertical feeder runs in systems where multiband capability is required or where elliptical waveguide cannot be used because of high attenuation. The axial symmetry of circular waveguide allows the simultaneous propagation of two orthogonal TE_{11} modes, which is particularly useful when two different polarizations on two different frequencies are to be transmitted from a dual polarized antenna. A single waveguide run can carry two polarizations with 30 dB minimum isolation. It also has the lowest loss of all other types of waveguide. The disadvantages of circular waveguide, may however, outweigh the advantages, because:

1. It is only practicable for straight runs,
2. It requires complicated and critical networks to make the transition from rectangular waveguide.
3. It may have moding problems, and support more than the fundamental mode when the frequency range in use is high enough.

The attenuation/frequency curves are shown in Figure 9.15. The maximum VSWR for a complete single-polarized system, of any length, including straight sections and transitions is expected to be 1.04 maximum, and typically, 1.03. For a dual-polarized system 1.06 is the maximum expected VSWR, with a typical value of 1.04.

RECONVERTED MODE LEVEL
The (RML) reconverted mode level is the level of higher-order mode energy reconverted to the dominant mode in a circular waveguide system. Higher-order modes become trapped within the circular waveguide because they cannot propagate in the connecting rectangular waveguide. Group delay distortion and echo noise result when a portion of this energy, delayed in time, is reconverted to the dominant mode. In order to minimize this RML, each circular-to-rectangular transition may include a non-linear tapered circular-to-circular transition section which minimizes the generation of unwanted modes and prevents their propagation into the circular-to-rectangular transition section. In addition to this non-linear taper, a mode filter may be installed at each circular-to-rectangular section to attenuate the unwanted modes that may be present. Table 9.4 provides information on the waveguide size and the frequency band it operates in.

Table 9.4 Circular Waveguide Frequency Bands

Waveguide size	Frequency band, GHz		
WC281	3.700 – 4.200		
WC269	3.700 – 4.200		
WC205	5.925 – 6.425	6.925 – 6.425	
WC166	5.925 – 6.425	6.425 – 7.125	7.125 – 7.750
	7.725 – 8.500	7.900 – 8.500	
WC109	10.700 – 11.700	12.200 – 12.700	12.700 – 13.250

Microwave waveguides should be maintained under dry air or dry nitrogen pressure to prevent moisture condensation. The pressure which may be used is typically 10 lb/in^2 or 70 kPa gauge.

9.5.2.3 Elliptical Waveguide
This waveguide is perhaps the optimum choice for most microwave antenna feeder systems. It is constructed from high conductivity copper tubing with an elliptical cross-section. The corrugated wall gives the waveguide some degree of immunity to

crushing, together with light weight and good flexibility. A range of waveguide sizes as given in Table 9.5 provides a frequency range from 1.9 to 15.35 GHz. This type of waveguide is available in long continuous lengths which can be easily cut into the required length for any waveguide run, thus eliminating the need for multiple joints, elbows, and flexible sections. It also minimizes detailed waveguide system planning and provides improved electrical performance, with lower material and installation costs, compared with other types of waveguide.

The elliptical cross-section propagates the TE_{11} dominant mode, which is similar to the TE_{10} mode in rectangular waveguide, and operates below the cut-off frequencies of higher-order modes. Operating in the frequency band where only the dominant mode can exist eliminates signal distortion due to mode conversion and minimizes VSWR. The waveguide is usually operated under dry air or dry nitrogen at a pressure of 10 lb/s in^2, or 70 kPa gauge. The attenuation differs very little from its rectangular equivalents. When carefully transported and installed, it can provide a good VSWR performance, but relatively small deformations made to it can introduce sufficient impedance changes as can the introduction of foreign matter into the guide – to produce severe echo distortion noise.

Table 9.5 Elliptical Waveguide Frequency Bands

Waveguide size	Maximum potential operating range, GHz	TE_{11} mode cut-off frequency, GHz
EW20	1.9 – 2.700	1.60
EW28	2.6 – 3.500	2.20
EW37	3.3 – 4.300	2.81
EW44	4.2 – 5.100	3.58
EWS44	4.2 – 5.100	3.51
EW52	4.6 – 6.425	3.63
EW63	5.85 – 7.125	3.96
EW64	5.3 – 7.750	4.36
EW77	6.1 – 8.500	4.72
EW85	7.7 – 10.000	6.55
EW90	8.3 – 11.700	6.50
EW122	10.0 – 13.250	8.46
EW132	11.0 – 15.350	9.33

Figure 9.15 provides information on the attenuation characteristics for different operating frequencies. For all bands the maximum value for VSWR is expected to be in the range 1.15 – 1.20 for standard assemblies, and 1.06 to 1.08 for low VSWR assemblies. Both assembly VSWRs are based on 90 m or 300 ft lengths.

9.5.2.4 VSWR and Impedance

While antenna gain cannot easily be measured in the field, the antenna VSWR is easy to check upon installation and routine maintenance. The true VSWR is that at the antenna terminals and the VSWR at the end of the transmission line will be lower by the amount of attenuation in the line. For example, if the line has an attenuation of 3 dB, a VSWR of 1.5 measured at the station end of the line will be a true VSWR of 2.3 at the antenna. Figure 9.16 shows a chart for converting VSWR at the end of the transmission line to VSWR at the antenna terminals.

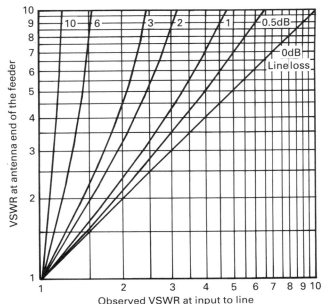

Graph to determine VSWR at antenna end of the feeder from
VSWR readings observed at the input to the feeder for various
line losses

Figure 9.16   VSWR at Antenna

9.5.3 Transmission Line Devices[4,5]

9.5.3.1 Circulators

A circulator is a three-port or four-port non-reciprocal device consisting basically of ferrite material, magnets, and three short lengths of transmission line terminated at a common junction. Power entering port 1 of a circulator is 'rotated' and emerges at port 2. Power entering port 2 emerges at port 3, and power entering port 3 emerges at port 1. Insertion loss in the forward direction (port 1 to 2, etc.) is approximately 0.5 dB, while the loss in the opposite direction is 25 – 30 dB or more. It is customary to use the circulator to couple two or three microwave radio equipments to a single antenna.

9.5.3.2 Isolators

An isolator is a three-port circulator, with one port terminated (port 3). When this network is employed a reduction in the echo-distortion noise results. It is the most effective solution to transmitter produced intermodulation – short of relocating transmitters – when the frequency separation between the desired and undesired signals is extremely close. This is because transmitter-produced intermodulation usually occurs by the interfering signal entering the transmitter from the antenna feeder, and mixing at the transmitter output stage. With an isolator placed between the transmitter and the antenna feeder, the backward path attenuates the interfering signal by 25 – 30 dB and reduces intermodulation products to acceptable levels.

Apart from interference from another frequency source, the isolator also insures that the noise contribution of the antenna system is as low as possible by isolating all high VSWR components (rigid bends, twists, etc.) from the transmission line. Any energy entering the output port (port 2) such as mentioned above, will be passed to the load at port 3 and absorbed. The load at port 3 must exactly match the impedance of the isolator to prevent energy from being reflected from port 3 back to the input port (port 1); otherwise the isolation between the input and output port would be reduced. In some cases an isolator may itself produce second-harmonic spurious energy and for this reason, a low-pass filter is often used between the isolator and the antenna.

9.5.3.3 Diplexers and Quadruplexers[1]

To combine both the transmitter output and receiver input, so they may be coupled onto one antenna, a bandpass filter is placed between the transmitter and an isolator, and the receiver and the same isolator before being coupled to the transmission line feeding the antenna. The transmitter band-pass filter is tuned to the center frequency of the transmit carrier, and has a bandwidth up to twice the expected RF bandwidth. Likewise for the receive filter. The combination of the two filters and the circulator is commonly known as a *diplexer*.

If two radio systems are to share the same antenna, the output of the diplexer of each system is fed into a separate isolator (optional) before being combined in the circulator which couples the two systems to the antenna transmission line. The combination of diplexer and circulators for combining two systems onto one antenna is called a quadruplexer.

9.6   TOWERS[2,3]

Microwave path engineering is greatly influenced by the availability of existing towers, and the costs of new towers. The height of the towers influences the length of the hop as can be seen from Sections 8.2, 8.3 and 8.4. In rural areas, heights of 350 ft may not be unreasonable since there is plenty of land which can be provided for supporting the guys of a guyed mast, or the base of a structured tower. In built-up areas land is not so freely available, and alternative means are often sought, such as the tops of buildings, etc. In addition to the lack of space in built-up areas, there are also the local building regulations to be contended with. Other codes which may limit the height of an antenna are those concerning aircraft. If the tower or mast is intended to be constructed near an airport it may be prohibited as soon as the idea is conceived. The planning of the siting of a tower must be co-ordinated with the relevant authorities, whether it is in a built-up or a rural area, if problems involving wasted time and money are not to be encountered. Two types of tower are normally used to support antennas. These are: the guyed mast; and the self-supporting tower.

The costs of the two different types are similar when the heights are small, but as the heights increase above 100 ft, the self-supporting tower's costs rise proportionally with the square of the height, whereas the guyed mast's costs rise linearly. Although it would appear that the guyed mast should be used for large heights, the limiting factor is the land area, which for example can be 55 times greater than a self-supporting structure when the height is 350 ft.

Tower foundations should be constructed of reinforced concrete with anchor bolts embedded before the concrete has set. Considerations which must be taken into account when a manufacturer is providing a quotation, are:

1. *Soil conditions*. This is because the soil bearing pressure is a major consideration in tower construction. Increasing the foundation area increases the soil bearing capacity or the equivalent design pressure. Unless accurate and specific information is given on the particular soil characteristics where the proposed tower is to be placed, the quoted costs from the manufacturer may be based on 'standard soil' as defined for example in the US EIA Standard RS–222A, i.e. the towers will be designed for a soil pressure of 4000 lb/ft^2 acting normal to any bearing area under specified loading. If the soil conditions are non-standard because the soil is granite and requires difficult excavation, or is sandy and requires extra large bases, the additional cost to accommodate these non-standard conditions may increase the cost of the tower by much more than expected.

2. *Wind loading*. EIA Standard RS–222A provides wind loading on steel towers, and RS–195A provides wind loading on antennae. The latter is normally used in specifying and determining wind loading. The figures used for wind loading are 30 lb/ft^2 under no ice, or normal conditions for flat surfaces, 40 lb/ft^2, and 50 lb/ft^2 for hurricane belt areas. Wind loading is the limit which is taken in design, and is not associated with operational requirements such as tower twist and sway. If a wind velocity which is taken as the design limit occurs, the tower may twist or sway enough to substantially degrade the signal, but it should return to normal when the wind stops.

In addition to the wind design loading, an 'operational loading' may be specified for which the twist and sway limits of the tower should not degrade the signal by more than 10 dB. Table 9.6 provides information taken from RS–222A on nominal twist and sway for a microwave tower. The wind loading expected for the area where the tower is to be sited should also be provided to the tower manufacturer; otherwise the manufacturer may quote on a minimum operational loading of 20 lb/ft^2.

Wind loading increases as the square of the wind velocity, where the constant K of proportionality is recommended by the EIA as 0.004 for pressures on the projected areas of flat surfaces, i.e.

$$\text{Pressure } P \text{ in lb/ft}^2 = K \times (\text{wind velocity } V \text{ in miles/h})^2 \qquad (9.28)$$

Hence

$$20 \text{ lb/ft}^2 \equiv 71.0 \text{ miles/h}$$
$$30 \text{ lb/ft}^2 \equiv 86.0 \text{ miles/h}$$
$$40 \text{ lb/ft}^2 \equiv 100.0 \text{ miles/h}$$
$$50 \text{ lb/ft}^2 \equiv 112.0 \text{ miles/h}$$

The loading on a tower in practice depends upon the sizes, shapes, locations and relative positions of all the anntennas, reflectors, waveguides and other devices which are mounted on it. The projected future use of the tower may also require consideration. The additional antennas together with their sizes may determine the present design.

3. *Building codes and regulations* pertinent to the tower site may have to be supplied to the manufacturer for, in some areas where the tower is above a certain height, a beacon needs to be installed and the tower painted to show that it is an obstruction. The ICAO (International Civil Aviation Organization) recommendations may also need to be complied with.

Table 9.6 Nominal Twist and Sway Values for Microwave Tower-Antenna-Reflector Systems*

A Total beam width of antenna or passive reflector between half-power points,°	Tower-mounted antenna		Tower-mounted passive reflector		
	B Limits of movement of antenna beam with respect to tower,°	C Limits of tower twist or sway at antenna mounting point,°	D Limits of movement of passive reflector with respect to tower,°	E Limits of tower twist at passive reflector mounting point,°	F Limits of tower sway at passive reflector mounting point,°
14	0.75	4.5	0.2	4.5	4.5
13	0.75	4.5	0.2	4.5	4.3
12	0.75	4.5	0.2	4.5	3.9
11	0.75	4.5	0.2	4.5	3.6
10	0.75	4.5	0.2	4.5	3.3
9	0.75	4.5	0.2	4.5	2.9
8	0.75	4.2	0.2	4.5	2.6
7	0.6	4.1	0.2	4.5	2.3
6	0.5	4.0	0.2	4.3	2.1
5	0.4	3.4	0.2	3.7	1.8
4	0.3	3.1	0.2	3.3	1.6
3.5	0.3	2.9	0.2	2.9	1.4
3	0.3	2.3	0.1	2.5	1.2
2.5	0.2	1.9	0.1	2.1	1.0
2	0.2	1.5	0.1	1.7	0.9
1.5	0.2	1.1	0.1	1.2	0.6
1.0	0.1	0.9	0.1	0.9	0.5
0.75	0.1	0.7	0.1	0.7	0.4
0.5	0.1	0.4	0.1	0.4	0.2

*Source:EIA RS-222A

All figures in the Table can be positive or negative, according to the direction of the movement, twist or sway.

Notes
1. Half-power beam width of the antenna to be provided by the purchaser of the tower.
2. (*a*) The limits of beam movement resulting from an antenna mounted on the tower are the sum of the appropriate figures in columns B and C.
(*b*) The limits of beam movement resulting from twist when passive reflectors are employed are the sum of the appropriate figures in columns D and E.
(*c*) The limits of the beam movement resulting from sway when passive reflectors are employed are twice the sum of the appropriate figures in columns D and F.
(*d*) The tabulated values in columns D, E, and F are based on vertical orientation of the antenna beam.
3. The maximum tower movement shown above (4.5°) will generally be greater than that actually experienced under conditions of 20 lb/ft^2 wind loading.
4. The problem of linear horizontal movement of a reflector-parabola combination has been considered. It is felt that in nearly all cases this will present no problem. According to tower manufacturers, no tower will be displaced horizontally at any point on its structure more than 0.5 ft per 100 ft of height under its designed wind load.
5. The values shown correspond to 10 dB gain degradation under the worst combination of wind forces at 20 lb/ft^2. This table is meant for use with standard antenna-reflector configurations.
6. Twist and sway limits apply to 20 lb/ft^2 wind load only, regardless of survival or operating specifications. If there is a requirement for these limits to be met under wind loads greater than 20 lb/ft^2, such requirements must be specified by the user.

9.6.1 Mast and Tower Land Areas

9.6.1.1 Guyed Mast

The guys may be spaced at 120°, for a triangular cross-section, with a length as projected onto the ground of 80 per cent of the mast height. With 80 per cent of the mast guyed, and additional allowance for guy anchors:

The *preferred area* is $1.39 \times 1.39 \times$ (tower height)^2
The *minimum area* is $1.39 \times 1.20 \times$ (tower height)^2

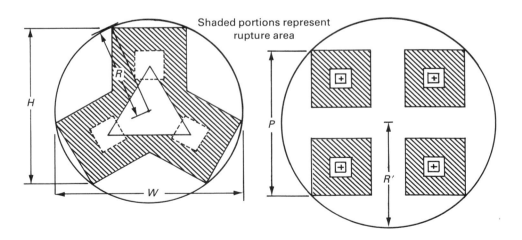

Figure 9.17   Approximate Land Area for Self-Supporting Towers

9.6.1.2 Self-Supporting Tower

Figure 9.17 shows the plan of a 3-leg and 4-leg self-supporting tower. The dimensions H, W, R, R' and F in feet for different tower heights are given in Table 9.7.

Table 9.7 Approximate Area Required for 3- and 4-leg Self-Supporting Tower

Tower height ft	R ft	W ft	H ft	R' ft	P ft
25	9.5	17.3	16.1	—	—
50	13.8	26.6	23.4	15.4	21.8
75	15.7	30.2	26.6	17.8	25.3
100	17.3	33.5	29.3	20.5	28.9
125	19.4	37.2	32.8	22.8	32.2
150	21.9	42.0	37.1	25.4	35.9
175	24.4	46.8	41.3	27.9	39.5
200	26.4	50.8	44.6	30.6	43.3
225	28.4	54.5	48.0	33.2	46.9
250	30.5	58.7	51.2	35.8	50.6
275	32.4	62.1	54.8	38.1	53.8
300	34.8	66.8	59.0	40.8	57.6
325	36.4	70.0	61.6	42.6	60.0
350	38.0	73.1	64.4	44.5	64.0

REFERENCES

1. Brodhage, H. and Hormuth, W., *Planning and Engineering of Radio Relay Links* (Siemens and Heyden, 1977).
2. White, R. F., *Engineering Considerations for Microwave Communications Systems* (GTE Lenkurt, San Carlos, California, 1975).
3. Freeman, R. L., *Telecommunication Transmission Handbook* (Wiley, 1975).
4. Antenna Systems, Catalogue 31, Andrew Corporation, 1981.
5. *Antennas/Transmission Lines/Duplexers/Filters* Catalogue 16, Decibel Products, 1978.
6. Raff, S. J., *Microwave System Engineering Principles* (Pergamon, 1977).
7. 'Economic and Technical Aspects of the Choice of Transmission Systems', Appendix to Section B.IV.3 of the handbook *Propagation* (ITU, 1971).

10 RADIO EQUIPMENT AND MODULATION

10.1 RADIO EQUIPMENT

Microwave links may extend from 10 km to 6000 km. Facilities for light routes may be minimal, providing only enough equipment to ensure point-to-point communications. On the other hand the facilities may be very heavy, requiring multi-radio channels and heavy route layout with sophisticated diversity switching. Systems may be constructed for nominally good service during certain limited hours of the day, thus providing considerable savings on capital equipment costs, or they may be built to provide a very high reliability and quality of service on a 24-hour-a-day basis for the worst month of the year.

For heavy traffic, and long-haul systems which require no, or very few, drop and insert facilities at the repeater stations, the end-to-end traffic is of the 'through' type. Other systems may require multiple access at some or all of the repeater stations, and thus have drop and insert facilities for supergroups, groups or channels. To provide for these different system types, two types of FM microwave equipment are in common use. These are the IF heterodyne, and the baseband or remodulating, type.

The IF heterodyne equipment receives the RF signal from the repeater receiver, and converts it to an IF. The IF signal is then passed into the repeater transmitter facing the next hop and converted to the new RF frequency for transmission at the required power level over the next hop. This type of radio equipment permits the elimination of the demodulation to baseband, and the remodulation from baseband up to RF stages, thus reducing considerably the distortion and noise which accumulates in the system. It is the preferred choice for systems handling exclusively, or almost exclusively, long-haul traffic, with little or no requirement for drop and insert along the route.

The baseband or remodulating type is preferable for short-haul or systems requiring multiple access, or for industrial long-haul systems. The received repeater RF signal is down-converted to the baseband, where supergroups, groups, or channels are dropped off, and/or the carrier blocks are inserted, and then modulated up to RF by the repeater transmitter facing the next hop, at a new frequency and power level.

Besides considering the IF heterodyne and the remodulating type of equipment, we note some other factors in the selection of the best radio equipment for a particular system:

1. The baseband bandwidth end-to-end, the baseband frequency response, loading and noise performance (*S/N, NPR*, crosstalk, etc.).
2. The value of the radio gain available, as determined from the transmitter output power and the receiver noise quieting.
3. The operating frequency band, and the required frequency spacing between radio channels, as determined by the transmitter deviation, receiver selectivity and frequency stability.
4. Primary power requirements and the options available.
5. The supervisory functions available, including the order wire, alarms and controls.
6. The equipment reliability, including the availability of redundant versions such as frequency diversity, 1-for-*N*, 2-for-*N* multi-line switching, hot stand-by, or hot-stand-by at transmitters and space diversity at the receivers.
7. Provisions for testing, spares, and maintenance.

With the rapidly changing nature of radio equipment, and the continuing development of new equipment and the upgrading of old ones, the user of the equipment should plan for the expected life of new equipment, and the provision for additional subsystems which may in the future be required to be added to the proposed system.

10.1.1 The FM Receiver

In most applications the radio link receiver shares the same antenna and feeder system as its companion transmitter facing in the same direction. Figure 10.1 is a simplified block diagram of a typical radio link transmitter and receiver in the ideal form. The receiver may or may not be connected to the common feeder by means of a circulator and preselector or band-pass filter. The circulator and filters are used to reduce the effects of the adjacent transmitter energy to a negligible amount in the receiver front end.

As mentioned in Section 9.5.3, an isolator may also be used in the transmitter

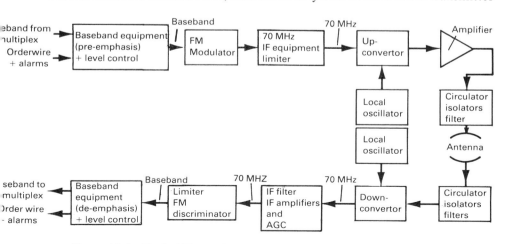

Figure 10.1   Typical Transmitter and Receiver: Baseband-to-Baseband

path to reduce the received signal entering the transmitter output stage and forming intermodulation products which would be then transmitted, and possibly cause interference in other receivers within or external to the system[3]. The isolator also reduces the effect of echo distortion, caused by feeder or antenna impedance mismatch.

Figure 10.2 shows a typical configuration of a transmitter and receiver, designed to use a single antenna, aided by a circulator and isolators. From the circulator and/or preselector filter, the incoming FM signal enters an amplifier (in some cases), and thence a mixer or down-convertor, where it is mixed with the local oscillator signal which is at a much higher level, to produce the IF spectrum. The resulting output of the down-convertor is amplified and filtered and amplified again. The IF band centered on a particular frequency is chosen by the filter. Most installations have standardized on a 70 MHz IF; however, the CCIR discusses a 140 MHz IF for systems designed to carry voice channels in excess of 1800 in a standard CCITT FDM configuration. Small-capacity systems may use an IF centre frequency of 35 MHz. The IF amplifier stages may contain a phase equalizer to correct the delay distortion introduced by the RF and IF filters, especially if they have been designed with their skirts close to the outside of the edges of the received information bandwidth. The RF filters are usually much wider in bandwidth than the received signal RF bandwidth, but the IF filters, whose purpose is to reject as much as possible of all unwanted frequencies, aim to have the IF bandwidth as close to the received signal RF bandwidth, with the consequence that delay distortion is more apparent the smaller the difference, and the sharper the IF filter cut-off (steep phase changes with increasing poles). As the received RF signal is of the order of −80 dBm, an RF amplifier of 7 − 10 dB does not raise the level anywhere near that required at the receiver output (see Table 3.1), which is expected to be around −30 dBm. This does not present a problem, because the mixer requires a low-level RF signal to be mixed with a high local oscillator signal, in order to keep the second- and higher-order harmonics at a low level. Most of the amplification occurs at the IF stage, because this is not so difficult at the lower intermediate frequencies. Amplifier gains in this stage may be of the order of 80 dB or so, depending on the AGC, and filter insertion losses. The IF is then fed into an FM demodulator, where the composite baseband appears at the demodulator output. Finally this composite baseband signal passes into a de-emphasis network; it is amplified, and sections of the baseband are filtered off to provide the supervisory alarms, order-wire, pilots, and voice-channel informa-

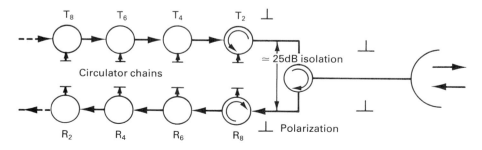

Figure 10.2   Common Antenna Operation

tion ready to be taken to the carrier or multiplex equipment.

10.1.2 The FM Transmitter

The composite baseband enters a pre-emphasis network and then a variable gain baseband amplifier whose output is used to frequency-modulate a local oscillator which may either be at an IF frequency, or at the final RF frequency. If the IF system is used, the IF is then up-converted to the output frequency, where it may be further amplified.

10.1.3 Diversity Combiners[1]

Section 8.5 considers the different types of diversity. The reduction in the outage time a diversity link is expected to have over a non-diversity link is shown in Fig. 8.32. Figure 10.3 shows the fading distribution for a non-diversity system with Rayleigh

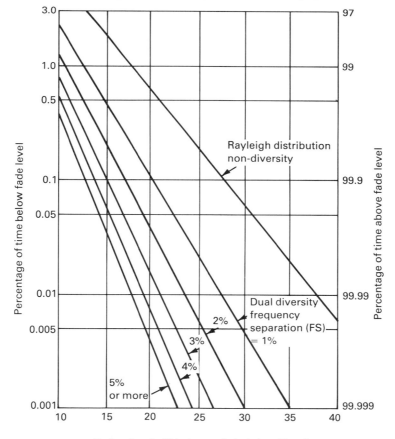

Figure 10.3 Diversity Fading Distributions (Courtesy of John Wiley, Reference 1)

fading versus frequency diversity systems for various percentages of frequency separation compared against the unfaded level.

When the different signals of the diversity system reach the receiving antenna or antennas, it is important that there is phase equality of the signals when they are combined. At RF or IF (or predetection), this can involve complicated circuitry, and is not possible at RF for the case of frequency diversity. Figure 10.4 is a block dia-

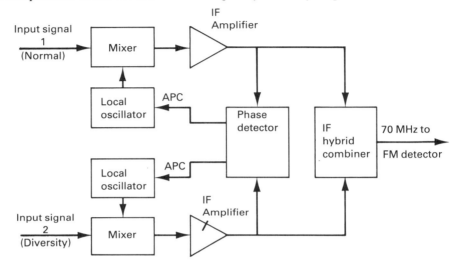

Figure 10.4 Pre-detection Combiner (Courtesy of John Wiley, Reference 1)

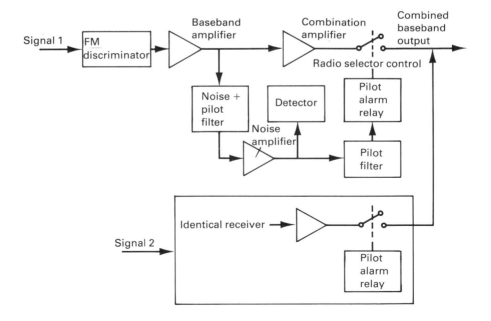

Figure 10.5 Post-detection Combiner (Courtesy of John Wiley, Reference 1)

gram of such an IF predetection combiner used in both frequency and space diversity.

Combining the received diversity is usually done after detection (post detection), i.e. in the baseband range, because the circuits are more simple. However, two system demodulators are required, thus increasing the equipment outlay over that where the switching is carried out at IF using the selection method. Figure 10.5 shows the post-detection combiner. Using the selection method, signal interference due to phase and level transients can be avoided if basic delays on the signal paths are made the same. This can be achieved by selecting identical equipments with the same electrical lengths and by insuring equal lengths of feeders, etc. The highest modulating frequency in the baseband for which a shift of less than 30° is required is decisive for the accuracy of adjustment at all frequency levels. In baseband combining and selection methods, the adjustment of the basic delays is achieved by the choice of suitable cable lengths.

10.1.3.1 Types of Combiner

Three types of combiner which are commonly used are:

> Selection combiner
> Equal gain combiner
> Maximal ratio combiner (ratio-squared)

SELECTION COMBINER
This combiner uses one receiver at a time. The output signal-to-noise ratio is equal to the signal-to-noise ratio into the combiner from the selected receiver.

EQUAL GAIN COMBINER
This combiner uses both receivers simultaneously, and simply adds the receiver output signals to give a signal-to-noise ratio at the combiner output as:

$$S_o/N_o = (S_1 + S_2)/(2N) \tag{10.1}$$

where N is the receiver noise power.

RATIO-SQUARED COMBINER
The signal-to-noise ratio at the output of this type of combiner is given by:

$$(S_o/N_o)^2 = (S_1/N)^2 + (S_2/N)^2 \tag{10.2}$$

For both equations 10.1 and 10.2 the following is assumed:

> All receivers have equal gain.
> Signals add linearly, whereas noise adds on an RMS basis.
> The noise is Gaussian.
> All receivers have equal noise power N at their output.
> The output signal-to-noise ratio at the combiner output is constant.

Figure 10.6 shows a comparison between the three types of combiners, where the signal-to-noise ratio at the output of each combiner, referred to that of a single

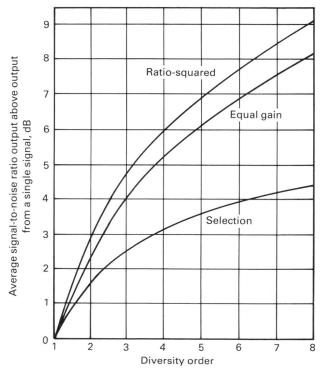

Figure 10.6   ***S/N Improvement plotted against Diversity Order***
(Courtesy of John Wiley, Reference 1)

signal only appearing at the output of each combiner, is plotted for various orders of diversity. The order of diversity refers to the number of independent diversity paths. Using space diversity would give a diversity order of two. Similarly, frequency diversity will also give a diversity order of two. If a system employed both frequency and space diversity together, the diversity order would be four, etc.

10.1.4 Pilot Tones

Pilot tones are used not only to monitor carrier equipment groups, supergroups, etc., but also to monitor the condition of the microwave link. These pilots are known as radio-relay continuity pilots. The pilot or pilots provided may be used, as discussed in Section 3.4.15, for example, in:

Gain regulating
Monitoring equipment performance
Frequency comparison, and carrier terminal frequency alignments
Measurements of level stability
The switching out of faulty links, and switching in of others
The control of diversity combiners

The control of diversity combiners involves the sensing by the combiner circuits of the presence of the continuity pilot. This informs the combiner that the path is still intact. The reason why this is important, over the obvious reason, is that most commonly used post-detection combiners use *noise* as the means to determine the signal level contribution from a particular path that should appear at the combiner output. The path with the least noise, as with the ratio-squared combiner, provides the greatest path contribution. Thus if a path were to fail, it would be comparatively noiseless, and would be made to contribute the most of the imagined signal to the combiner output. With a pilot continuity-sensing circuit, this situation is prevented from occurring.

Pilots are inserted before the transmitter modulator, and stopped at the baseband of the receiver. This is to prevent them from passing down the line to the next transmitter, etc. For the next section in the hop, the pilot is reinserted before the transmitter modulator, etc. Table 10.1 provides a list of the CCIR recommended* radio continuity pilots, together with other information which is relevant to the use of these pilots.

10.1.5 Pilot Deviations

Table 10.1 also shows the RMS deviation of the carrier produced by the pilot. The peak deviation produced by this pilot, as it is a sinusoid, will be the RMS value $\times \sqrt{2}$.

Table 10.1 Radio Continuity Pilots

System channel capacity N	Limits of band occupied by telephone channels, kHz	Frequency limits of baseband, kHz (1)	Continuity pilots frequency, kHz	RMS deviation produced by pilot, kHz
24	12 – 108	12 – 108	116 or 119	20 (2, 5)
60	12 – 552 60 – 300	12 – 552 60 – 300	304 or 331	25, 50, 100 (3)
120	12 – 552 60 – 552	12 – 552 60 – 552	607 (4)	25, 50, 100 (3)
300	60 – 1300	60 – 1364	1499, 3200, (6) or 8500 (6)	100 or 140
600	60 – 2540 64 – 2660	60 – 2792	3200 or 8500	140
960 900	60 – 4028 316 – 4188	60 – 4827	4715 or 8500	140
1260 1200	60 – 5636 60 – 5564 316 – 5564	60 – 5680	6199 8500	100/140 140

* CCIR Recommendation 401–2.

Table 10.1 Radio Continuity Pilots (cont.)

System channel capacity N	Limits of band occupied by telephone channels, kHz	Frequency limits of baseband, kHz (1)	Continuity pilots frequency, kHz	RMS deviation produced by pilot, kHz
1800	312 – 8204 316 – 8204	300 – 8248	9023	100
2700	312 – 12388 316 – 12388	308 – 12435	13627	100
TV			8500 9023 (7)	140 100

(1) Including pilot or other frequencies which might be transmitted to line.
(2) Other values may be used by agreement between administrations concerned.
(3) Alternative values dependent on whether the deviation of the signal is 50, 100 or 200 kHz (CCIR Recommendaqtion 404–2).
(4) Alternatively 304 kHz may be used by agreement between the administrations concerned.
(5) This deviation does not depend on whether or not a pre-emphasis network is used on the baseband.
(6) For compatibility in alternative use with 600-channel telephony systems and TV systems.
(7) The frequency 9023 kHz is used for compatibility purposes between 1800 channel telephone systems and TV systems, or when the establishment of multiple sound channels so indicates.

10.1.6 Interconnection of Baseband Frequencies

The CCIR Recommendation 380–3 provides information on the baseband impedances and levels for circuits with differing system channel capacity N. Table 10.2 reproduces this information.

Table 10.2 Interconnection at Baseband of Radio Relay Systems

Maximum number of telephone traffic channels (5)	Limits of band occupied by telephone channel, kHz	Frequency limits of baseband, kHz (4)	Nominal impedance at baseband, Ω	Relative power level per channel (dBr) (1, 2)			
				Radio relay system R (7)	Main repeater T	T'	Radio relay system input R' (7)
24	12 – 108 (3, 6)	12 – 108 (3, 6)	150 bal.	−15	−23	−36	−45
60	12 – 252 60 – 300	12 – 252 60 – 300	150 bal. 75 unbal.	−15	−23	−36	−45
120	12 – 552 60 – 552	12 – 552 60 – 552	150 bal. 75 unbal.	−15	−23	−36	−45
300	60 – 1300 64 – 1296	60 – 1364	75 unbal.	−18	−23	−36	−42

Table 10.2 Interconnection at Baseband of Radio Relay Systems (cont.)

Maximum number of telephone traffic channels (5)	Limits of band occupied by telephone channel, kHz	Frequency limits of baseband, kHz (4)	Nominal impedance at baseband, Ω	Relative power level per channel (dBr) (1, 2)			
				Radio relay system R (7)	Main repeater T	T'	Radio relay system input R' (7)
600	60 – 2540 64 – 2660	60 – 2792	75 unbal.	−20*	−23 −33	−36 −33	−45*
960	60 – 4028 316 – 4188	60 – 4287	75 unbal.	−20*	−23 −33	−36 −33	−45*
1260 †	60 – 5636 60 – 5564 316 – 5564	60 – 5680	75 unbal.	−28	−33	−33	−37
1800	312 – 8204 316 – 8204 312 – 8120	300 – 8248	75 unbal.	−28	−33	−33	−37
2700	312 – 12388 316 – 12388 312 – 12336	300 – 12435	75 unbal.	−28	−33	−33	−37

(1) The particular preferred values of the relative power level given in the table are agreed with the CCITT. These values apply to future systems.

(2) The level shown is referred to a point of zero relative level in the system, in accordance with the practice of the CCITT.

(3) For 12-channel systems, either of the basic groups A (12 – 60 kHz) or B (60 – 108 kHz) recommended by the CCITT may be accommodated in the band 12 – 108 kHz.

(4) Including pilots or frequencies which might be transmitted to line.

(5) Larger-capacity systems are not excluded by the table.

(6) A permissible alternative arrangement uses the frequency range 6 – 108 kHz. With this first alternative, it is possible to use only the noise measuring channel, situated above the baseband according to CCIR Recommendation 398–3. A further permissible alternative arrangement uses the frequency range 12 – 120 kHz. With this second alternative, it is possible to use only a continuity pilot situated below the baseband, according to CCIR Recommendation 381–2.

(7) The variation with frequency, over the range of baseband frequencies, of the equivalent loss of a homogeneous section of the hypothetical reference circuit from point R' to point R, should not exceed a limit of ±2 dB relative to the nominal value except under abnormal propagation conditions. This tolerance is similar to that accepted by the CCITT for line links by means of cable (see CCITT Recommendation M45).

10.1.6.1 Definition of the Points of International Connection at Baseband Frequencies

The points of international interconnection at baseband frequencies, called R' and R, form the input and output of a radio-relay system, conforming to CCITT Recom-

* Alternative levels $R = -23$ dBr and $R' = -42$ dBr can be used when the associated line transmission equipment is wholly of a type for which the CCITT recommends baseband interconnection levels $T = -33$ dBr and $T' = -33$ dBr (Main repeater station equipped with transistors).

† Other limits of band occupied by telephone channels may be used by agreement between the administrations concerned.

mendation G423 and CCITT Recommendation 380–3. At the output of the radio-relay system (point R), the following conditions are found in the baseband:

1. All the telephony groups (groups, supergroups, master-groups, etc.), and the pilots (line regulating, frequency comparison and monitoring pilots) included in the baseband are assembled in the position in which they are transmitted, as defined in the CCITT and CCIR Recommendations mentioned above.
2. All the continuity and switching pilots and other signals transmitted in a radio-relay system outside the telephony band, inherent to the radio equipment, are suppressed in accordance with CCIR Recommendation 381–2.
3. Any radio-relay protection switching is performed as part of the radio-relay system. With diversity reception, the combined output of the receivers used corresponds to point R.
4. Any de-emphasis networks are part of the radio equipment, so that the relative levels of the telephone channels are independent of frequency, within the limits of the tolerances stated in (7) of Table 10.2.

A similar point R' is defined for the baseband input of a radio-relay system, where similar conditions are to be met.

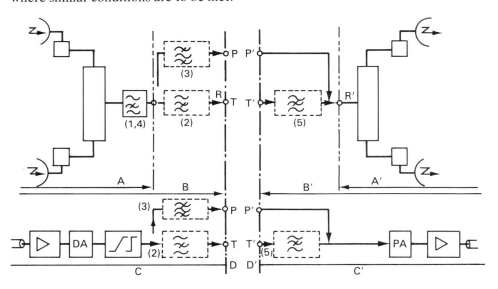

Figure 10.7   Baseband Interconnections (Courtesy of CCIR) Recommendation 380–3)

A, A'	Radio-relay system
B, B'	Line-link by means of radio-relay system
C, C'	Line-link by means of cable system
D, D'	Boundary of the high-frequency line equipment
R	Radio-relay system output
R'	Radio-relay system input
Point P'	Provided for possible injection of regulating pilots

Between T and T'	Telephony translating equipment and/or direct through-connection equipment
DA	De-emphasis network
PA	Pre-emphasis network
(1)	Blocking of continuity pilots, and if necessary, of regulating pilots
(2)	Blocking, if necessary, of regulating pilots, and pilots that must not go beyond the line link
(3)	Through-connection filter for regulating pilots, or for telephone groups can, if necessary, be inserted
(4)	Blocking of unspecified pilots or supervisory signals
(5)	Filter for blocking any unwanted frequency before injecting a pilot, insuring with (2) the requisite protection against a pilot (or other) frequency coming from another regulated line section (B or C, as the case may be).

10.2 MODULATION

Modulation may be defined as the process whereby a signal is transformed from its original form into a signal that is more suited for transmission over the medium between the transmitter and receiver. There are two major types of modulation:

Amplitude modulation
Angle modulation

In amplitude modulation, the phase angle of the carrier is kept constant and its amplitude is made proportional to the modulating signal. Similarly, in angle-modulation the amplitude is held constant and its phase angle varies with the modulating signal.

Many varieties of angle modulation are possible depending on the selection of the functional relationship between the angle and the modulating wave. Two of these varieties are important enough to be given the individual names of *phase modulation* (PM) and *frequency modulation* (FM). Phase modulation can be defined as angle modulation in which the instantaneous phase deviation is proportional to the modulating signal voltage. Similarly, frequency modulation is angle modulation in which the instantaneous frequency deviation is proportional to the modulating signal voltage. The similarity of the PM and FM waveforms show that for angle-modulated waves it is necessary to know the modulation function, because the waveforms appear the same and PM or FM cannot be distinguished from them alone.

10.2.1 Frequency Modulation

Because of numerous excellent texts on the different forms of modulation used in transmission, to detail each major type in this text would not be justified. The only form of modulation which will be treated briefly here is frequency modulation. Frequency modulation may be succinctly defined as follows:

Frequency modulation (FM) is the process in which amplitude changes of the modulating wave are used to vary the instantaneous frequency of the carrier wave from its unmodulated value.

The magnitude of the frequency change for a given amplitude of the modulating signal is called the *frequency shift*. While frequency shift is controlled by the amplitude of the modulating wave, the rate at which the carrier frequency is shifted is called the *deviation rate*, and this is controlled by the frequency of the modulating wave. For example, if a carrier of 5 GHz is modulated by a 200 kHz modulating signal of a given amplitude, the frequency swing of the carrier will depend on the modulation amplitude, but the rate of swing will be 200 kHz. Assume that the swing is 1 MHz for the given modulation peak amplitude, then the carrier will swing 1 MHz above 5 GHz and 1 MHz below 5 GHz, at a rate of 200 kHz. If a modulating signal with the same amplitude, but with, say, a new frequency of 400 kHz is used, the carrier will still swing ±1 MHz from the 5 GHz value, but the rate of this swing will now be 400 kHz, instead of the previous 200 kHz.

10.2.1.1 Bandwidth

Since the frequency shift is dependent upon the amplitude of the modulating wave, it would appear that the bandwidth required for FM transmission could be made considerably less than that of the modulating wave, by just reducing the amplitude of the modulating wave. However, the individual cycles of a modulated carrier wave are not sinusoidal, because of the instantaneous variations in frequency which occur during modulation. The result is that the modulated carrier wave contains not only an upper and lower sideband as for a single modulating sinusoid associated with amplitude modulation, but a large number of upper and lower symmetrically placed sidebands.

Where a single sinusoidal modulating wave is used, the spectrum of the modulated carrier wave is symmetrical with respect to the carrier frequency. In this case the sideband frequencies are displaced from the carrier by integral multiples of the modulating sinusoidal *frequency*. For example, a 200 kHz modulating wave, modulating a 5 GHz carrier, will produce a first-order pair of sidebands spaced 200 kHz from the 5 GHz carrier, a pair of second-order sidebands spaced 400 kHz from the 5 GHz carrier, etc.

Figure 10.8 shows the frequency spectrum of a modulated carrier for the cases where the modulating sinusoidal frequency is fixed, but its amplitude (or deviation – see Section 10.2.1.2) varies, and where the modulating sinusoidal frequency varies, but its amplitude (deviation) is fixed. Although the total bandwidth of a frequency-modulated wave is quite large, the higher-order sidebands often contain only a small portion of the total wave energy. In these cases, the actual bandwidth can therefore be reduced considerably without introducing excessive distortion.

10.2.1.2 Deviation

When the amplitude of a modulating signal produces a frequency shift in the carrier wave, this frequency shift produced away from the unmodulated carrier frequency is known as the *deviation*. A system is normally designed to have optimum perfor-

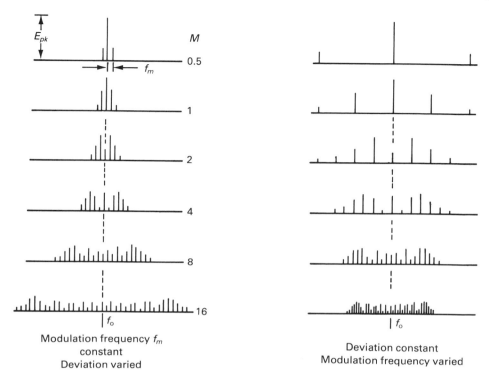

Figure 10.8   FM Spectra for Different Values of Modulation Index M

mance at a certain deviation, called the *nominal deviation*. Any change in the deviation will produce, in addition to a change in system baseband level, a change in the noise performance, as outlined in Sections 6.2 and 6.3. The most common method of adjusting the deviation of an FM system is based on suppressing the carrier. With no modulation, all of the energy from a transmitter is placed in the carrier, but as the modulation level increases, sidebands appear and their levels increase. This is followed by the carrier level falling to a minimum, and then as the modulation level continues to increase, rising again, etc.; these nulls in carrier levels are the 'Bessel zeros'.

The modulation index M, is defined as the ratio of the peak test-tone deviation Δf, to the highest modulating frequency f_m (if more than one frequency modulates the transmitter). The modulation index can be expressed as:

$$M = \Delta f / f_m \tag{10.3}$$

When a carrier ($\cos \omega_o t$) is frequency-modulated with a sinusoidal baseband signal the mathematical representation of the modulated carrier is given by:

$$e(t)/E_{pk} = J_o(M) \cos \omega_o t + J_1(M) [\cos(\omega_o + \omega_m)t - \cos(\omega_o - \omega_m)t] + \dots$$

$$\dots + J_n(M) [\cos(\omega_o + n\omega_m)t + (-1)^n \cos(\omega_o - n\omega_m)t] + \dots \tag{10.4}$$

where J_n = the Bessel functions of the first kind.
$e(t)$ = the instantaneous voltage of the modulated signal.
E_{pk} = the peak voltage of the modulated signal.
t = time in seconds.
ω = the carrier angular frequency in rad/s.
ω_m = the modulating angular frequency in rad/s.
n = number between 0 and infinity to complete the expansion of equation 10.4.
M = the modulation index.

It can be shown that

$$J_n(M) = M^n \sum_{p=0}^{\infty} \frac{(-1)^p . M^{2p}}{2^{2p+n} . p! \, (n+p)!} \tag{10.5}$$

Thus an angle-modulated wave with single-frequency sinusoidal modulation consists of a carrier plus an infinite number of side frequencies on both sides of the carrier at $f_o \pm nf_m$, where $n = 0, 1, 2, \ldots$ Figure 10.9 shows equation 10.5 plotted as

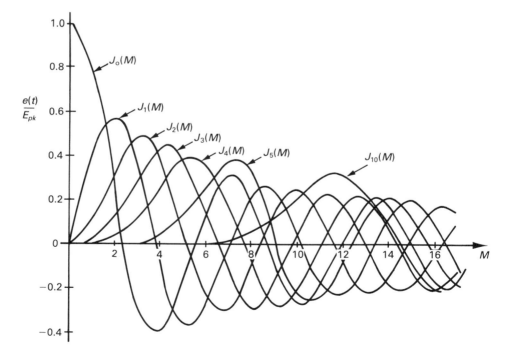

Figure 10.9   Bessel Functions

a function of the modulation index M, with the assistance of equation 10.4. When $M = 2.4$, there is no carrier. In practice this carrier does not disappear entirely, but goes through a minimum. There are other higher values of M which also produce this carrier null, but it is the first null of J_o, which is important. On FM systems without emphasis, the deviation is the same for all baseband frequencies, and the modulating index varies uniformly with the baseband frequency. On systems using emphasis, this relationship becomes more complex, for the deviation is not uniform across the band.

In aligning a transmitter the manufacturer will specify the modulating sinusoidal frequency, and its level (which is related to the peak deviation Δf), for the first Bessel zero, or alternatively for a modulation index of 2.4. This alignment is performed to permit the transmitter when operating with a non-sinusoidal input waveform from the baseband input to the transmitter modulator circuits to operate linearly, and to insure that either saturation of the transmitter does not occur, or that the RF bandwidth allocation is not exceeded. With the baseband composite waveform modulating the transmitter, the relative carrier and sideband amplitudes will vary with the amplitude and frequency of the modulating signal, but the total power contained in the modulated waveform remains constant.

10.2.1.3 Carlson's Rule

In principle, when an FM signal is modulated, the number of sidebands is infinite and the bandwidth required to encompass such a signal is similarly infinite. In practice, however, it turns out that for any bandwidth B, a large fraction of the total power in the spectrum is confined to sidebands which lie within some finite bandwidth which is larger than B, so that no serious distortion of the signal results if the sidebands outside this bandwidth are lost. Experimentally, it has been found that the distortion resulting from band-limiting an FM signal is tolerable as long as 98 per cent or more of the power is passed by the band-limiting filter. As 98 per cent of the signal power is contained in the first $(M + 1)$ sidebands, a filter which passes $(M + 1)$ upper sidebands from the carrier and also $(M + 1)$ lower sidebands from the carrier will therefore not cause significant distortion of the received FM signal. Thus for sinusoidal modulation, the bandwidth required to transmit or receive the *FM* signal is given by:

$$B = 2(M + 1) f_m \tag{10.6}$$

where M = the modulation index.
f_m = the sinusoidal frequency.

If a baseband composite signal is used, then in order to insure that the top frequency will pass through the filter, f_m is the highest baseband frequency. Equation 10.6 may be expressed in terms of the deviation by using equation 10.3, to give:

$$B_{RF} = 2 \, (\Delta f_{pk} + f_m) \text{ (Carlson's rule)} \tag{10.7}$$

where B_{RF} = the RF bandwidth of the transmitted signal, or the IF bandwidth of
            the received signal.
Δf_{pk} = the *peak* deviation.
f_m = the highest baseband modulating frequency.

FM signals are classified into two categories based on the value of M. If
$M \ll 1$, the FM signal is classed as a 'narrow-band' FM signal and the bandwidth of
the narrow-band signal is equal to $2f_m$, which is the same as the bandwidth of a
double-sideband AM signal. When $M \gg 1$, the FM signal is called a 'wide-band'
FM signal and its bandwidth is approximately $2\Delta f_{pk}$.

Section 6.3.6 deals with the RMS test-tone deviations as recommended by
CCIR Recommendation 404–2, which are used to align the transmitters carrying
various channel capacities.

10.2.1.4 Peak Deviation

Peak deviation is one of the main parameters used to determine the RF and IF
bandwidth of the transmitter and receiver. It is important to point out that the (per
channel) RMS test tone used for aligning the transmitter produces a sinusoidal peak
deviation calculated by multiplying the RMS value by $\sqrt{2}$. This sinusoidal peak value
in no way represents the multichannel peak deviation which comes about due to
many voice channels adding on the baseband. It is, however, possible to derive this
multichannel peak deviation from the sinusoidal peak deviation by adding a factor
which represents the difference in the statistical signal from that of a sinusoid representing the composite signal. Hence:

Peak sinusoidal deviation per channel

$= 2^{1/2}$ RMS test tone deviation per channel. $\qquad\qquad$ (10.8)

Peak deviation of multichannel signal
$=$ RMS test tone deviation/channel $\times$ G$(2.6 + 2 \log N)$ for $12 \leqslant N < 60$ $\quad$ (10.9)

Peak deviation of multichannel signal
$=$ RMS test tone deviation/channel $\times$ G$(-1 + 4 \log N)$
for $60 \leqslant N < 240$ $\qquad\qquad$ (10.10)

Peak deviation of multichannel signal
$=$ RMS test tone deviation/channel $\times$ G$(-15 + 10 \log N)$ for $240 \leqslant N$ $\quad$ (10.11)

where G$(\ldots) =$ antilog $[(\ldots)/20]$

By multiplying 3.76 by the right-hand side of equation 10.9, and 3.76 by the
right-hand sides of equations 10.10 and 10.11, the peak deviation of a multichannel
signal may be found. The multipliers 3.76 (11.5 dB) and 4.47 (13 dB) correspond to
the peak factors, which represent the difference in the statistical signal from that of
a sinusoid representing the composite signal. Thus we have:

Multichannel peak deviation
= RMS test tone deviation per channel $\times$ multiplying factor.   (10.12)

Table 10.3 shows the values of the multiplying factors for various system channel numbers N. This has been reproduced from that given in CCIR Report 418–1, pp. 332. Refer to Section 6.3.5 and Figure 6.10 for additional information.

Table 10.3 Peak Deviation Multiplying factors

Number of System Telephone Channels N	*Multiplying factor =* *(Peak factor)* $\times$ *antilog [(dB above modulation reference)/20]*
$3 < N < 12$	$4.47 \times$ antilog [(a value in dB specified by equipment manufacturer or station licensee subject to approval)/20]
$12 \leqslant N < 60$	$3.76 \times$ antilog [(2.6 + 2 log N)/20]
$60 \leqslant N < 240$	$3.76 \times$ antilog [($-1 + 4$ log N)/20]
$240 \leqslant N$	$3.76 \times$ antilog [($-15 + 10$ log N)/20]

10.2.1.5 Design of the RF and IF Bandwidth of the Radio

Having obtained the peak frequency deviation for N channels, we can determine the RF bandwidth of the transmitter by using Carlson's rule.

If the baseband signal causes the system to become overloaded it means that one or more of the following limits have been exceeded:

1. All the available or usable baseband spectrum is in use.
2. The point at which the total baseband signal power or system loading, if increased, would cause too much intermodulation.
3. System usage is such that any increase in either the top baseband frequency or the system loading would cause the RF bandwidth to exceed that legally allowed, for the particular band.

The first two of these limits often have some degree of flexibility, but the third, however, is a legal limitation which cannot be exceeded without violation of the law. The allowable maximum bandwidth or the necessary or occupied bandwidth (whichever is the greater) for a radio, can be derived from CCIR Report 418–1, Vol. 9. This report defines the various emission characteristics and provides formulae for calculating the 'necessary bandwidth'. The type of service and the allowable bandwidth for a particular service is formalized in an 'emission designator', which includes, at the start of the designator, the bandwidth in kHz; then follows a letter indicating the type of modulation (A for AM, F for FM, P for PM), and next a code number indicating the type of transmission (for composite transmission in the case of FDM it is the number 9).

Table 10.4 is a reproduction of Table IIB, CCIR Report 418–4.

Table 10.4 Frequency Modulation: Necessary Bandwidth (cont.)

Description and class of emission	Necessary bandwidth B_{n_1}, Hz	Examples Details	Designation of emission
Facsimile F4 (direct frequency modulation of carrier frequency by analogue half-tone picture signal)	$B_{n_1} = 2D + 0.855\,W$ for $0.14 \leqslant m < 0.77$ $B_{n_2} = 2D + 1.23W$ for $0.77 \leqslant m < 1.7$ $B_{n_3} = 2D + 1.5W$ for $1.7 < m \leqslant 3.45$	Diameter of cylinder 70 mm, number of lines per mm 5, speed of rotation 1 rev/s $W = 1100, D = 1500, m = 2.73$ Bandwidth $B_n = 4650$ Hz	4.65 F4
Four frequency diplex telegraphy F6	If the channels are not synchronized $B_n = 2.2D + 4Q$, where Q is the maximum speed in either channel	400 Hz spacing between frequencies $D = 600$ Hz, $Q = 50$ bauds Bandwidth $B_n = 1520$ Hz	1.52 F6
Composite transmission F9	$B_n = 2f_p + 2K\Delta f_{pk}$ $K = 1$	Microwave radio relay system specification: 60 telephone channels occupying baseband between 60 and 300 kHz; RMS per-channel deviation 200 kHz; continuity pilot at 331 kHz produces 100 kHz RMS deviation of main carrier, Computation of B_n: Δf_{pk} $= 200 \times 1000 \times 3.76 \times 2.02$ $= 1.52$ MHz $f_p = 0.331$ MHz $B_n = 3.702$ MHz	3700 F9
Composite transmission F9	$B_n = 2f_m + 2K\Delta f_{pk}$ $K = 1$	Microwave radio relay system specifications: 960 telephone channels occupying baseband between 60 and 4020 kHz; RMS per-channel deviation 200 kHz; continuity pilot at 4715 kHz produces 140 kHz RMS deviation of main carrier Computation of B_n: $\Delta f_{pk} = 200 \times 1000 \times 3.76 \times 5.5$ $= 4.13$ MHz	16 300 F9

Table 10.4 Frequency Modulation: Necessary Bandwidth (cont.)

Description and class of emission	Necessary bandwidth B_{n_1}, Hz	Details	Examples	Designation of emission
		f_m $= 4.028\,\text{MHz}\,f_p = 4.715\,\text{MHz}$ $(2f_m + 2K\Delta f_{pk}) > 2f_p$ $B_n = 16.32\,\text{MHz}$		
Composite transmission F9	$B_n = 2f_p$	Microwave radio relay system specification: 600 telephone channels occupying baseband between 60 and 2540 kHz; RMS per-channel deviation 200 kHz; continuity pilot at 8500 kHz produces 140 kHz RMS deviation of main carrier. Computation of B_n: Δf_{pk} $= 200 \times 1000 \times 3.76 \times 4.36$ $= 3.28\,\text{MHz}$ $f_m = 2.54\,\text{MHz}\ K = 1$ f_p $= 8.5\,\text{MHz}$, and so $(2fm + 2K\Delta f_{pk}) < 2f_p$ Bandwidth $B_n = 17\,\text{MHz}$, to incorporate the pilot	17 000 F9	
Composite transmission F9	$B_n = 2f_m + 2K\Delta f_{pk}$ $K = 1$	Stereophonic frequency modulation Broadcasting (pilot tone system) with multiplex subsidiary communications subcarrier. $f_m = 75\,000\,\text{Hz}\ \Delta f_{pk}$ $= 75\,000\,\text{Hz}$ Bandwidth $B_n = 300\,\text{kHz}$	300 F9	

In Table 10.4:

B_n = the necessary bandwidth, Hz

m = $\dfrac{2D}{W}$, frequency modulation index in facsimile

W = maximum number of black plus white elements to be transmitted per second, in facsimile and television.

D = half the difference between the maximum and minimum values of the instantaneous frequency, Hz.

The value of Δf_{pk} the peak frequency deviation is calculated by multiplying the RMS value of the per-channel deviation by the appropriate multiplying factor given in Table 10.3. For FM/FDM systems the necessary bandwidth B_n is given by:

$$B_n = 2f_m + 2\,K\Delta f_{pk} \tag{10.7}$$

where f_m = the maximum modulating frequency, Hz
       Δf_{pk} = the peak frequency deviation.
       K = a constant.
       f_p = the frequency of the continuity pilot.

Where a continuity pilot exists above the maximum modulation frequency f_m the general formula becomes:

$$B_n = 2f_p + 2K\Delta f_{pk} \tag{10.7A}$$

Where the modulation index of the main carrier produced by the pilot is less than 0.25, and the RMS frequency deviation of the main carrier by the pilot is less than or equal to 70 per cent of the RMS value of per-channel deviation, the general formula becomes either

$$B_n = 2f_p \tag{10.7B}$$

or $B_n = 2f_m + 2K\Delta f_{pk}$ whichever gives the greater value for B_n

10.3   NOISE EQUALIZATION BY EMPHASIS

Figure 10.10 shows that the noise in the higher baseband channels is considerably

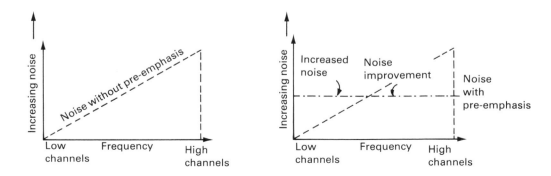

Figure 10.10   Effects of Pre-emphasis

worse than in the lower portion of the baseband. This can be verified by considering equation 5.49, where f_m is the baseband frequency. As the baseband frequency is raised to the highest frequency, the signal-to-noise ratio is reduced to a minimum, since S/N is inversely proportional to f_m^2. It would be desirable to alter the distribution of the noise by equalizing it as well as possible, over the baseband frequency range. This is accomplished by shaping the multichannel input signal level to the transmitter modulator, in terms of frequency, in order to raise or lower the test-tone deviation regions where the noise is higher or lower. The overall network characteristic is designed so that the total power of the composite baseband signal is essentially unaltered, i.e. just as much channel power is raised or boosted, as is decreased.

10.3.1 Pre-emphasis and De-emphasis

The characteristic which is most widely used, is defined in CCIR Recommendation 275–2. Figure 10.11 gives a plot of its baseband response. The pre-emphasis characteristic of the CCITT is such that the effective RMS deviation is not affected by it. In order to restore the baseband response back to a flat relationship, a de-emphasis net-

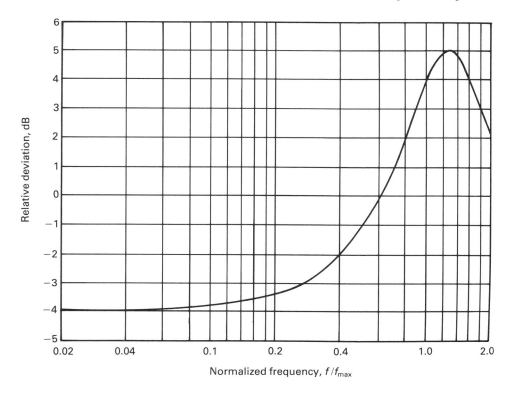

Figure 10.11   CCIR Pre-emphasis (Courtesy of CCIR, Recommendation 275–2, 1970)

work is employed at the system output. This de-emphasis network has a characteristic which is the exact opposite to that of the pre-emphasis network. The characteristic shown in Figure 10.11 is normalized so that for any particular channel capacity up to and including 2700 channels, the network response can be determined.

The characteristic shown in Figure 10.11 is given by the expression:

Relative deviation produced by the test-tone in dB

$$= 5 - 10 \log \left[1 + \cfrac{6.9}{1 + \cfrac{5.25}{\left(\cfrac{fr}{f} - \cfrac{f}{fr} \right)^2}} \right] \qquad (10.13)$$

where f_r = the resonant frequency of the network = $1.25 f_m$
　　　f_m = the highest telephone channel baseband frequency of the system.
　　　f = the baseband frequency.

Table 10.5 shows the values of f_m and f_T for FDM systems as described in CCIR Recommendation 380–3, in addition to the crossover frequency f_c, described below.

10.3.2 Crossover Frequency

The baseband frequency at which the test-tone level is unaltered is the *crossover frequency*, and is a very useful point for those measuring systems employing emphasis networks. It is used as the standard level reference frequency to specify test-tone levels for a system, and normally as the reference frequency when performing baseband frequency response tests (baseband to baseband frequency response). The crossover frequency f_c in terms of the highest baseband frequency f_m is given by:

$$f_c = 0.608 f_m \qquad (10.14)$$

The cross-over frequency is usually adopted when testing the loss between baseband terminal points of systems which are not in service.

The pre-emphasis and de-emphasis networks do introduce a loss into the baseband signal. This loss always totals 9 dB for the pre- and de-emphasis pair. The breakup of insertion loss is given by:

Insertion loss	Pre-emphasis loss, dB	De-emphasis loss, dB
I_{min}	9.0	0
I_c	5.0	4.0
I_{max}	1.0	8.0

The insertion loss of network is defined to be the loss in power between the output terminals and the input terminals of a network at a specific frequency or band of frequencies, i.e.

$$\text{Insertion loss } I = 10 \log (P_o/P_i) \qquad (10.15)$$

Table 10.5 Characteristic Frequencies for Pre-emphasis and De-emphasis

Maximum number of telephone traffic channels, N	Maximum frequency fm KHz	Resonant frequency fr KHz	Cross-over frequency fc KHz
24	108	135	66
60	300	375	182
120	552	690	336
300	1300	1625	792
600	2660	3325	1617
960	4188	5235	2546
1260	5636	7045	3427
1800	8204	10255	4998
2700	12388	15485	7532

REFERENCES

1. Freeman, R. L., *Telecommunications Transmission Handbook* (Wiley, 1975).
2. Brodhage, H. and Hormuth, W., *Planning and Engineering of Radio Relay Links* (Siemens and Heyden, 1977).

For Further Reading
3. CCIR Recommendations: 275–2, 381–2, 380–3, 398–3, 401–2, 404–2, 418–4, 1970/1974.
4. CCITT Recommendation G423, Orange Book, Vol. III–1 *Line Transmission*, VI Plenary Assembly, 1976.

11 TESTING OF MICROWAVE RADIO SYSTEMS

11.1 INTRODUCTION

The final objective to be set for any microwave radio system is to provide the best continuous distortion- and noise-free service to the customer that is possible within the economic constraints imposed. To attain this goal, the measurement of the performance parameters, and their adjustment to within specified limits, is of paramount importance. To meet the basic reliability and quality of service requirements as expected of a modern telecommunications network, initially and on a long term basis, a microwave radio system must be carefully engineered, installed and maintained. It is possible, of course, to over-design a system so that in any situation envisaged, the goal could be obtained, but on a more realistic level; economic considerations usually dictate the actual level of performance or grade of service provided. Under the economic constraints placed upon the system, the performance parameters such as the transmit and receive power levels, the deviation, frequency response, and the noise and distortion are carefully controlled so that they are within defined limits, and so that the system performs up to its specifications. To control the performance parameters implies the ability to measure them accurately. The measurements made in a microwave radio relay system are divided into two main categories which are: those concerned with equipment operating parameters, such as deviation, levels, etc., and those concerned with noise and its sources. These measurements are also divided up according to the logical sequence in which they occur. Thus we have:

1. *In-station tests*. These cover all of the measurements required at a single terminal, and all the alignments necessary at each station before progressing to the hop tests (see 2 below).
2. *Hop tests*. These permit the measurement of a single working link in the system, and the alignment of the hop parameters. These tests are required for each of the hops before the system tests (see 3 below).
3. *System tests*. These tests are performed to ensure that the total system is operating to specifications, and to complete fine alignment of system parameters.

In the measurement and alignment of a radio system, the equipments which are considered for the complete alignment are:

1. *The multiplex or carrier equipment.* Tests and adjustments are made at channel, group, and supergroup (if equipped) levels, and on the HF baseband. These tests have been described already in Section 3.10, and will not be included in this chapter.
2. *The baseband equipment.* This equipment is associated with the radio terminals, and may appear at intermediate or repeater locations if a return to the baseband level is required. This equipment serves as an interface between the multiplex equipment and the radio transmitter and receiver. It accepts the composite baseband signal and provides level co-ordination, impedance matching, amplitude and time delay equalization and pre- and de-emphasis. It also serves to monitor the radio continuity of the pilot signal, and may contain the baseband diversity combiners if so equipped. Because of the importance of its functions, the performance must be satisfactory at all times. Figure 10.1 shows the placing of the baseband equipment in a single heterodyne transmitter and receiver.
3. *The radio equipment.* Specific measurements and alignment procedures for the transmitter, and for the receiver, are necessary to ensure the correct deviation, transmitted power levels, AGC readings, noise quieting, etc. Usually included with the radio equipment is the antenna system VSWR or return loss measurements.

The rest of this chapter will be in two parts. These are on: In-station tests, and on hop tests. Each test will be described after the reason for the test and the information which it provides; (where necessary) an equipment description; and a reference to the preceding chapters where details of the theory may be found. After the test description a typical result to be expected will be given where possible and any further tests necessary as a follow-up will be referred to.

11.2 IN-STATION TESTS[1]

The tests dealt with in this section are:

> Antenna/transmission-line test
> Meter tests
> Transmitter tests – Frequency, RF output power, Deviation
> Receiver tests: AGC Calibration, Sensitivity (Quieting)
> Alarms: Receiver noise, Receiver pilot, AFC, Low power, Transmit pilot

11.2.1 Antenna/Transmission-Line Test

Reason for the Test
To verify that the minimum return-loss requirements across the operating frequency band are met, so that a minimum noise contribution is made to the baseband and the maximum RF signal is transmitted by the antenna.

Suggested Test Equipment
RF sweep oscillator
Frequency-response test set
Modulator
Directional coupler, 10 dB

Test Procedure
Note. Do not disconnect the transmission-line/antenna from the transmitter unless the transmitter is first turned off.

Method 1
1. Connect the test equipment by referring to Figure 11.1.

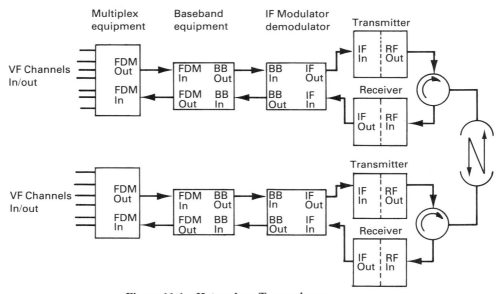

Figure 11.1   Heterodyne Transceivers

2. Calibrate the test set for the return loss listed in the system specification.
3. Connect the output of the directional coupler to the input of the antenna transmission line.
4. Observe the return loss display on the test set. It is required that the return loss of the antenna transmission line is not less than that listed in the system specification. If the requirement is met, proceed to step 8.
5. If the requirement of step 4 is not met:
   (*a*) Disconnect the antenna from the transmission line.
   (*b*) Terminate the transmission line at the antenna end.
   (*c*) Repeat steps 1 to 4.
6. If the requirement of step 4 is met after performing step 5, the fault is in the antenna. Use standard procedures to correct the fault.

7. If the requirement of step 4 is still not met after performing step 5, the fault is in the transmission line. Use standard procedures to correct the fault.

8. Photograph the return loss displayed on the test set and affix it to the equipment test record (ETR), for future reference.

9. Remove the test equipment, and connect the antenna transmission line to the radio equipment.

Method 2

The 'Bird–Thruline' RF directional wattmeter may be used for RF frequencies up to 1 GHz, with powers up to 1000 W. Procedure for use:

1. Turn off the transmitter power.

2. Connect the wattmeter to the transmitter end of the transmission line.

3. Turn on the transmitter power, and with the appropriate element inserted into the wattmeter read the value of the forward power P_f, by having the arrow on the element pointing in the direction of the antenna.

4. With the transmitter power still on, reverse the direction of the arrow, by turning the element through 180°. Read the power in the reverse direction, or the reflected power P_r.

5. The return loss is given by equation 7.36 or by RL = 10 log (P_f/P_r) dB. If the return loss is lower than that listed in the system specification (possibly RL < 20 or VSWR > 1.225), the fault is possibly in the antenna. To check this, first turn off the transmitter.

6. Terminate the transmission line in 50 Ω at the antenna end after disconnecting the antenna, and repeat steps 2 to 4.

7. If the return loss is the same as that measured in step 5, the fault is in the transmission line, and appropriate steps must be taken to rectify it. If the return loss is far less than that measured in step 5, the fault is in the antenna, and likewise the appropriate steps must be taken.

8. Turn off the transmitter power, and disconnect the wattmeter and reconnect the system feeder to the normal condition before restoring transmitter power.

Notes

1. To increase the sensitivity of the measurement of the power in the reflected direction, the wattmeter element may be changed with an element which reads a lower power for full scale deflection. If this change is to be made, first the transmitter must be turned off, and power reapplied only when the new element has been placed into the wattmeter, and the arrow pointing in the direction of the transmitter. Do not rotate a 10 W element when transmitter power is applied, as it may damage the wattmeter microammeter, or the element.

2. See also Sections 9.1.1 and 9.5.2.4 for reference.

11.2.2 Meter Readings

Reason for the Test

On commissioning a transmitter and receiver, using a multi-meter, voltages at various test points (TPs) are measured. This permits a reference to be set for all future maintenance work. If these TPs are monitored and recorded on a regular basis, any degradation in system performance can generally be quickly located by reference to the sets of TP readings which have changed in time. Also, from the preventive maintenance viewpoint, if there is a gradual change in the TP readings of a particular module in the transceiver, this module may be readjusted or replaced if necessary before problems arise.

Some commercial transceivers have a front panel meter with a multiposition switch which permits test point readings to be taken without the use of a multi-meter; others have a front panel meter with a multiposition switch to monitor some parameters, and require in addition multi-meter TP readings to complete the set. A small variation in a meter reading should not be interpreted as an indication of a defective unit or module, unless the equipment performance is known to have deteriorated as demonstrated by other tests, or a significant variation exists between those readings taken upon commissioning or module replacement and those just recorded. Two panel meter readings which have special significance are the receiver AGC, and the transmitter power. These readings indicate the strength of the received and transmitted signals respectively. If these readings are normal for a particular site, it can usually be assumed that the transceiver is performing satisfactorily. Whenever a unit is adjusted or replaced, a complete set of meter readings for the radio transmitter or receiver should be recorded, and, in addition, comments made on why there has been a change in the readings taken.

Suggested Test Equipment
Multi-meter
Appropriate test leads

Test Procedure
1. Using the front panel meter, record the readings taken for the various multi-position switch positions, together with the date.
2. Using the multi-meter, record the readings taken at the recommended TPs, together with the date.
3. Plot the readings taken on a graph, with the abscissa taken as the date.
4. Note any irregularities, and take appropriate action when necessary.

Equipment where Test Points Occur

TRANSMITTER
1. Output of the transmitter voltage regulators
2. AFC voltage after the loop filter
3. AFC voltage from the phase detector
4. AFC reference osillator rectified voltage
5. Driver current
6. Power amplifier current
7. Power amplifier power out
8. Transmitter power out
9. Transmitter reflected power

RECEIVER

10. Output of the receiver voltage regulator
11. Down-convertor mixer current
12. Down-convertor local oscillator level
13. IF filter output level
14. Discriminator output level
15. AGC

ALARMS

16. Transmitter pilot level
17. Receiver pilot level
18. NODAN or noise level (the NODAN circuit is explained later in this section)
19. NODAN alarm
20. Receiver and transmitter summed alarm test point
21. Low power

The receiver and transmitter summed alarm is sometimes provided to indicate on the front panel of either unit that a fault has occurred. Where the fault originates from is not necessarily indicated, but may be found knowing what alarms are being summed, and what the readings at the various test points for these alarms should be.

The NODAN circuit in the receiver permits continuous comparison of noise in a 3.1 kHz slot above the combined baseband and pilot signal to be compared against a standard noise level of say 58 dBrnC0. If the detected noise exceeds the reference, then the comparator triggers and an alarm is raised.

11.2.3 Transmitter Carrier-Frequency Test

In some transmitters, the local oscillator frequency used to up-convert the baseband information is derived from a lower and more stable crystal oscillator. The frequency multiplication required to attain the local oscillator frequency may be quite high and hence the use of straight-chain multipliers may be better replaced by phase-locked loop multipliers. If this is so it is usually represented as the automatic frequency control (AFC) circuit or circuits (if a double conversion is employed). Figure 11.2 shows such an arrangement in a single conversion transmitter. This test, with the use of a calibrated frequency counter, measures the RF frequency out of the up-convertor when the AFC circuit is enabled or disabled. It also permits the fine adjustment of the crystal oscillator to provide the up-convertor (oscillator/modulator) with an output RF frequency which is within specified limits when under the control of the crystal oscillator. When the AFC loop is switched to the manual position (MFC), a reading of the RF frequency allows the phase-detector output voltage to be set mid-range, so that the loop will stay in lock indefinitely (or for a long period of time), even though the crystal oscillator frequency will change due to the ageing of the crystal and variation of its temperature.

Because of frequency allocations, and interference problems which may arise, it is imperative that the transmitter RF frequency is adjusted correctly and remains

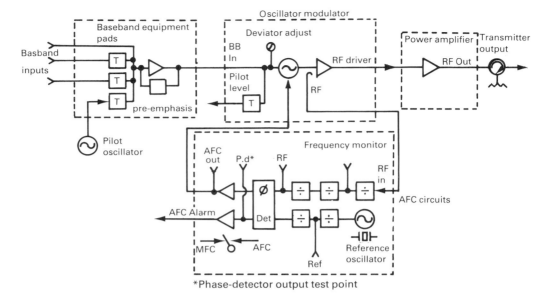

Figure 11.2   Baseband Modulating Transmitter

within specified limits during its operational lifetime. This test should be carried out monthly, or at worst every three months.

The other parameters which may require consideration are correct operation of the AFC action, the capture range, the sensitivity, the sense and the error frequency.

Suggested Test Equipment
1. Calibrated Temperature-Controlled-Crystal-Oscillator (TCXO) frequency counter which can measure up to or above the RF of the transmitter.
2. Test cords and adapters.

Test Procedure
1. Key on the transmitter if necessary.
2. Set the AFC/MFC switch on the AFC unit to the MFC position.
3. Connect a frequency counter to the frequency monitor jack on the AFC unit. The frequency monitor jack may be at the RF frequency or may be divided down by some of the loop dividers. The frequency is usually referred to as the 'monitored frequency'.
4. Note that the frequency displayed by the counter must be within a frequency range specified by the manufacturer.
5. If the requirement is not met, adjust the frequency control of the RF oscillator feeding the up-converter (this is not the lower-frequency crystal oscillator) for the required indication.
6. Set the MFC switch to the AFC position. It is required that the displayed frequency must be within the range specified by the licensing authority.

7. If the requirement is not met, adjust the fine frequency control on the lower frequency crystal oscillator for the required indication.
8. Record the displayed frequency.
9. Record the test point readings of Section 11.2.2.

11.2.4 Transmitter RF Output Power and Panel Meter Calibration

This test describes the method of accurately measuring the transmitter output power and, if necessary, calibrating the transmitter front panel meter to indicate the RF output power from the transmitter.

Suggested Test Equipment
1. Power meter with a thermistor mount
2. Coaxial attenuator, 50 Ω, with the required attenuation and power rating to reduce the RF power to a level that can be accepted by the power meter.

Test Procedure
RF OUTPUT POWER
1. Remove the transmitter from service.
2. Disconnect the feeder to the isolator and connect the power meter and thermistor mount to the transmitter output via the coaxial attenuator.
3. Key on the transmitter.
4. Record the power meter indication. It is required that the output power is according to the manufacturer's specification.
5. If the requirement is not met the power amplifier or driver stage may be defective. Take appropriate action to determine the cause of the problem, and rectify.

PANEL METER CALIBRATION
6. Rotate the multiposition switch on the front panel to the PWR position.
7. Key the transmitter and compare the reading against that of step 4. It is required that the panel meter and power meter indications should be the same.
8. If the requirement is not met adjust the power calibration control, or take the appropriate action as specified by the manufacturer.
9. Return the transmitter to service and proceed with other tests as required.

POWER METER CALIBRATION
To ensure that the reading from the power meter is an accurate representation of what is being measured, the power meter may be calibrated. The following is the procedure:
1. Couple the power meter to a signal source (an RF signal generator) through a calibrated fixed attenuator and a switchable attenuator.
2. Set the RF generator to a level which is the same or higher than the power of the transmitter, and which does not exceed the power meter input rating when the switchable attenuator is not providing any attenuation.
3. Make a calibration chart in steps of 1 dB, to a maximum of steps of 5 dB of meter readings versus input power to the power meter by switching in the attenuator.

4. If so required, connect the calibrated attenuator and power meter to the antenna end of the waveguide or coaxial run, and measure the output level, being certain to observe safety precautions, especially those related to possible eye damage (cataracts forming) due to looking into an energized waveguide.

5. Subtract the power reading obtained at the antenna end from that obtained in step 4 of the RF output power-test procedure above, to obtain the loss figures of the transmission line and isolator/circulator units.

6. Compare the figures against the design specifications and determine whether the loss figures so obtained are acceptable. If not, take remedial action.

For routine maintenance tests, the Bird–Thruline RF directional wattmeter may be used, as described in Section 11.2.1, Method 2.

Record all measurements and the dates on which the measurements are taken, preferably in the preventive maintenance log book.

11.2.5 Baseband Equipment Tests

The co-ordination of signal levels in conformity with channelizing and radio equipment interface requirements is performed by baseband pad, filter and amplifier circuits in both the transmitting and receiving paths. In order to verify that these circuits are functioning properly, the baseband equipment must be isolated from the multiplex and radio equipment, and the signal level must be measured in each direction of transmission. The measurement is most readily accomplished by connecting a signal generator to serve as the FDM input of the baseband equipment, and, with the pilot disabled, use a voltmeter connected through a test transformer to determine the signal level at the baseband output. The frequency and amplitude of the generator output, and the desired voltmeter reading, depend upon the application, and may be stated in the system design specifications.

The pilot level is measured in the transmitting path only, and may require that the baseband equipment is isolated. With the pilot generating circuits enabled, and with no signal at either the FDM or baseband input, the voltmeter is again connected to the baseband output. The desired output level – which is usually 10 dBm0 for microwave systems – is given in the design specifications for any particular system.

In remodulating or baseband radio systems, there is a direct interface of baseband and transmitter/receiver equipment, i.e. the baseband signal directly modulates the RF carrier, or a submultiple of it. A heterodyne system, however includes an IF modulator/demodulator interface as shown in Figure 11.3. The purpose of the modulator is to generate an IF carrier – in most instances, 70 MHz – that can be frequency-modulated by the transmit path baseband equipment output. It also provides amplification and IF level co-ordination. The demodulator provides amplitude limiting, demodulation, level co-ordination and when necessary, 'mop-up' group delay equalization.

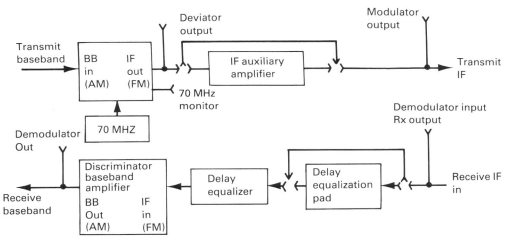

Figure 11.3   IF Modem Interface in Heterodyne System

11.2.6 IF Modem

As with baseband equipment, measurements of the IF modem performance characteristics are taken with the equipment isolated. Generation of the 70 MHz (or whatever the value is) signal must be among the first measurements taken in a heterodyne type of system. The unmodulated IF carrier frequency itself can be checked by connecting a frequency counter to the 70 MHz IF monitor jack or other test point provided by the manufacturer. The counter reading and the date of the test are recorded so that frequency drift can be detected, and corrected by subsequent measurements. The level of the IF carrier can be determined by connecting an RF power meter to either the deviator or the modulator test point. The desired reading depends upon the system application and configuration, and also upon interface considerations, but typical values are −5 or +1 dBm at the deviator (modulator) IF output, and +5 dBm at the discriminator (demodulator) IF input.

When the characteristics of the unmodulated 70 MHz carrier have been ascertained, and adjusted if necessary, the IF modem performance is tested in the presence of a modulating signal.

11.2.7 Transmitter Deviation

As mentioned previously, in a FM microwave radio system, changes in the baseband signal amplitude cause variations in the carrier frequency, with a higher-level input resulting in a greater frequency shift, or deviation from a centre frequency. The difference at any given time between the modulated and the unmodulated carrier frequencies is the deviation*; the maximum frequency shift permitted being the

* See Section 10.2.1.2

'peak deviation'. Deviation in a heterodyne system is a function of the IF modem, whilst in a remodulating or baseband system it is a function of the frequency-modulated oscillator (FM) or 'Osc. Mod.' In either case, the 'Bessel zero' method is employed to determine and adjust the amount of deviation, the Bessel zero being defined as the point at which all of the carrier energy has been distributed amongst the sidebands. At this point, the modulation index equals 2.405, and the carrier amplitude is zero. Since modulation varies in proportion to the level of the modulating frequency, and the required level-versus-frequency characteristic of the modulator clearly defines the point at which the Bessel zero will occur, a criterion may be established to which deviator sensitivity may be adjusted.

The sensitivity of the deviator, whether it is an IF modem, or an 'Osc. Mod', determines the degree of the frequency shift for a given input amplitude and frequency. To set the desired system deviation requires the use of a modulating source and a detector to monitor the carrier frequency component. This equipment is often combined into a 'deviation test set' usually supplied by the manufacturer. If for example, a heterodyne-type transmitter operating at 2 GHz and carrying 960 telephone channels is designed for a peak deviation of 283 kHz (200 kHz RMS × 1.414) with an input level of −33 dBm, the 117 kHz modulating source (283/2.405 = 117 kHz) would supply an equivalent input to the deviator. To allow an in-service measurement, the monitoring device would be connected to the 70 MHz monitor point. If the deviation were correct at this point, the spectrum analyzer would indicate a carrier zero level. To ensure that the modulation index is 2.405 and that there are no higher modulation indices, the deviation adjustment is normally set to its lowest gain at the start of the test and slowly increased until the first carrier null appears. This will then be the desired peak deviation for a given fixed-input test-tone level, and frequency.

The measurement and adjustment of the frequency deviation of a transmitter is an out-of-service test. It may be performed over the RF path between adjacent terminals, or on a loop-back basis at an individual site. In this case a 50 Ω attenuator of the power rating of the transmitter is used to represent the expected path loss between the transmitter and the receiver. For example, if the transmitter power is 10 W, and the expected receiver level is −80 dBm, the attenuator must have a minimum power rating of 10 W, and an attenuation of 120 dB, since 40 dBm + 80 dBm = 120 dB.

The 'Osc. Mod.' consists of a basic oscillator, oscillating at the proper RF frequency. The tank circuit of the oscillator is equipped with a varactor diode which acts as the modulator. The parameters which require consideration are: linearity, frequency stability, RF amplitude response, and level.

Suggested Test Equipment
1. Spectrum analyzer
2. Signal generator (top frequency 200 kHz)
3. Level meter for calibration of the signal generator
4. Attenuator (if required) to pad down the transmitter power to a level which is acceptable to the spectrum analyzer.

5. Frequency counter to set the frequency of the signal generator to that required for the carrier null.
6. Appropriate test leads and adaptors.

Test Procedure
1. Remove the transmitter from service.
2. Attach if necessary the attenuator from the transmitter output to the spectrum analyzer input.
3. Key on the transmitter.
4. Disable the pilot oscillator.
5. Set the signal generator to the specified frequency and level required to produce the Bessel zero.
6. Adjust the spectrum analyzer to display the carrier and sidebands of the transmitter RF signal.
7. If the carrier is not at a null, reduce the signal generator level to a minimum.
8. Increase slowly the level of the signal generator until the first Bessel zero is observed.
9. Note the level out of the signal generator. It is required that the signal generator output level must be within 0.2 dB of the appropriate test-tone level for the Bessel zero.
10. If the requirement of step 9 is not met, set the oscillator level to that specified.
11. Adjust the baseband-level control, or the deviator sensitivity control for the required carrier null as shown on the spectrum analyzer.
12. When the carrier null has been obtained, remove the test equipment and restore the transmitter back to service, or proceed with other tests as required.

Figure 11.4 shows typical displays of the spectrum analyzer.
For reference, see also Sections 10.2.1, 6.3.5, and 6.3.6.

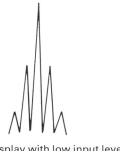
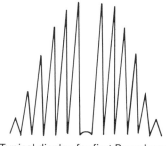

Typical display with low input level   Typical display for first Bessel zero
(carrier null)

Figure 11.4   Spectrum Analyzer Diagrams

11.2.8 The Receiver Equipment

Refer to Figure 10.2. If more than one receiver is attached to an antenna by means of the isolators and circulators shown in the diagram, there may also be filters to assist the relevant receiver in the selection of its signal. These filters act as open-circuits to the off-frequencies, and cause reflections and therefore additional echo noise. To overcome this there may be additional isolators required. In addition it becomes important that the frequency plan is carefully chosen to exclude any mixing products which may fall within the receiver's IF bandwidth. Section 9.5.3 provides additional details.

The parameters which may require consideration are the losses in the duplexer configuration, the amplitude response, the group delay, the (2A-B) type interfering signal, and the return losses.

The Mixer
Refer to Figure 10.1. The mixer converts the incoming RF signal to a lower IF frequency which is easier to work with. It is usually a ring hybrid balanced type at 2 GHz or lower. Above 2 GHz the use of a waveguide splitter is the most common method. The local oscillator, which is at a frequency of the RF ± the IF center frequency, provides the bias on the diodes, and its output power is chosen for the best operating point of the diodes for noise figure, harmonic and intermodulation distortion suppression, and conversion loss. The first transistor stage following the mixer operates at a level which may be 6 dB lower than that of the incoming RF signal. This means that for good noise performance, its noise figure should be very low. Following the mixer and this first transistor stage, there is the IF pre-amplifier section which amplifies the IF signal to the required working level. At this point there are the amplitude-response and gain-adjustment potentiometers.

Parameters to consider are: RF to IF amplitude response, noise figure, spurious oscillations and conversion gain.

Local Oscillator
The local oscillators are usually crystal-controlled, with the crystal sometimes in an oven which is temperature-controlled. The crystal frequency may be a fifth overtone in the 90 – 120 MHz range. After the basic oscillator there are amplification stages and multiplier stages which bring the local oscillator signal to the proper operating frequency and level. One important aspect of the local oscillator signal is the noise close to oscillator centre frequency. For a good oscillator this should be about -70 to -80 dBc/Hz, at the oscillator itself. At the local oscillator frequency this noise referred to the carrier (dBc) is increased by $10 \log M$, where M is the multiplication ratio.

The parameters to be considered are: the output frequency, the output level, the thermal noise, and spurious oscillations.

The IF Filter
The IF filter sets the operating bandwidth of the receiver. It is a unit which may have traps to reject unwanted signals created by the frequency plan, and is usually self-

equalized for group delay. The bandwidth is normally chosen to be larger than the RF bandwidth of the transmitter by a factor of $1.5 - 2$, in order to prevent excessive group-delay distortion.

The parameters to be considered are: return loss, amplitude response, group delay distortion, and linearity of differential gain.

The IF Amplifier

After the IF filter, the multistage IF amplifier provides a constant output level over a wide range of incoming signal. It accomplished this feature by having an automatic gain control working in conjunction with the IF amplifier. The IF amplifier is usually a broadband amplifier with a good noise figure.

The parameters to be considered are: the gain at the minimum incoming receiving level, the amplitude response, and the noise figure.

AGC

The automatic gain control circuit samples the output of the IF amplifier, detects it, and provides a voltage proportional to the IF amplifier output, which is then reapplied to the IF amplifier to bias the transistors so as to maintain a constant gain. The AGC also provides a means of monitoring the level of the received RF signal, since it is proportional to it. The speed of the AGC action should be fast enough to keep the IF amplifier output constant, but slow enough to prevent oscillation and suppression of fast level changes in the RF signal.

The parameters to be considered are: constant IF amplitude output, and the AGC metering circuit.

The IF Limiter

The purpose of the IF limiter is to remove any amplitude modulation that may be present in the receiver before demodulating the FM signal. The suppression of the AM is usually achieved by a pair of limiting diodes, or by transistor limiters. Each stage of limitation can provide up to 20 dB of limiting, and, if more than this is required, additional limiter stages may be employed. Since limiter stages are clipping stages, and therefore far from linear, harmonics and intermodulation products which are formed must be removed. The limiter is therefore followed by a filter before the signal enters the demodulator.

The parameters to be considered are: the percentage of AM limiting, the differential gain and the differential phase.

The Demodulator

The demodulator transfers the FM information back into its original composite baseband signal form. It comprises the discriminator and the baseband amplifier. The discriminator is usually a balanced type with two tuned circuits and a pair of diodes providing the 'S'-shaped curve – level output versus frequency input. The demodulator provides a means of monitoring the IF centre frequency by the discriminator zero.

The parameters to be considered are: sensitivity, linearity and response.

The baseband amplifier provides the proper output level, at the same time it has the facility for frequency-response adjustments, and provides the de-emphasis if the transmitted signal has been pre-emphasized.

11.2.9 Receiver AGC Calibration

In most equipment configurations, the receiver has an associated meter that indicates the received signal strength derived from the AFC circuit in the IF amplifier. These AGC meter readings are relative indications of the received carrier level. An accurate graphic record of AFC meter readings versus RF input level must be made to establish when the antenna systems are optimally aligned. This graph is also helpful in determining the condition of the transmission path during subsequent maintenance routines, and it may indicate fading depth below the no-fade level, especially if the fading duration is long enough for the meter to respond, and be read.

The procedure for developing an AFC curve requires that an RF signal generator is connected to the receiver input – at either the directional coupler or another calibrated RF test input point – to simulate a received signal. With no RF input, the AGC meter should indicate near zero. The frequency of the output generator is tuned to the centre frequency of the receiver under test, and its level precisely adjusted to the nominal input signal level, with compensation being made for losses in the directional coupler, hybrids and circulators (unless the test point calibration makes allowances for these factors). For example, if the required received signal level is −60 dBm and the loss through the coupler, etc., is 8 dB, the generator output level should be set for −52 dBm. The simulated received signal is varied in 5 dB steps to a point below the practical threshold or mute point of the receiver, and the observed meter readings are recorded. It is more suitable to plot the results directly as the measurements are taken, in a form as shown in Figure 11.5.

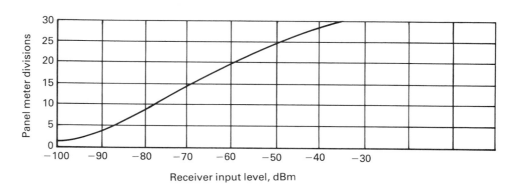

Figure 11.5   Typical AGC Curve

Most microwave radio receivers are equipped with circuits that indicate an alarm condition when the received signal drops to a predetermined level. During the AGC measurement procedure, therefore, the appropriate alarm indicator should be observed as the simulated received signal approaches this minimum level in order to check that the alarm is functioning, and that its threshold setting is the same as that specified by the manufacturer. Some receivers may cause the alarm to be automatically muted if the pilot is not received. If this is so the alarm cut-off (ACO) switch should be activated when performing this procedure, for otherwise no readings will be observed on the AGC panel meter. The RF signal generator normally simulates only a received unmodulated carrier for this test, and thus on demodulation there will be no pilot.

In remodulating, or baseband, receivers, an off-frequency alarm circuit may also be provided. When the difference between the received signal and the local oscillator frequencies exceeds a predetermined limit, this circuit causes an alarm indication. Operation of such an alarm may also be checked during the AGC procedure. With the signal generator set for the unfaded received signal level and the proper frequency, a counter is connected to an appropriate IF monitoring point. With the AFC function disabled, the local oscillator is tuned until the off-frequency alarm indicator appears. Typically, the counter should read either 69 or 71 MHz, ± 200 kHz, at that time. The local oscillator is then retuned to produce the proper IF output, which in most systems is 70 MHz ± 10 kHz.

Suggested Test Equipment
RF signal generator
RF frequency counter
Power meter with thermistor mount
Switchable attenuator in 1 dB steps minimum
Appropriate test cords and adaptors

Calibration of the Signal Generator
1. Attach to the output of the signal generator, the coaxial or waveguide lengths, together with the switchable attenuator between them, that will be used to feed directly into the directional coupler.
2. With the attenuator switched for zero dB attenuation, set the received carrier frequency on the RF signal generator at a level which permits the RF frequency counter to read the frequency.
3. Attach the RF frequency counter to the end of the cabling from the generator and set the signal generator to the received carrier frequency.
4. Remove the frequency counter, and attach the power meter thermistor mount. Set the output of the signal generator to a level of approximately −10 dBm as read from the generator meter indication, and the output of the coaxial or waveguide after switching in 30 dB attenuation. Using the generator fine level control, set to a level of −40 dBm.

Test Procedure
1. Arrange for the receiver panel meter to read AGC. If the panel meter indication is steady, note the indication and proceed to step 2.
2. If the panel meter indication is not steady do not proceed with this test unless there is no need to determine the unfaded received signal level.

3. Insure that there are no alarms up on the receiver to be tested.
4. Remove the transceiver from service.
5. Turn off the transmitter, leaving the receiver on.
6. Activate the alarm cut-off (ACO) circuit on the receiver if equipped.
7. Remove the power meter thermistor mount from the signal generator set-up, and connect the signal generator via the attenuator and cabling to the directional coupler.
8. Connect the frequency counter to the IF ouput and read the frequency.
9. If the frequency is out by more than ± 10 kHz for a 70 MHz IF, the local oscillator should be adjusted until the required condition is met.
10. Remove the frequency counter from the IF output and restore the IF circuits to normal.
11. Using the switchable attenuator, vary the signal into the directional coupler from −40 dBm to the receiver mute point, in steps of 5 dB, and record and plot the AGC panel meter readings.
12. Remove the test equipment and return the receiver and transmitter to service or proceed with other tests as required.

Note: The receiver quieting or sensitivity test described below in Section 11.2.10 may be completed at the same time as the AGC calibration, for the test set-up is the same except for an additional item of equipment. The advantage is the saving of time and additional unnecessary work in calibration, etc.

Test Results
1. By comparing the reading of AGC as taken in step 1 of the test procedures with the AGC calibration chart as determined in step 11, a measure of the unfaded received signal level can be obtained. This reading should be compared with that calculated in order to determine if excessive loss exists in the radio path between the radio receiver and the preceding transmitter.
2. If this unfaded received signal level reading drops by 3 dB at any time it means that the incoming signal has been reduced, which may have been caused by:
    (*a*) Fading
    (*b*) Low transmitter power
    (*c*) Antenna misalignment
    (*d*) Increased feeder loss

See also Section 8.4.4, including equation 8.77, on the calculation of unfaded received signal level.

11.2.10 Receiver Sensitivity or Quieting

This test is one of the more important tests, since it permits first a verification that the receiver is operating satisfactorily, and permits secondly the elimination of the receiver as the cause – out of all the possible causes – of excessive hop noise, should such a problem occur. The theoretical quieting curve was described in Section 7.14, and the theoretical quieting curve for a 600-channel system was derived as shown in Figure 7.7. It is again mentioned that for convenience this test should be performed alongside the AGC calibration.

This section describes the method to be followed when measuring the receiver noise performance. The per-channel noise is compared to a worst expected figure when measued for a known received signal level. The per-channel noise itself can be measured with a selective level meter, or with a white noise receiver. When a selective level meter is used, the noise should, however, be measured in a 3.1 kHz bandwidth; otherwise a correction factor must be made to the noise measurement.

Suggested Test Equipment
1. RF signal generator.
2. RF frequency counter.
3. Power meter with thermistor mount.
4. Switchable attenuator in 1 dB steps minimum.
5. Frequency selective level meter with a 3.1 kHz measuring slot or noise test set.
6. Appropriate test leads and adaptors.

Noise Test Set Requirements
Noise generator. Appropriate high-pass and low-pass filters for the system under test (see Section 7.11, and Tables 7.1, 7.3 & 7.4)
Noise receiver. Appropriate bandpass filter for the top slot of the system under test (see Section 7.12, and Tables 7.1, 7.3 & 7.4)

INITIAL PROCEDURE
1. Set the output of the noise generator to equal the system noise loading level plus the output level of the receiver under test (see Section 7.11), e.g. for a 300-channel system, the receiver baseband level is -23 dBr. Thus:

$$\text{System loading level from equation } 6.8 = -15 + 10\log 300 \quad = 9.8\,\text{dBm0}$$
$$\text{Receiver output level} \qquad\qquad = \qquad\qquad\qquad -23\,\text{dBr}$$
$$\text{Noise generator output} \qquad\qquad = +9.8 + (-23) \quad = -13.2\,\text{dBm}$$

2. On the noise receiver, select the top-slot band-pass filter using Table 7.1, which, for example, is 1248 for a 300-channel system.
3. Connect the output of the noise generator to the input of the noise receiver.
4. Calibrate the noise receiver to zero reference. With the noise receiver calibrated in this manner, the NPR readings taken during the receiver quieting test can be converted to dBrnC, or any other noise unit. Section 7.8 provides the relevant conversion equations.

Test Procedure
1. Insure that there are no alarms up on the receiver to be tested.
2. Remove the transceiver from service.
3. Turn off the transmitter, leaving the receiver on.
4. Activate the ACO circuit on the receiver if so equipped.

5. Calibrate the signal generator.
6. Connect the signal generator via the switchable attenuator and patch cord to the directional coupler.
7. Connect the selective level meter or the noise receiver to the baseband of the receiver-out jack.
8. With a level of −40 dBm into the receiver, measure the noise at the appropriate test frequency or noise slot according to Table 7.1. If a selective level meter is used, measure the signal-to-noise ratio.
9. Change the input level from −40 dBm in steps of 5 dB until the receiver mute point is reached. Record the noise or the signal-to-noise ratio and directly plot the results on the quieting proforma.
10. Compare the receiver quieting curve plotted in step 9 with the corresponding theoretical quieting curve as derived in Section 7.14. It is required that the quieting curve plotted in step 9 shall not exceed the noise corresponding to the theoretical curve as derived in Section 7.14, or as provided by the manufacturer.
11. If the requirement of step 10 is not met, compare the measured curve with that illustrated in Figure 7.7 and determine which part of the quieting curve is out-of-limits.
    If the noise recorded is higher for input levels below the FM improvement threshold, the receive unit assembly is at fault.
    If the noise recorded is higher for input levels in the linear section, again the receiver unit assembly is at fault.
    If the noise recorded is higher in the lower curved section of the curve the receiver baseband amplifier is at fault.
12. Remove the test equipment and return the receiver to service, or proceed with other tests as required.

Section 7.14 describes the possible problem areas and the items of equipment in the receiver which may be the cause of the problem.

11.2.11 Alarm Verification

This section provides details for verifying, and if necessary adjusting, the operation of the transceiver alarms. The alarms which may be present are:

TRANSMITTER
AFC alarm
Low power alarm
Pilot level and alarm

RECEIVER
AGC alarm
Pilot alarm
NODAN alarm

11.2.11.1 Transmitter Alarms
SUGGESTED TEST EQUIPMENT
1. AC voltmeter.
2. Various leads and adaptors.
3. Termination plugs.

TEST PROCEDURES
AFC alarm
1. Remove the transmitter from service.
2. Activate the alarm cut-off.
3. Verify that the AFC switch is in the AFC position, and that no alarms on the transmitter are up.
4. Switch the AFC switch to the MFC position. It is required that both the transmitter alarm and the AFC alarm should come up.
5. If the requirement is not met, take appropriate action.
6. Restore the AFC switch to the AFC position. Both alarms should cease.
7. Restore the transmitter to service, or proceed with other tests as required.

Low power alarm
1. Remove the transmitter from service.
2. Activate the alarm cut-off.
3. Switch the multiposition front panel switch to read power on the front panel.
4. Note the panel meter reading.
5. Either by simulation (as per manufacturer's instructions), or by reducing the output power of the transmitter by manual adjustment, reduce the power by 3 dB.
6. Verify that the low-power alarm comes up, as required.
7. If the requirement of step 6 is not met, the problem may be caused by a faulty alarm circuit; a defective diode detector in the oscillator-modulator or power amplifier; or an incorrect panel meter calibration.
8. Remove the test equipment and return the transmitter to service, or proceed with other tests as required.

Transmit pilot level
1. Remove the transmitter from service.
2. Activate the ACO.
3. Connect the AC voltmeter to the pilot monitor jack and note the pilot level. It is required that the pilot level should be −10 dBm0 if measured at a dBr point, or at a level specified by the manufacturer.
4. If the requirement of step 3 is not met, adjust the pilot level control for the correct level.
5. Remove the AC voltmeter from the pilot monitor jack or test point.

Transmit pilot alarm
1. Simulate a drop of 10 dB ± 1 dB in the pilot level. It is required that the transmitter pilot alarm and transmitter alarm should come up.

2. If the requirement is not met, adjust the pilot alarm threshold potentiometer until the alarm indicator just lights, or is raised.
3. Restore the pilot level to normal. It is required that the transmitter pilot alarm and transmitter alarm should clear.
4. Return the transmitter to service, or proceed with other tests as required.

11.2.11.2 Receiver Alarms
SUGGESTED TEST EQUIPMENT
1. RF signal generator.
2. Power meter with thermistor mount.
3. Frequency counter.
4. Noise receiver.
5. Switchable pad.
6. Test leads and adaptors as required.

TEST PROCEDURES
AGC alarm or NODAN alarm
The AGC alarm circuit is normally set to mute the receiver at a per-channel *S/N* of 30 dB (58.5 dBrnC0). However, the AGC alarm threshold may be set to a convenient level to meet the user's requirements. This alarm setting may be completed when the quieting curve is being compiled. A receiver may be equipped with either an AGC alarm or a NODAL alarm, but not usually both, since they each perform the same function.
1. Remove the receiver from service.
2. Activate the ACO.
3. Set the frequency of the RF signal generator to the assigned frequency of the receiver under test.
4. Calibrate the signal generator as per Section 11.2.9.
5. Refer to the quieting curve (Section 11.2.10) to see what the input RF signal level to the receiver must be for a *S/N* of 30 dB.
6. Set the RF signal generator to a level 5 dB higher than that determined in step 5.
7. Connect the generator and switchable pad and cabling to the directional coupler input.
8. Reduce the signal generator output to the level as determined by step 5. It is required that the NODAN alarm and receiver alarms should be raised.
9. If the alarms are raised with an input level ± 2 dB as determined by step 5, the requirement is met.
10. If the requirement of step 9 is not met, set the RF signal generator to the level as determined by step 5.
11. Adjust the AGC alarm threshold potentiometer until the AGC and receiver alarms are just raised.
12. Adjust the NODAN alarm threshold potentiometer until the NODAN and receiver alarms are just raised.
13. Repeat steps 6 to 11 to verify correct setting.
14. Remove test equipment and return the receiver to service, or proceed with other tests as required.

Receiver pilot alarm

This test may be completed as one of the hop tests, where the transmit pilot is reduced by 10 dB, and the far-end receiver alarm verified as correctly set, or is set accordingly.

1. On receipt of the reduced transmit pilot level by 10 dB below the normal, note that the receiver pilot and receiver alarms are raised.
2. If the alarms are not raised adjust the receiver pilot level potentiometer until the alarms are just raised.
3. If this adjustment does not permit the alarms to be raised, check the receiver pilot tuned circuit detector, since it may be tuned to a slightly different frequency.

11.3 TERMINAL-TO-TERMINAL TESTS

The tests dealt with in this section are:

> Baseband gain
> Frequency response
> White noise loading
> Spurious tones

11.3.1 Baseband Gain and Frequency Response

The procedure described is for measuring and, if necessary, adjusting the gain and frequency response of the HF baseband of a radio channel, from terminal to terminal. The baseband gain and frequency response of each channel should be as flat as possible relative to the normal channel gain. If they are not, service over the channel may be adversely affected. The requirements stated in these procedures are usual for a one-hop baseband-baseband system.

These tests may be performed over the RF path between adjacent channels, or on a loop-back basis at an individual site. If loop-back is preferred, the transmitter frequency must be changed to that of the receiver frequency, using a frequency-convertor and an attenuator to reduce the power level to that compatible with the receiver: alternatively, the transmitter and receiver IFs can be looped, ensuring that the levels are compatible. When these and other hop tests are performed over the RF path, it requires the co-operation of staff at both ends of the hop under test, These tests are out-of-service tests.

PREREQUISITE
The deviation of the transmit terminal must be correctly adjusted.

SUGGESTED TEST EQUIPMENT
At the transmitter site
1. Test oscillator with output impedance equal to that of the baseband

2. Variable input impedance level meter
3. Patch cords

At the receiver site
1. Variable-input impedance-level meter
2. Loop-back test set (optional)
3. Patch cords

11.3.1.1 Baseband Gain
TEST PROCEDURE
1. Remove the channel from service.
2. At both the transmit and receive sites activate the ACO switch.
3. Set the test oscillator to the crossover frequency as given in Table 10.5, or as recommended by the manufacturer.
4. Set the oscillator output frequency to the impedance of the baseband, and at the baseband transmit dBr level, as given in Table 10.2.
5. Connect the test oscillator to the baseband input of the transmitter.
6. At the receiver end, connect the level meter switched to the baseband impedance to the baseband output of the receiver.
7. Note the level meter indication. It is required that the level should be as listed in Table 10.2 for the system channel capacity.
8. If the requirement of step 7 is not met, adjust the *receiver* baseband level control for the correct indication.
9. If the baseband level control of step 8 is out of range, check the transmitter deviation.
10. Record the level in the equipment test records together with the date the test was performed.

11.3.1.2 Frequency Response
TEST PROCEDURE
1. At the transmit site, insert the appropriate frequencies one by one at the dBr level as given in Table 10.2 or by the manufacturer. The frequencies to be used may be either those recommended by the manufacturer, or selected across the baseband at intervals which provide at least ten readings.
2. The requirement may be ± 0.5 dB about the reference frequency, unless otherwise specified by the manufacturer.
3. The frequency to be used as the reference is the crossover frequency ($0.608 f_{max}$) again, unless otherwise specified by the manufacturer.
4. At the receiver site, measure the level of each of the frequencies transmitted, and record its deviation from the reference level in the equipment test records. It is required that the deviation is less than or equal to 0.5 dB, the reference frequency, unless otherwise specified.
5. If the requirement of step 4 is not met, it will be necessary to locate and correct a faulty or incorrectly adjusted unit at either the transmitter or receiver site.
6. At each site, remove the test equipment and return the transmitter and receiver to service, or proceed with other tests as required.

11.3.2 Noise Characteristics and Testing

Noise in a microwave radio system is derived from many sources. To achieve optimum quality and reliability of service, the effect of the noise from each source on the system must be determined and then minimized. With such equipment-dependent characteristics as operating frequency, power and deviation to be measured and adjusted, the system noise performance in every baseband-derived communication channel is made to give the lowest practical noise level. Within any VF channel, the total, or loaded, noise level results from the interaction of many contributors. All the individual contributors, however, can be placed in one of two general categories: those noise sources and types that are present at all times in the system, and those that exist only when a modulating signal is present. Thermal noise, for example, exists whether or not a modulating signal is applied to the system and may thus be grouped with the first category. This noise may be called the 'idle' noise contribution. Idle noise, as the term indicates, is that residual noise measured in a baseband-derived VF channel with no modulation present. In an interference-free environment, idle noise is composed of thermal and 'intrinsic' elements. The intrinsic, or basic, element is noise in the transmission path generated within the baseband, transmitter and later receiver stages. In most practical systems, however, the effects of noise from other forms of interference must be taken into consideration.

Idle noise may be considered to be excessive if it does not conform to the recommendations given in the system design specifications. In a remodulating-type radio, idle noise could be excessive if it were within 3 dB of the loaded noise requirement; for example, to meet a loaded noise requirement of 20 dBrnC0, the idle noise should not exceed 17 dBrnC0. The remainder of the loaded noise would be due to contributors whose effects are apparent only when the system is being modulated. These contributors all fall within the general category of *intermodulation distortion* producers. Echo distortion, for example, is a type of intermodulation noise that is created when delayed echo signals – frequently the result of discontinuities or moding within the waveguide run, or, less frequently, caused by path reflections – appear in the FM portion of the system.

In any radio system, tuned or active components have the capacity to degrade signal quality because they are all inherently non-linear, i.e. either their amplitude response or their rate of phase shift is not uniform over a band of frequencies. When a single frequency passes through such a non-linearity, it emerges as a fundamental frequency f and harmonics of that frequency. When more than one frequency passes through the same non-linearity, harmonics and intermodulation products between the fundamental and the harmonics are produced. Under busy-hour traffic conditions in a radio relay system, the number of individual frequencies in a given VF channel is very large, and many VF channels are simultaneously transmitted over the same facilities. When the number of channels approaches 240, the modulating spectrum becomes so uniform over the baseband that it has characteristics similar to that of random or white noise. A white noise signal is thus a convenient tool for conducting noise tests. Chapter 7 dealt extensively with the noise loading of a radio link, and with the principles of how to operate the noise test sets, etc.

11.3.2.1 Idle Noise Considerations

High levels of idle noise may be produced by high thermal contributions from a low RF received signal level or improper deviation. High intrinsic noise contributions may be produced from noisy radio or baseband equipment. Baseband level interference may be produced from adjacent relay sets, or by interference at or near the RF carrier which permits the IF filters to pass the difference frequencies into the lower stages of the receiver. If the test results indicate a high idle noise condition, the flow chart given in Figure 11.6 can provide a logical progression of steps for isolating the cause or causes. Depending upon where in the spectrum the excess idle noise appears, one or more contributing mechanisms may be found. For example, thermal noise generated in the front end of a receiver will increase in direct proportion to the received signal level in the higher slots, but will be negligible in the lower slots. With an adequate receiver RF input signal level and negligible interference, low-slot idle noise may be isolated by substituting such components as modulation amplifiers, frequency-modulated oscillators, IF amplifiers, discriminators, and power supplies. Radio frequency interference (RFI) may result from interaction with systems external to that under test (inter-system RFI), from sources within the system itself (intra-system RFI), or from within one particular facility (infra-station RFI). When the frequency of the interfering signal lies near one of the in-band test slots, it will add to the idle noise contributor. If the interference frequency appears in the IF but is outside the baseband spectrum of the radio, it will interact with the transmitted baseband traffic and produce not idle noise but excessive intermodulation distortion, due to the excessive phase delay at the filter edges. This distortion will appear as a high intermodulation noise level, usually in the high, but sometimes in the low, slots.

Inter-system RFI can be detected by disabling the transmitter of the system under test and noting any idle noise swings that would indicate the presence of a foreign signal. This may also be done by using a selective level meter slowly swept in frequency over the baseband of the receiver, and noting any high level frequencies observed. It may also be suspected when other transmitters are operating nearby, particularly if they are operating as a submultiple of the system's receiver frequency (a 3 GHz radar transmitter interfering with a 6 GHz receiver for example), and can be readily identified through bucket curve analysis. Inter-system RFI can be reduced in many cases by careful antenna orientation (polarization), and by shielding or using shrouded antennas.

Intra-system RFI sources can be identified by disabling suspected transmitters within the building or area. If the interfering signal is still present with the receiver waveguide terminated to eliminate external signals, the cause could possibly be inter-rack interference, and proper grounding and shielding may correct the problem. In addition to this, the power supply to the receiver may be monitored with an oscilloscope to determine if high-level spikes appear due to surge operation of nearby equipment. If intra-system RFI is not found, and yet the interference contribution to idle noise continues, such intra-system conditions as antenna coupling may exist.

11.3.2.2 Intermodulation Noise Considerations

If idle noise is acceptable, but the measured loaded noise level still exceeds the

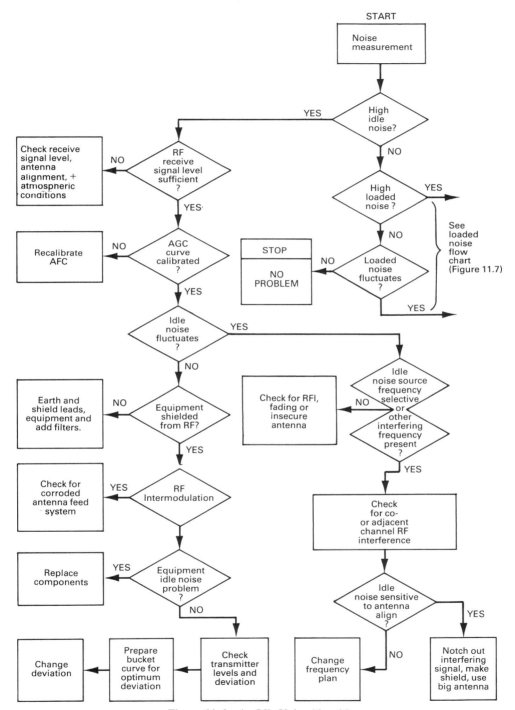

Figure 11.6   An Idle Noise Algorithm

system's loaded noise requirement, excessive intermodulation distortion is the problem. Again, a flow chart is given in Figure 11.7 to assist in isolating contributors. Excessive intermodulation results from poor baseband linearity or from delay distortion including waveguide echo distortion. Poor baseband linearity produces noise which usually occurs in all of the test slots, but is particularly noticeable in the lower slots. Delay distortion influences only the higher slots, unless an extremely long echo path is involved. Poor linearity, as reflected in high low-slot loaded noise, is generally related to the amplitude/frequency characteristics of the deviator–discriminator pair. A non-linear deviator paired with a mirror-image non-linear discriminator may result in excellent linearity and good low-slot loaded noise performance; however, such dedicated pairing is frequently not possible. Most manufacturers linearize each component against a factory standard or a linearity test set. It is possible by careful adjustment in the field to obtain reasonably good results.

Excessive low-slot loaded noise may also be caused by baseband amplifier input or output levels that are too high, or by unbalanced baseband amplifiers, and so balance adjustments and maximum levels are normally specified by manufacturers for these active devices. Distortion due to unequal transit delay time across the passband of the IF signal results in excessive loaded noise in the high-test slot. If such group delay distortion is present, it can be corrected by an equalization technique. Delay equalization is best accomplished with a test set-up which displays the delay characteristic on an oscilloscope, as shown in Figure 1.4. These test sets normally have the capability of measuring from baseband to IF (for heterodyne repeater adjustments), IF to baseband (for remodulating equipment) and section/system mop-up adjustments. When such a test set is not available, delay adjustments can be made to achieve the best noise in the higher baseband slots as indicated by the noise receiver.

Another intermodulation contributor that produces excessive high-slot loaded noise is waveguide echoes. These are commonly the result of impedance and aperture discontinuities, waveguide moding, and interference between the desired transmission path and a secondary advanced or delayed path. Echo distortion, as described in Section 7.9, causes delay ripples that, in modern high-performance microwave links, are often the major noise contributors. Path intermodulation, generated by secondary reflections from buildings or other terrain features, may be suspected if the high-slot idle noise is stable but the loaded noise fluctuates rapidly and is sensitive to slight adjustments in antenna orientation.

11.3.2.3 Noise Loading Test

This section outlines the steps to be taken in making a noise loading (NPR) test on the HF baseband of a radio channel from terminal-to-terminal. It is not applicable to systems with a channel capacity less than six voice channels. In a factory, it is usual for each system to be optimized for NPR on a back-to-back basis. Once installed in the field, the antennas and transmission lines introduce their own distortion, which usually causes the intermodulation noise to degenerate somewhat over the back-to-back value. The requirements stated in this section refer to a one-top baseband-to-baseband system. This test is more meaningful if performed over the RF path between two adjacent terminals, rather than on a loop-back basis at an individual

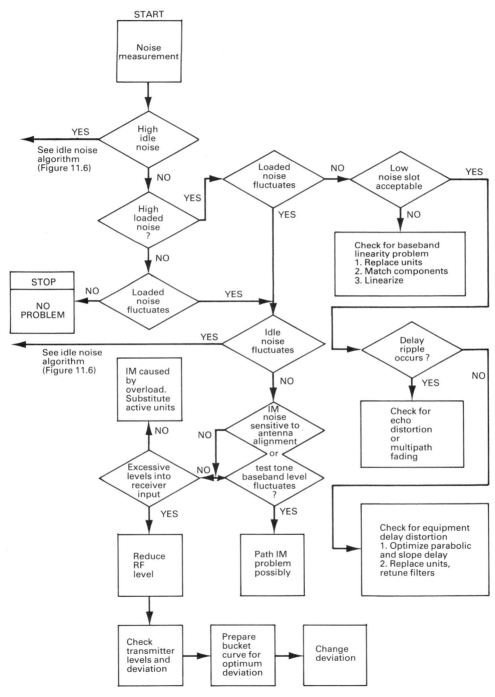

Figure 11.7   A Loaded Noise Algorithm

site. If performed over an RF path, it does however, require the co-operation of staff at both ends of the hop to be tested. The system usually is performed by taking the system out of service, but the CCIR have made provisions for these tests to be completed over a working link which is carrying live traffic (see Section 7.13).

PREREQUISITES
1. The requirements given in this section assume that the test is performed under normal propagation conditions for the hop under test. Testing must not be performed during fading conditions.
2. A normal condition can be verified by monitoring the AGC of the receiver under test with the front panel meter. The reading is then compared with the AGC curve and the incoming signal level determined. The incoming signal must be steady.
3. The terminal-to-terminal baseband and frequency response requirements should be met before performing this test.

SUGGESTED TEST EQUIPMENT
At the transmitter terminal
1. Noise generator
2. Patch cords and adaptors

At the receiver terminal
1. Noise receiver
2. Patch cords and adaptors

   Note that the noise generator and receiver must be equipped with compatible filters, as per Table 7.1, 7.3 or 7.4.

TEST PROCEDURE
1. Remove the radio channel from service.
2. At both transmitter and receiver sites activate the ACO.
3. Disconnect the hop transmit pilot.
4. At the transmitter and receiver sites set up the noise generator and receiver according to the instructions given in Section 7.11 and 7.12.
5. Using the white noise generator and receiver, measure and record the total noise [intermodulation + residual $(I + R)$], and the residual noise R in each of the 3.1 kHz baseband slots which are available on the noise generator. It is required that the total noise $(I + R)$ should not exceed the installed performance by more than 2 dB or the user's maintenance requirements for the system under test.
Note. The absolute accuracy of the test equipment is only about $\pm$ 1 dB. In those instances where the requirements are close to the capability of the equipment, great care must be taken to maintain the test equipment in good working condition and to perform the tests with as much care and precision as possible. Comparison of the results with those obtained on other radio channels should assist in evaluating the performance of a marginal system.

6. If the requirement of step 5 is not met, compare the total and residual noise measurements and refer to Section 11.3.2.4 or Figures 11.6 and 11.7.
7. With high intermodulation noise in the bottom slot only, adjust the linearity control on the receiver discriminator for minimum total noise.
8. If the requirement of step 5 is not met, compare the total and residual noise measurements and refer to the noise causes and localization below or to Figures 11.6 and 11.7.
9. Remove the test equipment and return the channel to service, or proceed with other tests as required.

11.3.2.4 Noise Causes and Localization

1. When all noise measurements do not meet specifications, check the noise loading test equipment for the following:
   (*a*) Correct loading level
   (*b*) Possibility of ground loop interference
2. Also consider the possibility that the test equipment used in earlier sections of the alignment procedures may have been out of calibration.
3. If both the residual and total noise are excessive, and are not the fault of the test equipment, check for:
   (*a*) Low transmitter output power
   (*b*) Low receiver RF input signal
   (*c*) Low receiver IF level
   (*d*) Incorrect receiver baseband output level
Note. In most cases, residual noise will be the controlling factor.
4. High total noise in all slots may also indicate a problem with:
   (*a*) A damaged antenna
   (*b*) The transmission line between the radio and the antenna
   (*c*) Multipath distortion
5. High noise in the top slot only is most probably caused by a problem in:
   (*a*) The transmitter modulator
   (*b*) The receiver demodulator
   (*c*) The IF filter
6. High noise in the low slot is most probably caused by a problem in:
   (*a*) Either baseband amplifier
   (*b*) The transmitter modulator
   (*c*) The receiver demodulator

11.3.3 Residual Noise and Spurious Tones

This section outlines the steps to be taken to measure residual noise and to detect and measure spurious tones in the HF baseband of a radio channel from terminal to terminal. The test measures the general noise condition of the radio channel and serves to check the total effect of the noise from the individual radios in the system under test. It is normally performed upon completion of the frequency-response and white-noise-loading tests.

Prerequisite
The tests should be performed under normal propagating conditions. Testing should not be performed while fading is occurring as indicated when the residual noise is varying rapidly.

Residual Noise
Residual noise is defined as the total noise power present in a radio channel measured at the output of the receiver with the input to the transmitter of the transmitting station terminated. It is the power summation of the residual noise powers generated by the intermediate radio bays including the noise generated by the terminal equipment. For residual noise measurements, it is recommended that a level meter with a 3.1 kHz band be used. When using a level meter with a bandwidth other than 3.1 kHz, a correction factor must be added to the residual noise calculations

$$\text{Correction factor} = 10 \log [3100 \text{ Hz}/(\text{actual bandwidth in Hz})] \qquad (11.1)$$

In practice, the residual noise performance may differ from the predicted residual noise due to the following:

> Variations in transmitter power
> Variation in propagation conditions
> Radio channel baseband gain frequency roll-off
> Measurement tolerances

Spurious Tones
Spurious tones are any unwanted signal which may be present in the baseband output signal of the FM receiver. Spurious tones may be introduced by interference from external sources, or may be generated in the radio equipment. Each tone may cause an audible note in a telephone or data circuit. Spurious tone measurements should be made with the bandwidth selector switch set to the narrow bandwidth position (80 Hz). It is necessary to restrict the bandwidth of the level meter to reduce the residual noise component in the measurement, since the residual noise of the radio system measured in the wide bandwidth position may be almost equal to the spurious tone requirement.

Suggested Test Equipment
1. Selective level meter
2. 75 Ω terminating resistor
3. Patch cords and adaptors as required

Test Procedure
1. Remove the channel from service and disable the hop pilot.
2. Connect the selective level meter to the receiver baseband output, insuring that the impedance of the SLM is chosen to match that of the baseband.

3. Insure that all inputs to the transmitter at the far end of the hop are terminated with a resistor which matches the baseband impedance (in this case 75 Ω is chosen).
4. Set the bandwidth of the level meter to 3.1 kHz.
5. Tune the level meter to frequencies evenly spaced across the baseband frequency spectrum so that the readings are 50 kHz apart. The residual noise levels must not exceed −70 dBm0 for systems with a deviation of 35 kHz RMS. For other deviations, the corresponding noise can be found from:

$$\text{Maximum residual noise} = -70 - 20 \log (\text{deviation in kHz}/35) \text{ dBm0} \quad (11.2)$$

Note. Do not measure spurious tones encountered at the above frequencies. If a tone is encountered, tune slightly to one side and record.

6. Place the SLM into the smallest bandwidth using the bandwidth selector switch.
7. Manually, slowly change the receiving frequency of the SLM over the entire baseband range. If any spurious tones are found then measure and record:
   The frequency of the tone
   The level of the tone greater than −65 dBm0
   The noise level directly adjacent to the tone greater than −50 dBm0
   There are two requirements:
   (*a*) Spurious tones: maximum level is −65 dBm0 using an 80 Hz bandwidth.

   If another bandwidth is used the correction factor is:
   10 log [80/(actual bandwidth in Hz)]   (11.3)

   (*b*) Noise adjacent to tone: maximum level is −81 dBm0, using an 80 Hz bandwidth. If another bandwidth is used, equation 11.3 applies here also.
8. If the requirement of step 7(*a*) is not met, it is an indication that the radio channel contains a noisy transmitter or receiver. This equipment may need to be replaced and the system re-aligned.
9. If the requirement of step 7(*b*) is not met, the source of the spurious tone or tones should be isolated within the system. When the equipment generating a tone is located, it should be realigned and the measurement repeated.
10. Disconnect the test equipment and proceed with other tests as required.

11.3.4 Multiplex Tests

See Section 3.10 for the procedures relating to these tests. The tests are usually done over a hop or hops, depending on where the drop-and-insert facilities are. If the tests are not done between adjacent hops, but between terminals in which there may be several hops, each intervening link transceivers must first be fully aligned.

11.3.5 Near-end Baseband Crosstalk Tests

It is necessary to perform crosstalk measurements only when other methods of locating sources of high noise have failed. This test is an out-of-service test which requires the removal of all traffic between the adjacent terminals and the station under test. Near-end crosstalk measurements are performed by transmitting an HF baseband signal in one direction and observing the level of the same signal in the receiving direction at the same site.

Crosstalk Causes and Localization

Crosstalk is mainly caused by direct radiation of signals from one channel to another at baseband frequency, or at the RF transmitting and receiving levels. It normally occurs if the cables carrying different baseband signals for different systems or spurs are physically too close to each other, especially if improper shielding and grounding of the cabling in the jack-field and the equipment rack occurs. Crosstalk at the RF transmitting and receiving levels may occur if the RF channels are insufficiently separated.

Locating the crosstalk may be done by an elimination process. For example, if the crosstalk is excessive the following procedures may apply:

1. To determine if the crosstalk is due to the baseband equipment at the transmit terminal, switch off the transmitter at this terminal.
2. If the crosstalk remains, the source of the crosstalk is at the near-end terminal. Check the jack-field and all cables carrying the baseband for direct radiation.
3. If the crosstalk disappears by the switching off of the transmit terminal, the source of the crosstalk is at the far-end terminal.

SUGGESTED TEST EQUIPMENT
1. Test oscillator
2. Selective level meter
3. Baseband terminating resistors
4. Cords and adaptors as required

TEST PROCEDURE
The test set-up is to terminate both the far-end transmitter and receiver basebands with the terminating resistors. The test oscillator is fed into the transmitter baseband, and the selective level meter placed on the near-end receiver output.

1. Remove each channel from service in both directions.
2. At the far-end or adjacent site only connect the baseband terminating resistors.
3. At the measuring terminal connect the test oscillator.
4. Set the test oscillator output to the appropriate test tone and level and enter on the equipment test record.
5. Connect the selective level meter to the measuring terminal receiver baseband output.
6. Set the level meter bandwidth switch to the 80 Hz position.

MEASUREMENT
7. Tune the test oscillator to a series of specific frequencies across the baseband and measure, and record, the crosstalk at each frequency with the level meter. It is required that the measured crosstalk must not exceed -65 dBm0 for an 80 Hz bandwidth. If another bandwidth is used the correction factor of equation 11.3 must be applied.

Note. When the crosstalk is measured at specific frequencies, the basic noise should also be measured with the level meter, by removing the input test signal from the test oscillator.

8. If the requirement of step 7 is not met, refer to the system block and level diagram and choose intermediate insertion and monitor points, in a process of elimination, to localize and correct which area of the HF section is causing the crosstalk.

9. Remove the test equipment and return the radio system to service, or proceed with other tests as required.

11.3.6 Far-end Crosstalk Tests

Far-end crosstalk measurements are performed by transmitting an HF baseband signal on one channel and observing the level of the same signal on a diversity channel at the other site. The test equipment and preliminary discussion given in Section 11.3.5 applies to this section as well.

Test Procedure
1. Remove both channels from service between the terminals in the direction under test.
2. Connect the level meter set to an 80 Hz bandwidth to receiver B at the receiver terminal (measuring terminal).
3. Connect the test oscillator to transmitter A at the transmitter terminal.

Measurement
4. Set the level oscillator to the appropriate test tone and level and record in the equipment test record.
5. Tune the level meter to the test tone frequency and measure and record the crosstalk level. It is required that the measured crosstalk must not exceed -65 dBm0 at 80 Hz bandwidth. If another bandwidth is chosen the correction factor given in equation 11.3 must be applied.
6. If the requirement of step 5 is not met, refer to the system block and level diagram and choose intermediate insertion and monitor points, in a process of elimination, to localize and correct which area of the HF section is causing the crosstalk.
7. Repeat steps 4 and 5 for a series of other specific frequencies across the baseband.
8. Repeat steps 4, 5, 6, and 7 for the other diversity channel.
9. At both terminals remove the test equipment and return the channels to service.

11.3.7 System Tests

The system tests are exactly the same as the hop tests, except that:
The staff are at either end of the system.
Different pilots are used.
System alarms are different.
Crosstalk tests may not be required.

11.3.8 Other Tests

If a diversity system is employed, there may be further tests required on the protection unit, and on the system itself – where confirmation of the correct switching operation is sought. The equipment manufacturer should provide details of the necessary tests required on each of the additional items of equipment, and on the system as a whole.

The engineers' order-wire also may require alignment. This is usually done by inserting a 1000 Hz – 800 Hz test tone, and measuring and adjusting the levels through to the baseband. In addition there may be supervisory equipment which carries the alarms from different sites on a polling basis. This may also employ a section of the sub-baseband. Similarly, the checking of a tone through to baseband and adjustment of pads and amplifiers at the center frequency of the supervisory band may suffice.

REFERENCE

1.   Farinon Canada LR and TR Test Procedures, Farinon Electric, 1691 Bayport Avenue, San Carlos, California, 94070.

INDEX